T0325397

Condensed Matter Physics: A Modern Perspective

Online at: https://doi.org/10.1088/978-0-7503-3031-2

Condensed Matter Physics: A Modern Perspective

Saurabh Basu
Department of Physics, Indian Institute of Technology Guwahati, Guwahati, Assam, India

IOP Publishing, Bristol, UK

ISBN 978-0-7503-3031-2 (ebook)
ISBN 978-0-7503-3029-9 (print)
ISBN 978-0-7503-3032-9 (myPrint)
ISBN 978-0-7503-3030-5 (mobi)

DOI 10.1088/978-0-7503-3031-2

Version: 20221201

IOP ebooks

British Library Cataloguing-in-Publication Data: A catalogue record for this book is available from the British Library.

Published by IOP Publishing, wholly owned by The Institute of Physics, London

IOP Publishing, No.2 The Distillery, Glassfields, Avon Street, Bristol, BS2 0GR, UK

US Office: IOP Publishing, Inc., 190 North Independence Mall West, Suite 601, Philadelphia, PA 19106, USA

Dedicated to my family…

Contents

Preface

The study of condensed matter physics and its applications to the physical properties of various materials have found a place in the undergraduate curriculum for a century or even more. The perspective on teaching of condensed matter has remained unchanged for most of this period. However, the developments in condensed matter over the last few decades require a new perspective of teaching and learning of the subject. Quantum Hall effect is one such discovery that has influenced the way condensed matter physics is taught to undergraduate students. The role of topology in condensed matter systems and the fashion in which it is interwoven with physical observables needs to be understood by a student for deeper appreciation of the subject. However, the role of inter-particle interactions in shaping the properties of materials cannot be ignored. Thus, to have a quintessential presentation for undergraduate students, in this book, we have addressed selected topics that comprehensively contribute to the learning of condensed matter physics that emerged in the not-so-distant past, as well as those topics that have firmly laid the foundation of conventional condensed matter physics.

A little elucidation of the content will aid to a better understanding of the spirit of this text. We start with an electronic system in chapter 1 that transcends the free electron theory applied to metals. Interparticle interactions are invoked, and the consequences on the dielectric function that encodes interesting physical properties are discussed. In continuation of the ongoing discussion, experimental methods, such as the de Haas–van Alphen effect that traces out the Fermi surface of metals is discussed. The chapter is wrapped up with a brief discussion of the Fermi liquid theory, which was constructed by Landau, and re-emphasizes the shortcomings of the free electron theory in a most comprehensive manner.

In chapter 2 a follow-up of material properties is taken up in the form of studying magnetism. Different types of magnetic order and materials are introduced to students. The upshot of the discussion is that the electronic interactions drive magnetic order, and hence they need to be incorporated for a comprehensive understanding of magnetic properties. In this connection, we introduce spin models, such as the Ising model, XY model, Heisenberg model, etc. We solve these via controlled approximations. Furthermore, magnetism is shown to originate from itinerate electronic models, such as the Hubbard model. A self-consistent Hartree–Fock solution of the Hamiltonian yields a reasonable description for both the ferromagnetic and antiferromagnetic correlations.

Next, in chapter 3 we embark on the transport properties of 2D electronic systems and focus solely on the role of a constant magnetic field therein. This brings us to the topic of quantum Hall effect which is one of the main verticals of the book. Origin of the Landau levels and the passage of the Hall current through edge modes are discussed. The latter establishes a quantum Hall sample to be the first example of a topological insulator. Having discussed 2D electron gas, it is of topical interest to discuss the corresponding scenario in graphene.

Our subsequent focus in chapter 4 is the other vertical, namely, topology. Introducing the subject from a formal standpoint, we discuss the band structure and topological invariants in 1D and 2D. In particular, we talk about the Su–Schrieffer–Heeger (SSH) model, which, apart from being a possible realization for a polyacetylene chain, has emerged as a paradigmatic tool to study topology in 1D. In 2D, the usual hobbyhorse, namely graphene is taken. We dwell upon the possibility pointed out by Haldane whether graphene can become a topological insulator. Addition of the spin of electrons to the ongoing discussion emerged as a unique possibility to yield another version of the topological insulator, namely, the quantum spin Hall insulator, which may lie at the heart of the next generation spintronic devices.

In chapter 5, we introduce second quantization and discuss how it takes care of the indistinguishability of particles. Hence, we discuss representations in quantum mechanics, and subsequently study Green's functions at zero temperature. Wick's theorem is studied with a view to denote scattering processes using Feynman diagrams. Finally, the finite temperature Green's functions are studied, where we show how to compute the Mutsubara frequency sums in simple situations.

We go on to discuss superconductivity in chapter 6. We sequentially follow the historical developments of the field, phenomenological understanding of different phases, magnetic properties of superconductors, etc. Importantly, we present the Bardeen, Cooperand Schrieffer (BCS) theory, in as much detail as possible, in an effort to provide a microscopic description of the thermodynamic and the electromagnetic phenomena. Ginzburg–Landau theory, a phenomenological description of a superconductor to a metal transition, has been introduced in a nutshell. We finally wind up with a description of the experiments that determine the superconducting energy gap, and a very brief note of the unconventional cuprate superconductors.

Lastly, we describe the superfluid state in chapter 7. We describe fundamental concepts of Bose–Einstein condensation and the Gross–Pitaevskii equation. Next, we go on to describe bosons as a tool to comprehend the complicated many body physics in the presence of strong inter-particle interactions. This leads us to optical lattice, Bose–Hubbard model, and the phase diagram that encodes a transition from a superfluid phase to an insulating phase. Such a phase transition has been accessed experimentally. The ramifications of the presence of random disorder are shown to yield a glassy phase, namely the Bose glass phase. Finally, the ultracold gases with spin degrees of freedom are shown to demonstrate an even richer phase diagram.

All the while during the course of the book, we have included rigorous mathematical derivations wherever required, presented experimental details to connect with the ongoing discussions, and tried to be as lucid as possible in our presentation of topics and concepts. A whole lot of schematic diagrams are presented for clarity as well. We hope that students gain from the essence of this book, and it aids their understanding of both topical as well as traditional condensed matter physics. I shall be available and happy to answer queries, clarifications by students and researchers, and welcome comments for improvement.

Foreword

The book *Condensed Matter Physics: A Modern Perspective* by Professor S Basu presents a refreshingly modern take on the subject. It presents careful, detailed, analysis on several model topics that are at the heart of present day condensed matter physics. Some of these include the discussion of the role of topology in condensed matter, a treatise on superfluidity, and a detailed exposition on Green's function theory. The more conventional topics such as magnetism, superconductivity, and transport theory are also discussed in separate chapters; for these the author has carefully chosen a range of topics from the vast literature to bring out the essential points of the subject.

The book is expected to be useful for advanced undergraduate students and also beginning graduate students who aim to take up condensed matter for research in the future. The topics are lucidly described and each chapter provides a basic introduction to the subject which may be useful while delving deeper into it at a later time.

Overall, I find this book to be a useful addition to others on the subject and would recommend it for advanced undergraduate and graduate students.

Dr Krishnendu Sengupta
Senior Professor
School of Physical Sciences
Indian Association for the Cultivation of Science
Jadavpur, Kolkata
India

Acknowledgement

It is a proud privilege to acknowledge a lot of people who have actively or passively contributed in bringing up the book in the current form. First and foremost, I owe a lot to my current graduate students, Ms Shilpi Roy, Mr Sayan Mandal, Mr Dipendu Halder and Ms Srijata Lahiri. I am also thankful to my erstwhile students, namely, Dr Sudin Ganguly, Dr Sk Noor Nabi, Dr Sourav Chattopadhyay, Dr Priyadarshini Kapri, and Dr Priyanka Sinha who are currently pursuing their jobs or post-doctoral careers across the world. It is also a privilege to acknowledge Professor Gaurav Dar (BITS Pilani, Goa campus) who as has been an inspiration at all times. I am also thankful to my PhD supervisor, Professor Avinash Singh (IIT Kanpur), my collaborator, Professor S Murakami (TokyoTech), Professor B Tanatar (Bilkent), Professor A Perumal (HOD, Physics, IIT Guwahati), and a number of colleagues at the department of Physics, IIT Guwahati.

Finally, it is a sheer pleasure to acknowledge my lovely wife, Sanghamitra, and my angels, Shreya and Shreemoyee for being there all the while with me during writing of the book.

Author biography

Saurabh Basu

Saurabh Basu was born and raised in Kolkata (then Calcutta), India. His school and undergraduate colleges were in St. Lawrence High School and St. Xavier's College, respectively. He completed a Master's in Physics from IIT Bombay, PhD from IIT Kanpur, and after a couple of post-doctoral studies at TIFR, Mumbai and Queens University, Kingston, Canada, Saurabh had joined the Department of Physics, IIT Guwahati, India in 2003 and is currently a professor there. His broad research interests are different fields of theoretical condensed matter physics, with focus on topological materials, higher order topological insulators, ultracold physics, non-Hermitian systems, charge and thermal transport in mesoscale and nanoscale devices, Floquet dynamics, critical phenomena, etc. Dr Basu has about 90 research publications in different peer reviewed journals and a few book chapters. 12 PhD students have so far graduated with his guidance, and currently there are six students working for their PhD degree. Several Bachelors, Master's students, and intern students from different institutes have received his guidance. Apart from research and teaching at IIT Guwahati, he is passionate about improving the status of school and college education in the country. He has visited several institutions to upgrade hands-on training programmes there, and is committed to making science education more enjoyable for young students.

Saurabh Basu

Chapter 1

Electron liquid

1.1 Introduction

It is somewhat implicit that readers are familiar with the first course on quantum mechanics which mainly deals with the properties of a single and non-relativistic particle in the presence of a given potential that usually has a simple form. We give a brief recap on some of these problems below. Readers are encouraged to look at the classic texts on quantum physics [1–4]. For example, the eigensolutions (ϵ_n, ψ_n) of a free particle confined in an infinite potential well are found to be $\epsilon_n \sim n^2$, and $\phi_n \sim \sin\frac{n\pi x}{L}/\cos\frac{n\pi x}{L}$[1] or, in an infinite space, $\phi_n \sim e^{in\pi x/L}$ (n being the quantum number for the problem). On the other hand, a parabolic potential (simple harmonic oscillator) yields an equidistant energy spectrum of the form, $\epsilon_n = (n + \frac{1}{2})\hbar\omega$, and the eigenfunctions are denoted by a Gaussian ($\sim e^{-x^2}$) multiplied by a polynomial (Hermite polynomial) which possesses an even or odd parity depending on whether n is even or odd. Such an even–odd nature of the eigenfunctions is an artefact of the symmetric potential[2] which allows both even and odd solutions. Hence the Hilbert space gets fragmented into one half for even n, and the other half for odd n. Further, in a three-dimensional case, which is more complicated than its one-dimensional counterpart, in the presence of a Coulomb potential, appropriate for hydrogen (H) atom which has only one electron, the energy spectrum of an electron takes a form, $\epsilon_n = -\frac{13.6}{n^2}$ eV, where n denotes the principal quantum number. The eigenfunctions are denoted by the variables of the spherical polar coordinate system, where the radial function is a polynomial in the radial variable, r (Laguerre polynomial) multiplied by an exponentially decreasing function, namely, $e^{-\alpha r}$ ($\alpha = a_0^{-1}$, a_0 being the Bohr radius). The angular variables appear via the spherical harmonics, that is, $Y_{lm}(\theta, \phi)$. Interestingly, the Y_{lm} functions denote the eigenfunctions of the square of

[1] The cos solution comes for odd n in a symmetric well.

[2] Symmetric potential implies $V(x) = V(-x)$.

doi:10.1088/978-0-7503-3031-2ch1

the angular momentum operator, $\mathbf{L}^2$ and the angular part of the Hamiltonian for the electron in a H-atom is the same as that of $\mathbf{L}^2$ (apart from a constant), which implies that the commutator $[\mathcal{H}, \mathbf{L}^2] = 0$, and hence $\mathcal{H}$ and $\mathbf{L}^2$ share the same eigenfunctions.

The barrier transmission problems in quantum mechanics may find potential applications in the transport properties of semiconducting heterostructures. There are a variety of barriers, such as a finite step, finite well, or an infinite potential discontinuity (δ-function), etc where the formulae of finding the reflection and the transmission coefficients are prescribed by matching the boundary conditions of the wavefunctions and their derivatives across the boundary[3]. Interesting consequences occur when the energy of the particle is less than the barrier height, etc. An infinite sequence of potential profile results in the Kronig–Penny model for solids which describes the behaviour of the electrons in a crystalline solid, and yields the band structure which is indispensable for describing electronic properties of materials. Importantly, the results yield a classification of metals, semiconductors and insulators.

Exact solutions (or solutions in a closed analytical form) of the Schrödinger equation for a large variety of potential functions do not exist. Thus, in most cases of practical importance, one resorts to approximate methods for the solution of the problem. The reliability of the approximate methods depends on two things, the strength of the potential function has to be necessarily much smaller than the original Hamiltonian (that is, without the potential term(s)), and secondly, the eigensolutions of the original Hamiltonian should be known. Owing to the smallness of the former, it is referred to as the perturbation term (call it $\mathcal{H}'$), while the latter is called the unperturbed Hamiltonian (call it $\mathcal{H}_0$). The corrections to the unperturbed energies are computed with the eigenstates of the $\mathcal{H}_0$. In most cases, obtaining the effects of the perturbation term up to one or two orders of $\mathcal{H}'$ are usually sufficient, unless they turn out to be zero owing to some symmetries of the eigenfunctions, which appear as the selection rules for a given problem. For example, the degeneracies of the $n \neq 1$ levels of the H-atom may be lifted (totally or partially) by the presence of a weak electric field or a magnetic field. The electric field case is known as the Stark effect, where the dipole moment of the atom couples to the electric field to yield the perturbation term. The presence of the magnetic field gives rise to Zeeman effect where the spin of the electron couples with the external magnetic field.

Further, a time dependent electromagnetic field is considered to investigate the transition between different quantum states. The interaction of radiation with matter yields the transition probabilities between the ground state and the excited states, which aid in calculating Einstein's A and B coefficients and is an important area of research in the field of LASER. A reader must have also come across a few other forms of the potential, such as a linear potential (in the context of a slowly varying potential in the Wentzel–Kramers–Brillouin (WKB) approximation [8]) or a

[3] The derivative of the wavefunction is not continuous across an infinite potential discontinuity, nonetheless it differs by a known constant.

relatively weak cubic potential (x^3) along with a dominant harmonic term (x^2) to explain melting of solids, etc.

The time evolution of quantum systems is in general interesting owing to its applications in the fields of optics, condensed matter physics, dynamical systems, etc. For example, a periodic driving of a quantum system as time progresses (may not be necessarily a weak driving potential) has formal similarities with a periodic crystal potential in a solid. This leads to Floquet formalism and facilitates discussion of the stability of periodic orbits.

The formalism of quantum mechanics is based on matrix methods and each of the operators, for example, the Hamiltonian, momentum, etc can be expressed as matrices as the bases that are suitable for their description. Thus, quantum mechanics requires a sound background of linear algebra which formulates the key findings, such as the expectation values of the operators. These expectation values yield physical observables that are relevant to experimental scenarios. The method has facilitated reformulation of the simple harmonic oscillator problem (stated earlier), where an occupation number basis that counts the number of oscillators in a given quantum state, and thus yields a simpler, yet a more insightful description. The transition from one quantum state to the next higher or lower one which differ in energy by an amount $\frac{1}{2}\hbar\omega$, can be expressed as creation or annihilation of oscillators, respectively. This not only introduces a, $a^\dagger$ (annihilation, creation) operators in this particular problem, but also lays the foundation of treating many particle problems.

The description of quantum systems crucially needs a basis in which the operators are written down as matrices. In some problems, certain basis states are found to be more convenient than others, The change of basis or a relationship between two bases have important consequences in different problems in quantum mechanics. For example, two non-equivalent descriptions (bases) in a spin–orbit coupled system beget a unitary transformation between them, which are known as Clebsch–Gordon coefficients and are important topics to learn in atomic systems.

Finally, an undergraduate course on quantum mechanics should conclude with an introduction to relativistic quantum mechanics. The necessity for the relativistic formulation can be understood as follows. The time dependent Schrödinger equation relates a double derivative of the wavefunction with respect to the spatial coordinates (the kinetic energy term) to the first derivative with respect to time (energy). This is clearly not Lorentz invariant, as the Lorentz transformation equations put space and time at the same footing. Upgrading time to its double derivative (like the wave equation and hence relates $\frac{d^2}{dt^2}$ to $\vec{\nabla}^2$) yields the Klein–Gordon equation, applicable for spin-0 particles. For spin-$\frac{1}{2}$ particles, a downgrading of the space derivative, and relating $\vec{\nabla}$ to $\frac{\partial}{\partial t}$ preserves the Lorentz invariance, which yields Dirac equations. Among other important artefacts, negative energy solutions are found to be valid and applicable to 'holes' (or anti-particles), besides spin of the electrons naturally arising from the solution of the Dirac equation. One should remember that spin has been introduced by 'hand' in the Schrödinger equation. In addition, readers at the advanced level are encouraged to read the solution of the Dirac equation in the presence of a variety of potential functions, for example, a Coulomb potential, etc.

In condensed matter physics, the relativistic nature of particles finds an application in graphene, which shows relativistic dispersion for the valence electrons close to the Fermi energy. This gives rise to a vast field of two-dimensional (and even three-dimensional generalizations exist) Dirac materials, where the electronic dispersion is linear in the wave vector, just like that of a photon. However, the electrons have a much lower velocity (of the order of 10^5–10^6 m s^{-1}) than that of light, thus earning the name, *pseudo relativistic* dispersion. The linear dispersion has important ramifications on the transport properties of Dirac materials.

Technically speaking, physics deals only with one-body and many-body problems, because a two-body problem reduces to a one-body problem and a three-body problem is unsolvable. However, physicists and chemists have to routinely worry about $\sim 10^{23}$ number of particles. With this many of them, the density is such that the particles spend enough time within a few de Broglie wavelengths from one another, and hence we need to go beyond the single-particle description, that is, there is necessity to formulate a quantum many- body theory. The basic idea behind the approach is that instead of keeping track of a large number of strongly interacting particles, we can get away with a relatively smaller number of weakly interacting particles, called quasiparticles, or elementary excitations.

An elementary excitation occurs in a quantum system due to the application of an external perturbation. It is like a ripple on a pond, but in quantum theory these ripples are quantized. In crystal lattices, these ripples are caused by thermal effects and they are known as phonons, which carry both energy and quasi-momentum and weakly interact with each other. Phonons have finite (and reasonably large) lifetime, unlike the carriers, namely the electrons. However, if they decay faster than they are created, then the 'quasiparticle' description will not make much sense.

1.2 Jellium model

The electronic description of metals due to Bloch, Bethe and others in the 1930s neglects the electron–electron interaction, and in a vast number of cases, such a simplified description works. Even the distinction between a metal and an insulator can reliably be done by band-filling calculations, that is, without invoking the electron–electron interaction. Nevertheless, the band gaps in semiconductors and insulators are somewhat difficult to calculate quantitatively. However, modifications of the single-particle band structure, in the form Hartree–Fock corrections are computed, which is equivalent to computing the average energy shift in the single-particle energies in the presence of an average density due to all other electrons having been known. This yields the essence of the mean field theory. Such description has, by and large, been successful in explaining a host of material properties. Yet there are reasons for one to deliberate upon, and include the electron–electron interaction in studying the physics of materials. The salient ones are as follows:

1. the average inter-electron distance in a typical metal (which goes as $n^{-1/6}$, n being the density) is about 1 nm. For a small inter-particle distance, inclusion of interaction effects is indispensable.
2. There are physical examples, such as complex materials where significant deviations from the single-particle band theory are noted. The familiar examples are, transition metal oxides, cuprate superconductors, etc.

An extension of the free electron gas can be thought of via a jellium model where the electrons interact via Coulomb potential, and the overall charge neutrality of the system is maintained by a homogeneous positively charged background. We shall mainly concentrate on obtaining the dielectric function via standard many-body approaches, such as, Hartree and Hartree–Fock approximations, random phase approximation (RPA), etc. In the following we consider an electron liquid (because of the involvement of inter-particle interactions, we do not call it an electron gas any more) in 3D at $T = 0$, the so-called jellium model. Jellium is a prototype model for metals which is a uniform electron gas with positively charged background. Since the many-electron wavefunction can be solved using computational techniques, it is considered to be a convenient model for testing the characteristics of the density functionals.

1.2.1 The Hamiltonian

The Hamiltonian consists of terms that correspond to the kinetic energy of the electrons (H_{kin}), Coulomb interaction among the electrons ($H_{\text{e-e}}$), the interaction between the electrons and the background positive charge ($H_{\text{e-b}}$) which originates from the electrons interacting with the lattice vibrations, and finally a background energy ($H_{\text{b-b}}$) which is basically the Coulomb interaction between the positive (background) charges. An enumeration of different terms in the second quantized notation can be written as [5–7],

$$
\begin{aligned}
\mathcal{H}_{\text{kin}} &= \sum_{\mathbf{k},\sigma} \xi_{\mathbf{k}} c^{\dagger}_{\mathbf{k}\sigma} c_{\mathbf{k}\sigma} \\
\mathcal{H}_{\text{e-e}} &= \frac{1}{2\mathcal{V}} \sum_{\mathbf{k},\,\mathbf{k}',\,\mathbf{q}} V(\mathbf{q}) c^{\dagger}_{\mathbf{k}\sigma} c^{\dagger}_{\mathbf{k}'+\mathbf{q}\sigma'} c_{\mathbf{k}'\sigma'} c_{\mathbf{k}+\mathbf{q}\sigma} \\
\mathcal{H}_{\text{e-b}} &= e^2 \int d^3\mathbf{r}\, d^3\mathbf{r}' \frac{n(\mathbf{r})}{4\pi\epsilon_0 |\mathbf{r} - \mathbf{r}'|} \\
\mathcal{H}_{\text{b-b}} &= \frac{e^2}{2} \int d^3\mathbf{r}\, d^3\mathbf{r}' \frac{n^2(\mathbf{r})}{4\pi\epsilon_0 |\mathbf{r} - \mathbf{r}'|} \\
\mathcal{H} &= \mathcal{H}_{\text{kin}} + \mathcal{H}_{\text{e-e}} + \mathcal{H}_{\text{e-b}} + \mathcal{H}_{\text{b-b}}
\end{aligned}
\tag{1.1}
$$

where $\mathcal{V}$ is the volume, $c_{\mathbf{k}}$ ($c^{\dagger}_{\mathbf{k}}$) are the single-particle annihilation (creation) operators corresponding to momentum, $\mathbf{k}$, $\xi_{\mathbf{k}} = \epsilon_{\mathbf{k}} - \mu$, where ϵ_k denotes single-particle energies and μ being the chemical potential, the factor $\frac{1}{2}$ takes into double counting, and $n(\mathbf{r})$ is the density term whose Fourier transform is defined as,

$$n_{\mathbf{q}} = \frac{1}{\sqrt{\mathcal{V}}} \int d^3\mathbf{r} e^{-i\mathbf{q}.\mathbf{r}} n(\mathbf{r}) \\ = \frac{1}{\sqrt{\mathcal{V}}} \sum_{\mathbf{k}\sigma} c^{\dagger}_{\mathbf{k}\sigma} c_{\mathbf{k}+\mathbf{q}\sigma}. \tag{1.2}$$

It is somewhat odd to write some of the terms of the Hamiltonian $\mathcal{H}$ in equation (1.1) in real space, and others in momentum space. But with the Fourier transformation for the density term shown above, $\mathcal{H}$ can be fully written in the momentum space (by which we also implicitly assume that the electron liquid is homogeneous, and hence translational invariance holds). However, this will not hinder the discussion that is going to follow.

There is a subtle point that deserves a mention. Strictly speaking, a Fourier transform of the Coulomb potential cannot be done owing to a divergence of the integral[4]. However a screening term (which is also physically relevant owing to the screening effect from all other charges) of the form, $\frac{e^{-\lambda r}}{\epsilon_0 r}$ kills the divergence and yields $V(q) = \frac{e^2}{\epsilon_0 q^2}$ in the limit of zero screening ($\lambda \to 0$).

1.2.2 Hartree–Fock approximation

It may be noticed that all other terms excepting the $\mathcal{H}_{\text{e-e}}$ are quadratic in c-operators and hence can be diagonalized in the single-particle basis, however, $\mathcal{H}_{\text{e-e}}$ is quartic in c and hence needs an approximation to be at par with other terms for arriving at the solution. In a mean field approximation[5], the quartic term is decoupled as a sum of all possible quadratic terms, namely,

$$c^{\dagger}_p c^{\dagger}_q c_r c_s \simeq -\langle c^{\dagger}_p c_r\rangle c^{\dagger}_q c_s - \langle c^{\dagger}_q c_s\rangle c^{\dagger}_p c_r + \langle c^{\dagger}_p c_s\rangle c^{\dagger}_q c_r + \langle c^{\dagger}_q c_r\rangle c^{\dagger}_p c_s$$

where the signs are governed by anticommutation relation of the fermions. The above decoupling scheme replaces the interaction term $\mathcal{H}_{\text{e-e}}$ by,

$$\mathcal{H}^{\text{HF}}_{\text{e-e}} = \frac{1}{2\mathcal{V}} \sum_{\mathbf{k},\,\mathbf{k}',\,\mathbf{q}} \Big[- \langle c^{\dagger}_{\mathbf{k}\sigma} c_{\mathbf{k}'\sigma'}\rangle c^{\dagger}_{\mathbf{k}'+\mathbf{q}\sigma'} c_{\mathbf{k}+\mathbf{q}\sigma} - \langle c^{\dagger}_{\mathbf{k}'+\mathbf{q}\sigma'} c_{\mathbf{k}+\mathbf{q}\sigma}\rangle c^{\dagger}_{\mathbf{k}\sigma} c_{\mathbf{k}'\sigma'} \\ + \langle c^{\dagger}_{\mathbf{k}\sigma} c_{\mathbf{k}+\mathbf{q}\sigma}\rangle c^{\dagger}_{\mathbf{k}'+\mathbf{q}\sigma'} c_{\mathbf{k}'\sigma'} + \langle c^{\dagger}_{\mathbf{k}'+\mathbf{q}\sigma'} c_{\mathbf{k}'\sigma'}\rangle c^{\dagger}_{\mathbf{k}\sigma} c_{\mathbf{k}+\mathbf{q}\sigma} \Big]. \tag{1.3}$$

The above equation (1.3) can be split into two terms, one for $\mathbf{q} = 0$ and other for $\mathbf{q} \neq 0$, namely,

$$\mathcal{H}^{\text{HF}}_{\text{e-e}} = \mathcal{H}^{\text{HF}}_{\text{e-e}}(\mathbf{q} = 0) + \mathcal{H}^{\text{HF}}_{\text{e-e}}(\mathbf{q} \neq 0) \tag{1.4}$$

[4] $V(\mathbf{q}) = \frac{e^2}{\epsilon_0} \int_0^{\infty} \frac{1}{r} e^{i\mathbf{q}\cdot\mathbf{r}}$ is a divergent integral.

[5] This type of mean field decoupling was introduced by Weiss who replaced the magnetic exchange interaction of the form $\mathbf{S}_i \cdot \mathbf{S}_j$ by $\langle \mathbf{S}_i\rangle \mathbf{S}_j + \langle \mathbf{S}_j\rangle \mathbf{S}_i - \langle \mathbf{S}_i\rangle\langle \mathbf{S}_j\rangle$.

where,

$$
\begin{aligned}
\mathcal{H}_{\text{e-e}}^{\text{HF}}(\mathbf{q}=0) &= V(\mathbf{q}=0) \sum_{\mathbf{k}',\mathbf{k}'\sigma\sigma'} \langle c_{\mathbf{k}\sigma}^{\dagger} c_{\mathbf{k}\sigma} \rangle c_{\mathbf{k}'\sigma'}^{\dagger} c_{\mathbf{k}'\sigma'} \\
\mathcal{H}_{\text{e-e}}^{\text{HF}}(\mathbf{q}\neq 0) &= \sum_{\mathbf{k},\mathbf{q}\sigma} \left[V(\mathbf{q}) n_{\mathbf{k}-\mathbf{q}\sigma} \right].
\end{aligned}
\tag{1.5}
$$

The third and the fourth terms can be written as,

$$
\begin{aligned}
\mathcal{H}_{\text{e-b}} &= -V(\mathbf{q}\neq 0) n \mathcal{N} \\
\mathcal{H}_{\text{b-b}} &= \frac{\mathcal{V}}{2} V(\mathbf{q}\neq 0) n^2
\end{aligned}
\tag{1.6}
$$

where $n = \frac{\mathcal{N}}{\mathcal{V}}$ is the electron density. It may be noted that $V(\mathbf{q}=0)$ diverges since $V(\mathrm{q}) \sim 1/q^2$, and because of which, both $\mathcal{H}_{\text{e-b}}$ and $\mathcal{H}_{\text{b-b}}$ will diverge. However, that is not too much of a concern, as the charge neutrality implies that various infinities would cancel each other to ensure the finiteness of the total energy. Let us combine $\mathcal{H}_{\text{e-b}}$ and $\mathcal{H}_{\text{b-b}}$, which yields,

$$
\mathcal{H}_{\text{eb}} = \mathcal{H}_{\text{e-b}} + \mathcal{H}_{\text{b-b}} = V(\mathbf{q}=0)\left[\frac{\mathcal{V}}{2} n^2 - n\mathcal{N}\right] = \frac{V(\mathbf{q}=0)}{\mathcal{V}} \frac{\mathcal{N}^2}{2} = \frac{e^2}{2\lambda^2 \epsilon_0} \frac{\mathcal{N}^2}{\mathcal{V}}. \tag{1.7}
$$

Thus, $\mathcal{H}_{\text{eb}}$ vanishes in the thermodynamic limit owing the presence of $\mathcal{V}$ in the denominator. There is another interesting cancelation that is shown below.

Let us explicitly calculate the energy due to the background Hamiltonian, $\mathcal{H}_{\text{b-b}}$ (it is implicitly assumed that the eigenstates are known so that we can compute the energies), that is,

$$
E_{\text{b-b}} = \frac{e^2}{2} n^2 \int d^3\mathbf{r} d^3\mathbf{r}' \frac{1}{|\mathbf{r}-\mathbf{r}'|} e^{-\lambda|\mathbf{r}-\mathbf{r}'|}. \tag{1.8}
$$

Let us assume $\mathbf{r} - \mathbf{r}' = \mathbf{z}$, which upon substituting yields,

$$
E_{\text{b-b}} = \frac{e^2}{2} n^2 \int d^3z d^3\mathbf{r} \frac{1}{|z|} e^{-\lambda|z|} = \frac{e^2}{2} n^2 \mathcal{V} \int d^3z \frac{e^{-\lambda z}}{z} = \frac{e^2}{2} n^2 \mathcal{V} \frac{4\pi}{\lambda^2}. \tag{1.9}
$$

Thus the energy per particle is,

$$
\frac{E_{\text{b-b}}}{\mathcal{N}} = \frac{e^2}{2} \frac{4\pi}{\lambda^2} n. \tag{1.10}
$$

A similar calculation for the $E_{\text{e-b}}$ yields the form,

$$
\frac{E_{\text{e-b}}}{\mathcal{N}} = -e^2 \frac{4\pi}{\lambda^2} n. \tag{1.11}
$$

Also the $\mathbf{q} = 0$ term of the $\mathcal{H}_{\text{e-e}}$ term yields[6],

[6] This is called the **Hartree** term or the direct term (because of $q = 0$) in the electron–electron interaction. Also see chapter 5 for further discussions.

$$\frac{E_{\text{e-e}}(\mathbf{q}=0)}{\mathcal{N}} = \frac{e^2}{2}\frac{4\pi}{\lambda^2}n. \tag{1.12}$$

Thus, we have,

$$\frac{E_{\text{b-b}}}{\mathcal{N}} + \frac{E_{\text{e-b}}}{\mathcal{N}} + \frac{E_{\text{e-e}}(\mathbf{q}=0)}{\mathcal{N}} = 0. \tag{1.13}$$

Hence we are left with only the $\mathbf{q} \neq 0$ in the electron–electron interaction term which is called the *Fock* term or the *exchange* term (as opposed to the direct interaction, called the Hartree term), apart from the kinetic energy of the electrons.

Thus we get a Hartree–Fock Hamiltonian, $\mathcal{H}_{\text{HF}}$ (since both the direct and the exchange terms are included) which can be written as,

$$\mathcal{H}_{\text{HF}} = \sum_{\mathbf{k}\sigma}\left(\frac{\hbar^2 k^2}{2m} + \Sigma_{\text{HF}}(\mathbf{k})\right)c^{\dagger}_{\mathbf{k}\sigma}c_{\mathbf{k}\sigma} \tag{1.14}$$

where $\Sigma_{\text{HF}}(\mathbf{k})$ the '*self energy*' at the HF level and is written as,

$$\Sigma_{\text{HF}}(\mathbf{k}) = -\sum_{\mathbf{q}} n_{\mathbf{k}-\mathbf{q}\sigma}V(\mathbf{q}). \tag{1.15}$$

This is the only term that is relevant to us as it provides a momentum shift, that is, a $\mathbf{k}$-dependent correction to the non-interacting energy.

Here we have computed the self energy at the HF level. The self energy is in general a complex quantity, where the real part of it contributes to the total energy, while the imaginary part denotes the lifetime of the quasiparticles. Once the self energy (Σ) is computed, the total energy can be written as,

$$E_k = \epsilon_k + Re\ \Sigma(\mathbf{k}). \tag{1.16}$$

In general $\Sigma(\mathbf{k})$ may include contributions from the exchange, correlation, disorder, electron–phonon scattering, etc, we have only restricted ourselves to the exchange energy.

1.2.3 Hartree–Fock energy

Next we have to solve the Hartree–Fock (HF) Hamiltonian and obtain its eigensolutions. For this purpose, let us write down the Hamiltonian and the corresponding Schrödinger equation in real space with $(\psi_i(\mathbf{r}), \epsilon_i)$, namely,

$$\frac{\hbar^2}{2m}\nabla^2\psi_i(\mathbf{r}) - \frac{e^2}{2\mathcal{V}}\sum_j \int d^3\mathbf{r}' \frac{1}{|\mathbf{r}-\mathbf{r}'|}\psi_j^*(\mathbf{r}')\psi_i(\mathbf{r})\psi_j(\mathbf{r}) = \epsilon_i\psi_i(\mathbf{r}). \tag{1.17}$$

In the absence of a better estimate for the wavefunction, we assume a plane wave solution of the form, $\psi_i(\mathbf{r}) = \frac{1}{\sqrt{\mathcal{V}}}e^{i\mathbf{k}.\mathbf{r}}$, where a box normalization of the plane wave is assumed in a volume $\mathcal{V}$. Plugging this plane wave solution into the above equation (let us only concentrate on the interaction term as the kinetic energy term will trivially yield $\frac{\hbar^2 k^2}{2m}$),

$$\frac{e^2}{2\mathcal{V}}\sum_{\mathbf{k}_j}\int d^3\mathbf{r}'\frac{e^{-i(\mathbf{k}_j-\mathbf{k}_i).\mathbf{r}'}}{|\mathbf{r}-\mathbf{r}'|}\frac{1}{\sqrt{\mathcal{V}}}e^{i\mathbf{k}.\mathbf{r}}.$$

Replace $\mathbf{r} - \mathbf{r}' = \mathbf{z}$, which allows us to write,

$$\frac{e^2}{2\mathcal{V}}\sum_{\mathbf{k}_j}\int d^3\mathbf{z}\frac{e^{-i(\mathbf{k}_j-\mathbf{k}_i).\mathbf{z}}}{|\mathbf{z}|}\frac{1}{\sqrt{\mathcal{V}}}e^{i\mathbf{k}.\mathbf{r}}.$$

This confirms that the plane wave states indeed are the solutions of the HF Hamiltonian as just a coefficient consisting of

$$\frac{e^2}{2\mathcal{V}}\sum_{\mathbf{k}_j}\int d^3\mathbf{z}\frac{e^{-i(\mathbf{k}_j-\mathbf{k}_i).\mathbf{z}}}{|\mathbf{z}|}$$

gets multiplied with the plane wave states. This allows us to solve for the energy, ϵ_k as,

$$\begin{aligned}\epsilon_k &= \frac{\hbar^2 k^2}{2m} - \frac{e^2}{\mathcal{V}}\sum_{\mathbf{k}_j}\int d^3\mathbf{z}\frac{e^{-i(\mathbf{k}_j-\mathbf{k}_i).\mathbf{z}}}{|\mathbf{z}|}\\ &= \frac{\hbar^2 k^2}{2m} - e^2\sum_{\mathbf{k}'<\mathbf{k}_\mathrm{F}}\frac{4\pi}{|\mathbf{k}-\mathbf{k}'|^2}\\ &= \epsilon_k^0 + \Sigma_\mathrm{ex}(\mathbf{k}).\end{aligned} \tag{1.18}$$

Thus the single-particle energies are getting renormalized by the second term, which now can be solved by converting the sum into an integral,

$$\begin{aligned}\Sigma^\mathrm{ex}(\mathbf{k}) &= -\int\frac{d^3\mathbf{k}}{(2\pi)^3)}\frac{4\pi e^2}{|\mathbf{k}-\mathbf{k}'|^2}\\ &= -\frac{e^2}{\pi}\int_0^{k_\mathrm{F}}k'^2dk'\int_{-1}^{+1}\frac{d(\cos\theta)}{k^2+k'^2-2kk'\cos\theta}\\ &= -\frac{e^2}{\pi k}\int_0^{k_\mathrm{F}}k'dk'\ ln\left|\frac{k+k'}{k-k'}\right|\\ &= -\frac{e^2 k_\mathrm{F}}{\pi}\left[1+\frac{1-\alpha^2}{2\alpha}ln\left|\frac{1+\alpha}{1-\alpha}\right|\right] = -\frac{e^2 k_\mathrm{F}}{\pi}F(\alpha)\end{aligned} \tag{1.19}$$

where k_F is the Fermi wave vector and $\alpha = \frac{k}{k_\mathrm{F}}$. $F(\alpha)$ contains the wave vector dependence of the self energy. Here

$$F(\alpha) = 1 + \frac{1-\alpha^2}{2\alpha}ln\left|\frac{1+\alpha}{1-\alpha}\right|.$$

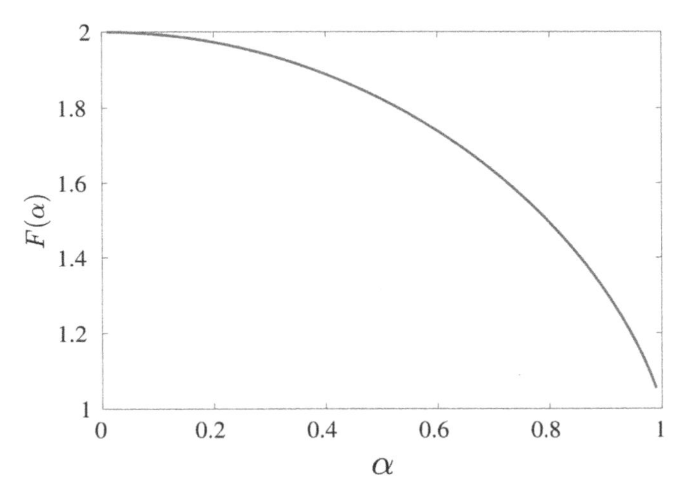

Figure 1.1. $F(\alpha)$ is plotted as a function of α. As $k \to k_{\rm F}$, $F(\alpha)$ smoothly decreases.

We plot $F(\alpha)$ as a function of α in figure 1.1. As k approaches $k_{\rm F}$, $F(\alpha)$ decreases indicating an increase in the value of the self energy, $\Sigma^{\rm ex}$ for electrons near the Fermi level.

Hence the ground state energy per particle is obtained after a straightforward calculation as,

$$E_g^{\rm ex} = \sum_k n_k \Sigma^{\rm ex}(\mathbf{k}) = \frac{(ek_{\rm F}^2/\pi)^2}{2\pi} \int_0^1 F(\alpha)d\alpha = -\frac{3}{4}\frac{e^2 k_{\rm F}}{\pi}. \tag{1.20}$$

1.3 Properties of the electron liquid

Let us introduce more familiar notations used to express the energies of an electron liquid. All the physical (measurable) properties are expressed in terms of a length scale, r_s which denotes the radius (in atomic units) of a sphere that encloses one unit of the electronic charge. The density of the system is written as,

$$n = \frac{3}{4\pi a_0^3 r_s^3}; \qquad a_0 \text{ is the Bohr radius.} \tag{1.21}$$

Also the density, n is related to the Fermi wave vector, $k_{\rm F}$ (for a non-interacting and non-relativistic electron gas in 3D) via,

$$n = \frac{k_{\rm F}^3}{3\pi^2} \tag{1.22}$$

thereby relating $k_{\rm F}$ to r_s via,

$$k_{\rm F} = \left(\frac{9\pi}{4}\right)^{1/3} \frac{1}{r_s} = \frac{1.9192}{r_s}. \tag{1.23}$$

This also gives the Fermi energy, ϵ_F as,

$$\epsilon_F = \frac{\hbar^2 k_F^2}{2m} = (k_F a_0)^2 \frac{\hbar^2}{2ma_0^2} = \frac{3.6832}{r_s} E_H \tag{1.24}$$

where E_H corresponds to the magnitude of the ground state energy of the hydrogen atom,

$$E_H = \frac{\hbar^2}{2ma_0^2} = 13.6 \text{ eV}.$$

Further the kinetic energy per particle can be denoted as,

$$\frac{E_{kin}}{N} = \frac{\langle H_{kin} \rangle}{N} = \frac{3}{5}\epsilon_F = \frac{3}{5}\left(\frac{9\pi}{4}\right)^{2/3} \frac{1}{r_s^2} = \frac{2.2099}{r_s^2} E_H. \tag{1.25}$$

Finally, the plasma frequency, ω_p (discussed later) which is related to the density of the electron liquid, n is written as,

$$\omega_p = \left(\frac{4\pi n e^2}{m}\right)^{1/2} = \left(\frac{12}{r_s^3}\right)^{1/2} E_H = \frac{3.4641}{r_s^{3/2}} E_H. \tag{1.26}$$

In the same spirit, we replace the Fermi wave vector, k_F by r_s in the expression for exchange energy, E_g^{ex}, so that it becomes,

$$E_g^{ex} = -\frac{0.9163}{r_s}. \tag{1.27}$$

This yields the total energy in the HF approximation in equation (1.20) as (see E_{kin} from above),

$$E_k = \frac{2.2099}{r_s^2} - \frac{0.9163}{r_s}. \tag{1.28}$$

Thus the total energy of the electron liquid is expressed in terms of the power series in r_s. A couple of more terms which account for the correlation energy, E_c (not discussed here) have been obtained by Gell-Mann and Brueckner [9] who showed E_c to have the form,

$$E_c = A \ ln r_s + C + \mathcal{O}(r_s) \tag{1.29}$$

where A and C are constants. If we include this, the total energy becomes,

$$E_g = \frac{2.2099}{r_s^2} - \frac{0.9163}{r_s} + E_c. \tag{1.30}$$

The above expression is quite reliable in the limit $r_s \to 0$, that is at large densities.

1.3.1 Effective mass

We write the parabolic energy dispersion for the non-interacting electrons as,

$$\epsilon_k = \frac{\hbar^2 k^2}{2m} \tag{1.31}$$

where m denotes the mass of the electrons. In some cases, such as in semiconductors, the inter-particle interaction effects (which cannot be calculated exactly any way) are included in the effective mass, m^*, such as,

$$E_k = \frac{\hbar^2 k^2}{2m^*}. \tag{1.32}$$

Comparing equations (1.31) and (1.32), one can write an expression for the ratio of the bare mass to the effective mass as,

$$\frac{m}{m^*} = \frac{\partial E_k}{\partial \epsilon_k}. \tag{1.33}$$

Hence taking the derivative of equation 1.16,

$$\frac{\partial E_k}{\partial \epsilon_k} = \lim_{\epsilon_k \to 0}\left[1 + \frac{\partial}{\partial \epsilon_k} Re\ \Sigma(\mathbf{k}) + \frac{\partial}{\partial E_k} Re\ \Sigma(\mathbf{k}) \frac{\partial E_k}{\partial \epsilon_k}\right]. \tag{1.34}$$

Collecting the terms containing $\frac{\partial E_k}{\partial \epsilon_k}\left(= \left(\frac{m}{m^*}\right)\right)$, one gets,

$$\frac{m}{m^*} = \lim_{\epsilon_k \to 0} \frac{1 + \frac{\partial}{\partial \epsilon_k} Re\ \Sigma(\mathbf{k})}{1 + \frac{\partial}{\partial E_k} Re\ \Sigma(\mathbf{k})}. \tag{1.35}$$

Thus the effective mass is obtained as the derivative of the self energy. In the present case, that is at the HF level,

$$\begin{aligned}
\frac{m}{m^*} &= 1 + \frac{\partial}{\partial \epsilon_k} Re\ \Sigma^{\text{ex}}(\mathbf{k}) \\
&= \frac{me^2}{\pi k} \frac{d}{d\alpha} F(\alpha) \\
&= \frac{me^2}{2\pi k_{\text{F}} \alpha^2} \left(\frac{1 + \alpha^2}{\alpha} ln \left| \frac{1 + \alpha}{1 - \alpha} \right| - 2 \right).
\end{aligned} \tag{1.36}$$

The effective mass diverges at the Fermi energy, that is, for $\alpha \to 1$ (see figure 1.2). This implies that the liquid becomes unstable, and has observable consequences at low temperatures. For example, it is well established that at low temperatures, the ground state becomes unstable to a long-range order, such as a magnetic order, or develop superconducting correlations. In fact there are also instances where such instability is absent, which implies that the divergence of the $\frac{m}{m^*}$ gets regularized.

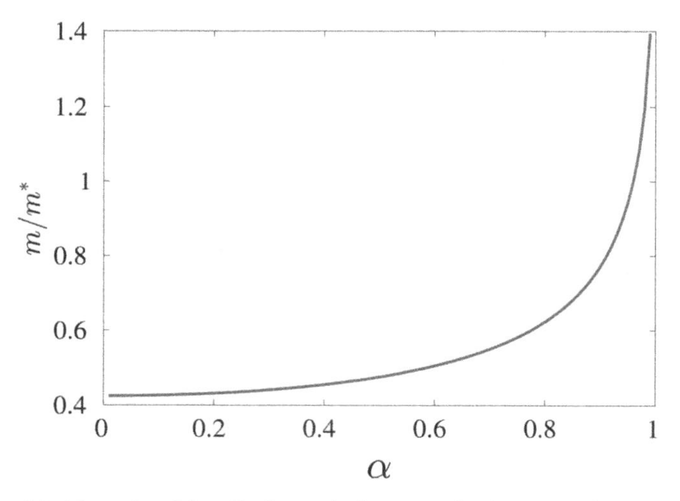

Figure 1.2. The ratio of the effective to the bare mass is plotted as a function of α.

This in turn means the regularization of the specific heat (remember specific heat is the second derivative of the energy with respect to temperature) via the presence of the correlation energies.

1.3.2 Magnetic properties

It is known that magnetism does not arise from interaction between the magnetic moments, but from the exchange part of the Coulomb interaction. In fact the exchange interaction (Heitler–London type) led us to the spin-only Heisenberg model or the itinerant Hubbard model which yields a good semi-quantitative basis for describing ferromagnetism in rare earth compounds. However, to describe the itinerant ferromagnetism in the transition metals, it is appropriate to resort to the electron liquid, that is, the carriers are moving in the background of positively charged ions, or the jellium model. In the non-interacting version of it, the kinetic energy of the carriers compete with the ordering energy, which is needed to align (or anti-align) the spins in presence of an external field. This results in a net effect which is weak and vanishes in the limit of zero external field. Such a scenario where the magnetic phenomena are absent is referred to as a non-magnetic or a Pauli paramagnetic behaviour of the electron liquid.

Now when electron–electron interaction is included, the spin alignment is mediated by the exchange interaction and can even be strong enough to sustain the order without any external magnetic field. To remind ourselves, the HF energy per electron (without taking into spins) is,

$$\left(\frac{E}{\mathcal{N}}\right)_{\mathrm{HF}} = \frac{30.1}{(r_s/a_0)^2}\ \mathrm{eV} - \frac{12.5}{(r_s/a_0)}\ \mathrm{eV}. \tag{1.37}$$

To discuss magnetic behaviour, we need to invoke the role of spins. A straightforward and efficient way to do this is writing down the HF energy in terms of the spin-dependent particle densities. Introducing the total density, $n\left(=\frac{\mathcal{N}}{\mathcal{V}}\right)$ explicitly in the above expression,

$$\left(\frac{E}{\mathcal{N}}\right)_{\rm HF} = \mathcal{N}\left[78.2\left(\frac{\mathcal{N}}{\mathcal{V}}\right)^{2/3}{\rm eV} - 20.1\left(\frac{\mathcal{N}}{\mathcal{V}}\right)^{1/3}{\rm eV}\right] \tag{1.38}$$

where we have used,

$$\frac{r_s}{a_0} = \left[\frac{3}{4\pi}\left(\frac{\mathcal{V}}{\mathcal{N}}\right)\right]^{1/3}$$

Introducing the number of $\uparrow$- ($\mathcal{N}_\uparrow$) and $\downarrow$- ($\mathcal{N}_\downarrow$) spins,

$$\begin{aligned}\left(\frac{E}{\mathcal{N}}\right)_{\rm HF}(\mathcal{N}_\uparrow, \mathcal{N}_\downarrow) &= \left(\frac{E}{\mathcal{N}}\right)_{\rm HF}\mathcal{N}_\uparrow + \left(\frac{E}{\mathcal{N}}\right)_{\rm HF}\mathcal{N}_\downarrow \\ &= \left[78.2{\rm eV}\left(\frac{\mathcal{N}_\uparrow}{\mathcal{V}}\right)^{2/3} + \left(\frac{\mathcal{N}_\downarrow}{\mathcal{V}}\right)^{2/3}\right] - \left[20.1{\rm eV}\left(\frac{\mathcal{N}_\uparrow}{\mathcal{V}}\right)^{1/3} + \left(\frac{\mathcal{N}_\downarrow}{\mathcal{V}}\right)^{2/3}\right]\end{aligned} \tag{1.39}$$

with the condition that $\mathcal{N}_\uparrow + \mathcal{N}_\downarrow = \mathcal{N}$. It is customary to introduce the spin polarization operator, namely,

$$P = \frac{\mathcal{N}_\uparrow - \mathcal{N}_\downarrow}{\mathcal{N}} \tag{1.40}$$

which allows us to write the number of spin-polarized particles as,

$$\mathcal{N}_\uparrow = \frac{\mathcal{N}}{2}(1 + P); \qquad \mathcal{N}_\downarrow = \frac{\mathcal{N}}{2}(1 - P). \tag{1.41}$$

This facilitates writing down the HF energy in terms of P as,

$$\begin{aligned}\left(\frac{E}{\mathcal{N}}\right)_{\rm HF} = \mathcal{N}\Bigg[&\frac{1}{2}\{(1 + P)^{5/3} + (1 - P)^{5/3}\} \\ &- \frac{1}{8}\left(\frac{\mathcal{V}}{\mathcal{N}}\right)^{1/3}\{(1 + P)^{4/3} + (1 - P)^{4/3}\}\Bigg].\end{aligned} \tag{1.42}$$

Let us now explore non-zero polarization ($P \neq 0$), or in other words non-vanishing spontaneous magnetization. Such a scenario will depend upon the sign of $E(\mathcal{N}, P) - E(\mathcal{N}, 0)$ (say $= \Delta E$). If $\Delta E < 0$, then spontaneous magnetization exists, and hence a ferromagnetic state is stabilized. Thus,

$$\frac{\Delta E(P)}{\mathcal{N}} = \frac{1}{2}\left[\{(1 + P)^{5/3} + (1 - P)^{5/3}\} - \frac{1}{8}\beta\{(1 + P)^{4/3} + (1 - P)^{4/3}\} - \mathcal{N}(1 - \frac{\beta}{4})\right] \tag{1.43}$$

where β is related to the inverse of density, via, $\beta = \left(\frac{\mathcal{V}}{\mathcal{N}}\right)^{1/3}$. $\frac{\Delta E(P)}{\mathcal{N}}$ plotted as a function of P is shown below (see figure 1.3) for a few values of β which shows a

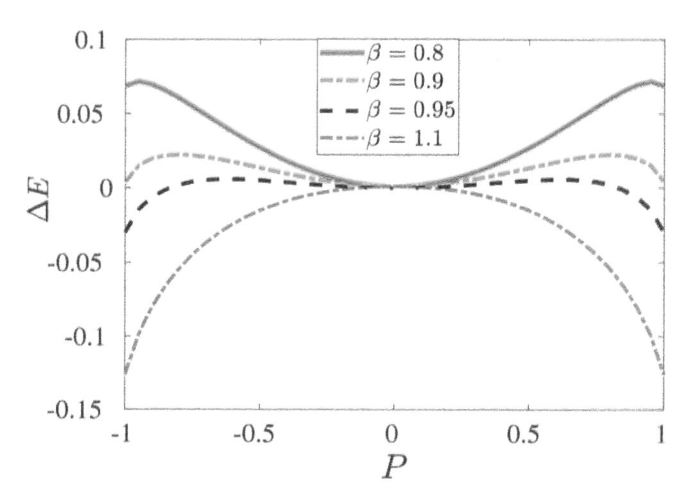

Figure 1.3. The change in energy per particle is shown as a function of the polarization, P. The magnetic state stabilizes for large β.

change in sign at a critical density given by $\beta_c = \frac{2}{5}(1 + \sqrt[3]{2}) \simeq 0.9$. This value of β_c translates to a density that corresponds to $r_s/a_0 \geqslant 5.5$. Clearly this number is related to the density of the electron liquid (or rather the inverse of it) and hence the critical density for ferromagnetic order to occur should be low enough such that the Wigner–Seitz radii (r_s should at least be $\sim 5.5a_0$. However, as we shall see, this is a too stringent criterion, and is only successfully cleared by cesium (Cs). Even the best known ferromagnets, such as Fe, Ni and Co do not meet the criterion.

Evidently, the premise on which these inferences are based is that the HF approximation is disconnected from reality. In fact the maximum density at which ferromagnetism can occur given by $r_s/a_0 \geqslant 5.5$ is short by one order of magnitude. It has been found by Zhong *et al* [13] that the upper limit of density corresponds to $r_s/a_0 \simeq 50$. The reason for such a large discrepancy lies in the jellium model itself. While the jellium model correctly assesses that completely free electrons (as in metals) cannot result in ferromagnetic correlations, it fails to account for the best known ferromagnets, such as, Fe, Ni and Co to have ferromagnetic ordering. A very important fact that the jellium model misses is the combination of the band structure effects along with the exchange interaction.

1.3.3 Screening and dielectric function

The development of the theory of an interacting electron gas occurred in close analogy with classical plasma, which is a highly ionized gas of electrons and positive charges formed at large temperatures and usually at low densities. Plasma can be produced in labs by irradiating metal surfaces by lasers. It is also found in the ionosphere that surrounds the Earth. A typical model suitable for investigating the properties of an electron gas is plasma where the positive charges are replaced by a uniform positive background. The only difference is that we are at the opposite limit, that is, in the regime of low temperature and large density. Thus, along with enforcing Pauli exclusion principle for the electrons, we are indeed talking about a quantum plasma which demonstrates collective behaviour that might have been expected from the long-range Coulomb interaction between the electrons. Such a collective behaviour gets manifested mainly through screening and collective oscillations.

To understand the concept of screening in an electron liquid, we may place a potential impurity, that is, an infinite scatterer in an electron liquid. Considering this as a perturbation, the electrons will tend to gather around the scatterer if it is positively charged, and will be repelled if it is negatively charged. Further, the electron charge density will oscillate far away from the scatterer. This is precisely a model for impurities in a solid. The mobile electrons view the impurities as potential scatterers. The impurities may have different valencies from that of the host, which will induce a localized charge, although the overall charge neutrality for the solid is retained.

As an example, consider the fluctuations in the local density due to the presence of the impurity denoted by $\delta n(\mathbf{r}', t')$, where $\mathbf{r}'$ and t' denote the space–time coordinate at any point other than at the location of the impurity. The density is given by, $n(\mathbf{r}, t) = \langle \psi^\dagger(\mathbf{r}, t)\psi(\mathbf{r}, t)\rangle$. Suppose the perturbation (due to the impurity potential) is turned on at $t = 0$. As a well-known technique in perturbation theory, the time dependence of the expectation values is obtained in terms of the matrix elements of the unperturbed Hamiltonian, $\mathcal{H}_0$ which, in principle can include inter-particle interactions (and does not include the potential scatterer).

Let us focus on the time evolution of the density operator $\rho(t)$[7], whose equation of motion (EOM) is given by,

$$i\frac{\partial \rho(t)}{\partial t} = \left[\mathcal{H}, \rho(t)\right], \quad (\hbar = 1), \tag{1.44}$$

where,

$$\mathcal{H} = \mathcal{H}_0 + \mathcal{H}'.$$

In order to make sure that the operators are correctly written in the Heisenberg representation (so that the time dependence of $\mathcal{H}'(t)$ is properly addressed), one needs a canonical transformation of the form,

$$\tilde{\rho}(t) = U(t)\rho(t)U^\dagger(t) \tag{1.45}$$

where $U(t) = e^{i\mathcal{H}_0 t}$ is a unitary transformation. Replacing equation (1.44) with $\tilde{\rho}$ yields,

$$i\frac{\partial \tilde{\rho}(t)}{\partial t} = \left[\tilde{\mathcal{H}}', \tilde{\rho}(t)\right] \tag{1.46}$$

where $\tilde{\mathcal{H}}' = U\mathcal{H}'U^\dagger$. The iterative solution can be written as,

[7] $\rho(t)$ should not be mistaken with the density, ρ. You may use $\hat{\rho}(t)$ to distinguish it from density.

$$\tilde{\rho}(t) = \tilde{\rho}(0) - i\int_0^t \left[\tilde{\mathcal{H}}, \tilde{\rho}(0)\right] dt' \tag{1.47}$$

where $\tilde{\rho}(0) = U(0)\rho(0)U^{\dagger}(0)$. Using equation (1.45),

$$\rho(t) = \rho(0) - iU^{\dagger}(t)\left\{\int_0^t \left[\tilde{\mathcal{H}}', \tilde{\rho}(0)\right] dt'\right\} U(t). \tag{1.48}$$

We now take an average using,

$$\langle A \rangle = Tr(\rho(t)A) \tag{1.49}$$

where ρ plays the role of the density matrix and A is an arbitrary operator. Here Tr denotes the trace of the matrix $\rho(t)A$. Using the above definition of $\rho(t)$, one can write,

$$\begin{aligned} \langle A \rangle &= Tr(\rho(0)A) - i\left\{\int_0^t dt' U^{\dagger}(t)[\tilde{\mathcal{H}}', \rho(0)]U(t)A\right\} \\ &= Tr(\rho(0)A) - i\left\{\int_0^t dt' U^{\dagger}(t)[\tilde{\mathcal{H}}', A_{\mathrm{H}}(t)]\right\} \end{aligned} \tag{1.50}$$

where A_{H} is the operator in the Heisenberg representation and is related to A by, $A_{\mathrm{H}} = UAU^{\dagger}$. Thus, the second term in the RHS of equation (1.50) yielding the renormalization of the expectation value due to the perturbation (the impurity potential) is,

$$\delta\langle A(\mathbf{r}, t)\rangle = -i\int_0^t dt' Tr(\rho(0)[\tilde{\mathcal{H}}', A_{\mathrm{H}}(\mathbf{r}', t')]). \tag{1.51}$$

In the linear response regime, that is when the perturbation is not too strong and it can be written as,

$$\mathcal{H}'(t) = \int d^3\mathbf{r} \; O(\mathbf{r})\phi(\mathbf{r}, t) \tag{1.52}$$

such that,

$$\tilde{\mathcal{H}}'(t) = \int d^3\mathbf{r} \; O_{\mathrm{H}}(\mathbf{r})\phi(\mathbf{r}, t). \tag{1.53}$$

This allows us to write,

$$\delta\langle A(\mathbf{r}, t)\rangle = -i\int_0^t dt' \int d^3\mathbf{r}' Tr(\rho(0)[O_{\mathrm{H}}(\mathbf{r}, t), A_{\mathrm{H}}(\mathbf{r}', t')])\phi(\mathbf{r}', t'). \tag{1.54}$$

For the Hamiltonian, one may consider $\mathcal{H} = \mathcal{H}_0 + \mathcal{H}'$, where $\mathcal{H}_0$ is the host Hamiltonian and $\mathcal{H}'$ is the perturbation term due to the presence of the impurity. The response of the liquid to this perturbation depicted by the impurity potential is given by,

$$\chi(\mathbf{r}, t, \mathbf{r}', t') = -i\langle[n(\mathbf{r}, t), n(\mathbf{r}', t')]\rangle_{\mathcal{H}'=0}\theta(t - t') \tag{1.55}$$

where the expectation value is taken with respect to the host material ($\mathcal{H} = \mathcal{H}_0$) and we are interested in the response of the system for $t > t'$.

A simple way of handling this problem is to seek the form of the potential function (potential energy per unit charge) at large distances from the impurity in the (interacting) medium. This potential is the superposition of the potential due to the impurity charge in vacuum, and the potential that is induced by the polarization charges in the medium. Let us denote the former by ϕ^{ext} and the latter by ϕ^{ind}. Also, the total potential that arises due to the superposition of the two is ϕ^{total}, that is,

$$\phi^{\text{total}}(\mathbf{r}, t) = \phi^{\text{ext}}(\mathbf{r}, t) + \phi^{\text{ind}}(\mathbf{r}, t).$$

Classical electrodynamics dictates that ϕ^{ext} is proportional to the ϕ^{total} with the dielectric constant, ϵ being a proportionality constant. In the Fourier space, this proportionality is written as,

$$\phi^{\text{ext}}(\mathbf{q}, \omega) = \epsilon(\mathbf{q}, \omega)\phi^{\text{total}}(\mathbf{q}, \omega) \tag{1.56}$$

$\epsilon(\mathbf{q}, \omega)$ captures the response of the system to the perturbation created by the impurity potential. Further, the induced potential can be obtained by employing Poisson's equations[8],

$$\nabla^2\phi^{\text{ext}} = 4\pi\rho^{\text{ind}} \quad \text{(in Gaussian unit)} \tag{1.57}$$

where ρ^{ind} is the charge density induced by the impurity potential.

Equipped with all of the above we can write,

$$\rho^{\text{ind}}(\mathbf{r}, t) = ie^2 \int_0^t dt' \int d^3\mathbf{r}' \,\text{Tr}\,(\rho(0)[n_{\text{H}}(\mathbf{r}, t), n_{\text{H}}(\mathbf{r}', t')])\phi^{\text{ext}}(\mathbf{r}', t) \tag{1.58}$$

where $n_{\text{H}}(\mathbf{r}, t)$ is the density operator in the Heisenberg representation which can be expressed as, $n_{\text{H}}(\mathbf{r}, t) = U(t)n(\mathbf{r}, t)U^{\dagger}(t)$. Introducing the susceptibility, $\chi(\mathbf{r}, t, \mathbf{r}', t')$ which denotes the density–density correlations via,

$$\chi(\mathbf{r}, t, \mathbf{r}', t') = -iTr(\rho(0)[n_{\text{H}}(\mathbf{r}, t), n_{\text{H}}(\mathbf{r}', t')])\theta(t - t') \tag{1.59}$$

which yields,

$$\rho^{\text{ind}}(\mathbf{r}, t) = e^2 \int \chi(\mathbf{r}, t, \mathbf{r}', t')\phi^{\text{ext}}(\mathbf{r}, t)d^3\mathbf{r}dt. \tag{1.60}$$

For a uniform system, Fourier transform of the above equations yields,

$$\rho^{\text{ind}}(\mathbf{q}, \omega) = e^2\chi(\mathbf{q}, \omega)\phi^{\text{ext}}(\mathbf{q}, \omega). \tag{1.61}$$

[8] The available texts on the subject usually deal in Gaussian unit. Thus to be compatible with these and a benefit to readers, we shall stick to Gaussian unit here. In SI unit, this equation can be written as, $\nabla^2\phi^{\text{ext}} = \frac{\rho^{\text{ind}}}{\epsilon}$.

Comparing it with Poisson's equation,

$$\phi^{\text{ind}} = \frac{4\pi\rho^{\text{ind}}}{q^2}. \tag{1.62}$$

Hence,

$$\begin{aligned}\phi^{\text{ind}}(\mathbf{q}\omega) &= \frac{4\pi\rho^{\text{ind}}}{q^2}\chi(\mathbf{q}, \omega)\phi^{\text{ext}}(\mathbf{q}, \omega) \\ &= V(\mathbf{q})\chi(\mathbf{q}, \omega)\phi^{\text{ext}}(\mathbf{q}, \omega).\end{aligned} \tag{1.63}$$

Thus the dielectric response of the system is obtained as,

$$\epsilon(\mathbf{q}, \omega) = \frac{\phi^{\text{ext}}(\mathbf{q}, \omega)}{\phi^{\text{ind}}(\mathbf{q}, \omega) + \phi^{\text{ext}}(\mathbf{q}, \omega)} = \frac{1}{1 + V(\mathbf{q}, \omega)\chi(\mathbf{q}, \omega)}. \tag{1.64}$$

It is the formal definition of the dielectric function which still requires the knowledge of the susceptibility, $\chi(\mathbf{q}, \omega)$ arising out of the density–density correlations.

Calculation of $\chi(\mathbf{q}, \omega)$ is indeed a daunting task for a fully interacting system because of the difficulty in calculating density–density correlations at two different space–time points. However, the problem is not too hard for a free Fermi gas, Since we are interested in investigating the effects of an impurity potential, whose effect has been included in ϕ^{ext}, ϕ^{ind} and $V(\mathbf{q})$, we can go ahead and compute $\chi(\mathbf{q}, \omega)$ in the non-interacting limit, that is, $\chi_0(\mathbf{q}, \omega)$. To obtain $\chi_0(\mathbf{q}, \omega)$ we have to calculate the expectation value of the commutator,

$$\langle[n_{\text{H}}(\mathbf{r}, t), n_{\text{H}}(\mathbf{r}', t)]\rangle = Tr(\rho(0)[n_{\text{H}}(\mathbf{r}, t), n_{\text{H}}(\mathbf{r}', t)]).$$

The subscript H denotes Heisenberg representation. $n_{\text{H}}(\mathbf{r}, t) = \psi^\dagger(\mathbf{r}, t)\psi(\mathbf{r}, t)$ where $\psi(\mathbf{r}, t)$ is the field operator. In the Fourier space, single-particle operators can be obtained using,

$$c_{\mathbf{k}}(t) = \int d^3\mathbf{r} e^{-i\mathbf{k}.\mathbf{r}}\psi(\mathbf{r}, t), \quad c_{\mathbf{k}}^\dagger(t) = \int d^3\mathbf{r} e^{i\mathbf{k}.\mathbf{r}}\psi^\dagger(\mathbf{r}, t). \tag{1.65}$$

Writing in terms of these c-operators one gets,

$$\begin{aligned}\text{Tr}\,(\rho(0)[n_{\text{H}}(\mathbf{r}, t), n_{\text{H}}(\mathbf{r}', t)]) &= \sum_{\mathbf{k}_1,\mathbf{k}_2,\mathbf{k}_3,\mathbf{k}_4} e^{i(\mathbf{k}_1-\mathbf{k}_2).\mathbf{r}} e^{i(\mathbf{k}_3-\mathbf{k}_4).\mathbf{r}'} \\ &\quad \left\langle\left[c_{\mathbf{k}_1}^\dagger(t)c_{\mathbf{k}_2}(t), c_{\mathbf{k}_3}^\dagger(t')c_{\mathbf{k}_4}(t')\right]\right\rangle \\ &= \sum_{\mathbf{k}_1,\mathbf{k}_2,\mathbf{k}_3,\mathbf{k}_4} e^{i(\mathbf{k}_1-\mathbf{k}_2).\mathbf{r}} e^{i(\mathbf{k}_3-\mathbf{k}_4).\mathbf{r}'} e^{i(\xi_{\mathbf{k}_3}-\xi_{\mathbf{k}_4}).t'} e^{i(\xi_{\mathbf{k}_1}-\xi_{\mathbf{k}_2}).t} \\ &\quad \left\langle\left[c_{\mathbf{k}_1}^\dagger c_{\mathbf{k}_2}, c_{\mathbf{k}_3}^\dagger c_{\mathbf{k}_4}\right]\right\rangle\end{aligned} \tag{1.66}$$

where $c_{\mathbf{k}}(t) = c_{\mathbf{k}} e^{-i\xi_{\mathbf{k}}t}$ and $c_{\mathbf{k}}^\dagger(t) = c_{\mathbf{k}}^\dagger e^{+i\xi_{\mathbf{k}}t}$.

Let us explicitly show the calculation of the commutator, namely,

$$\left[c_{\mathbf{k}_1}^\dagger c_{\mathbf{k}_2}, c_{\mathbf{k}_3}^\dagger c_{\mathbf{k}_4}\right] = c_{\mathbf{k}_1}^\dagger c_{\mathbf{k}_2} c_{\mathbf{k}_3}^\dagger c_{\mathbf{k}_4} - c_{\mathbf{k}_3}^\dagger c_{\mathbf{k}_4} c_{\mathbf{k}_1}^\dagger c_{\mathbf{k}_2}.$$

The above commutator has terms with equal number of creation and annihilation operators. Using Wick's theorem (see chapter 5 for details),

$$\left[c_{\mathbf{k}_1}^{\dagger}c_{\mathbf{k}_2}, c_{\mathbf{k}_3}^{\dagger}c_{\mathbf{k}_4}\right] = \delta_{\mathbf{k}_2,\mathbf{k}_3}c_{\mathbf{k}_1}^{\dagger}c_{\mathbf{k}_4}^{\dagger} - \delta_{\mathbf{k}_1,\mathbf{k}_4}c_{\mathbf{k}_3}c_{\mathbf{k}_2}.$$

Thus the commutator yields,

$$\sum_{\mathbf{k}_1,k_2} e^{i\{\mathbf{k}_1.(r-r')+\mathbf{k}_2.(r-r')\}}e^{i(\xi_{\mathbf{k}_2}-\xi_{\mathbf{k}_1})(t-t')}\langle n_{\mathbf{k}_2}\rangle - \sum_{\mathbf{k}_3,k_4} e^{i\{\mathbf{k}_4.(r-r')+\mathbf{k}_3.(r-r')\}}e^{i(\xi_{\mathbf{k}_3}-\xi_{\mathbf{k}_4})(t-t')}\langle n_{\mathbf{k}_4}\rangle.$$

Using the integral expression for the Θ-function,

$$\Theta(t - t') = -\int \frac{d\omega}{2\pi}\frac{e^{-i\omega(t-t')}}{\omega + i\eta}$$

where η is a positive definite quantity. In the Fourier space,

$$\chi_0(\mathbf{q}, \omega) = \int \chi_0(\mathbf{r}, t, \mathbf{r}', t')e^{i\mathbf{q}\cdot(\mathbf{r}-\mathbf{r}')}e^{i\omega(t-t')}d^3\mathbf{r}d^3\mathbf{r}'dtdt'$$

which yields using $\langle n_{\mathbf{k}}\rangle = n(\xi_{\mathbf{k}})$ ($n(\xi_{\mathbf{k}})$ is the Fermi-distribution function),

$$\chi_0(\mathbf{q}, \omega) = \sum_{\mathbf{k}} \frac{n(\xi_{\mathbf{k}}) - n(\xi_{\mathbf{k}+\mathbf{q}})}{\omega - (\xi_{\mathbf{k}+\mathbf{q}} - \xi_{\mathbf{k}}) + i\eta}. \tag{1.67}$$

This is called the Lindhard susceptibility of a free Fermi gas. It describes particle–hole excitations where a particle at an energy $\xi_{\mathbf{k}}$ from below the Fermi energy ($\xi = 0$) is promoted as a hole with energy $\xi_{\mathbf{k}+\mathbf{q}}$ by supplying an energy ω. χ_0 has a pole at $\omega = \xi_{\mathbf{k}+\mathbf{q}} - \xi_{\mathbf{k}}$.

However, usage of the free susceptibility, $\chi_0(\mathbf{q}, \omega)$ in place of $\chi(\mathbf{q}, \omega)$ ignores Coulomb interaction. However, there is one technique which can incorporate interaction effects by replacing the induced charges without interaction by the screened charge, which implies replacing $\chi_0 \to \frac{\chi_0}{\epsilon}$. Putting it in equation (1.67),

$$\epsilon(\mathbf{q}, \omega) = \frac{1}{1 + V(\mathbf{q})\dfrac{\chi_0(\mathbf{q}, \omega)}{\epsilon(\mathbf{q}, \omega)}}. \tag{1.68}$$

This yields an equation for the dielectric constant, $\epsilon(\mathbf{q}, \omega)$,

$$\epsilon(\mathbf{q}, \omega) = 1 - V(\mathbf{q}, \omega)\chi_0(\mathbf{q}, \omega). \tag{1.69}$$

The above approximation is known as the self-consistent field (SCF) approximation. It is also known by other names, such as, Lindhard approximation for the dielectric function and random phase approximation (RPA) for reasons discussed below.

Let us summarize the discussion held so far. Physically, the susceptibility $\chi(\mathbf{q}, \omega)$ denotes the change in density of a system in response to an external space and time dependent potential, $V(\mathbf{r}, t)$ which may be (and in this case is) from an impurity potential. In interacting systems, the screening reduces the effective interaction between a pair of charges due to the motion of other charges. In this context it is useful to invoke the dielectric constant and relate it to the change in density as a reaction to a change in the screened potential, which comprises the external potential and the modification of the potential due to other electrons.

The relationship between $\chi(\mathbf{q}, \omega)$ and the non-interacting version of it, that is, $\chi_0(\mathbf{q}, \omega)$ can be conveniently demonstrated by the diagrammatic perturbation theory for which we have given an exposure elsewhere (see chapter 5). An expansion of $\chi(\mathbf{q}, \omega)$ in terms of the connected particle–hole diagrams can be shown as in figure 1.4.

Here the continuous lines denote single-particle propagators, the wiggly line stands for the Coulomb interaction, namely, $V(q) = \frac{4\pi e^2}{q^2}$ and the dots where the lines meet denote particle–hole being created and destroyed, that is, they denote density fluctuations. Ignoring all diagrams on the RHS, except the first one yields $\chi(\mathbf{q}, \omega) \simeq \chi_0(\mathbf{q}, \omega)$, which is the RPA.

Let us now examine the two extreme limits for the non-interacting susceptibility, namely,

(i) $\omega \ll |\xi_{\mathbf{k}+\mathbf{q}} - \xi_{\mathbf{k}}|$,

(ii) $\omega \gg |\xi_{\mathbf{k}+\mathbf{q}} - \xi_{\mathbf{k}}|$.

Let us discuss (i). For small $\mathbf{q}$, $|\xi_{\mathbf{k}+\mathbf{q}} - \xi_{\mathbf{k}}| \simeq v_F \mathbf{k}.\,\mathbf{q}$ ($\hbar = 1$) where the $\mathbf{q}^2$ term has been neglected and $v_F = k_F/m$ is the Fermi velocity. Rewriting $\chi_0(\mathbf{q}, \omega)$ from equation (1.67),

$$\chi_0(\mathbf{q}, \omega) = \sum_{\mathbf{k}} n(\xi_{\mathbf{k}}) \left[\frac{1}{\omega - \xi_{\mathbf{k}+\mathbf{q}} + \xi_{\mathbf{k}} + i\eta} - \frac{1}{\omega + \xi_{\mathbf{k}+\mathbf{q}} - \xi_{\mathbf{k}} + i\eta} \right] \tag{1.70}$$

where $f(\xi_{\mathbf{k}+\mathbf{q}})$ is written in terms of $f(\xi_{\mathbf{k}})$ where $\mathbf{k} \to -(\mathbf{k} + \mathbf{q})$ has been replaced in the former. Also the sum over $\mathbf{k}$ implies the occupied $\mathbf{k}$-states, that is $|\mathbf{k}| < k_F$. This yields,

$$\begin{aligned} \chi_0(\mathbf{q}, 0) &= \sum_{|\mathbf{k}|<k_F} \left[\frac{1}{\omega - \xi_{\mathbf{k}+\mathbf{q}} + \xi_{\mathbf{k}} + i\eta} - \frac{1}{\omega + \xi_{\mathbf{k}+\mathbf{q}} - \xi_{\mathbf{k}} + i\eta} \right] \\ &= 2\mathcal{P} \sum_{|\mathbf{k}|<k_F} \frac{1}{\xi_{\mathbf{k}} - \xi_{\mathbf{k}+\mathbf{q}}} \end{aligned} \tag{1.71}$$

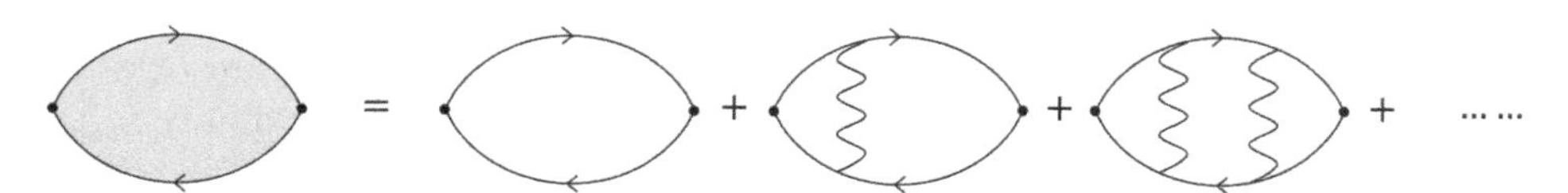

Figure 1.4. The renormalized propagator is shown as a sum of particle–hole diagrams with interaction lines between them shown via the wavy lines. The filled bubble on the left is the sum of infinite number of diagrams as shown on the right.

where $\mathcal{P}$ refers to the principal value[9]. In the limit $\mathbf{q} \to 0$ and $\omega = 0$,

$$\chi_0(\mathbf{q} \to 0, 0) = \sum_{\mathbf{k}} \frac{f(\xi_{\mathbf{k}}) - f(\xi_{\mathbf{k+q}})}{\xi_{\mathbf{k}} - \xi_{\mathbf{k+q}}}. \tag{1.72}$$

Changing the sum over $\mathbf{k}$ to an integral the density of states (DOS), $\rho(\xi)$,

$$\begin{aligned} \chi_0(\mathbf{q} \to 0, 0) &= \int_0^\infty d\xi \rho(\xi_{\mathbf{k}}) \left(\frac{\partial f}{\partial \xi_{\mathbf{k}}} \right) \\ &= -\rho(\epsilon_{\mathrm{F}}) \int_0^\infty d\xi \delta(\xi_{\mathbf{k}}) = \rho(\epsilon_{\mathrm{F}}). \end{aligned} \tag{1.73}$$

So the dielectric constant in this limit becomes,

$$\epsilon(\mathbf{q} \to 0, 0) = 1 + \frac{4\pi e^2 \rho(\epsilon_{\mathrm{F}})}{q^2}. \tag{1.74}$$

Now we use the standard result for the DOS of a free Fermi gas, namely, $\rho(\epsilon_{\mathrm{F}}) = \frac{3n}{2\epsilon_{\mathrm{F}}}$,

$$\epsilon(\mathbf{q} \to 0, 0) = 1 + \frac{6\pi n e^2}{q^2(\epsilon_{\mathrm{F}})} q^2 = 1 + \frac{\kappa^2}{q^2} \tag{1.75}$$

where $\kappa^2 = \frac{6\pi n e^2}{q^2}$. κ is called the Thomas–Fermi inverse scattering length.

It is now easy to calculate the effect of a charge placed in vacuum which would have created a potential, $\phi^{\mathrm{ext}}(\mathbf{r}) = \frac{e}{r}$. In the Fourier space, $\phi^{\mathrm{ext}}(\mathbf{q}) = \frac{4\pi e}{q^2}$. The actual potential $\phi(\mathbf{q})$ experienced by the medium is,

$$\phi(\mathbf{q}) = \frac{\phi^{\mathrm{ext}}(\mathbf{q})}{\epsilon(\mathbf{q})} = \frac{4\pi e}{\kappa^2 + q^2}. \tag{1.76}$$

In real space, one gets,

$$\phi(\mathbf{r}) = \frac{e}{r} e^{-\kappa r}. \tag{1.77}$$

Thus the same Thomas–Fermi inverse scattering length appears as a screening parameter in the Coulomb potential which is effectively screened at distances larger than $1/\kappa$.

Finally, solving for the susceptibility at small but finite q,

$$\chi_0(\mathbf{q}, 0) = -\frac{4m}{(2\pi)^3} \int_0^{2\pi} d\phi \int_0^{k_{\mathrm{F}}} k^2 dk \quad \mathcal{P} \int_0^{\pi} \frac{\sin\theta d\theta}{q^2 + 2qk\cos\theta}. \tag{1.78}$$

[9] $\frac{1}{x \pm i\eta} = \mathcal{P}(\frac{1}{x}) \mp i\pi\delta(x)$.

Performing the integral one gets,

$$\chi_0(\mathbf{q}, 0) = -\frac{m}{16\pi^2 q}\left\{4k_F q + [(2k_F)^2 - q^2]\, ln\left[\left|\frac{q + 2k_F}{q - 2k_F}\right|\right]\right\}. \tag{1.79}$$

The Thomas–Fermi approximation presented above is quite intuitive as it incorporates how the long-range Coulomb interaction is screened. However, since it is a long wavelength approximation, it is unable to capture the effects of local perturbations. Let us introduce the DOS at the Fermi level once more, namely, $\rho(\epsilon_F) = \frac{m}{k_F}$ which allows $\chi_0(\mathbf{q}, 0)$ to be written as,

$$\chi_0(\mathbf{q}, 0) = -\rho(\epsilon_F)F(y) \tag{1.80}$$

where

$$F(y) = \frac{1}{2} + \frac{1}{4y}(1 - y^2) ln\left|\frac{1 + y}{1 - y}\right|$$

with $y = \frac{q}{2k_F}$. $F(y)$ as a function of y is shown in figure 1.5. For small q, that is $y \to 0$, $F(y)$ can be expanded as,

$$F(y) = 1 - \frac{y^2}{3} + \mathcal{O}(y^4).$$

On the other hand, for small wavelengths (large q), that is for $y \to 1$ or $q \to 2k_F$, $F(y)$ diverges[10] due to the divergence of the natural logarithm. In the limit of

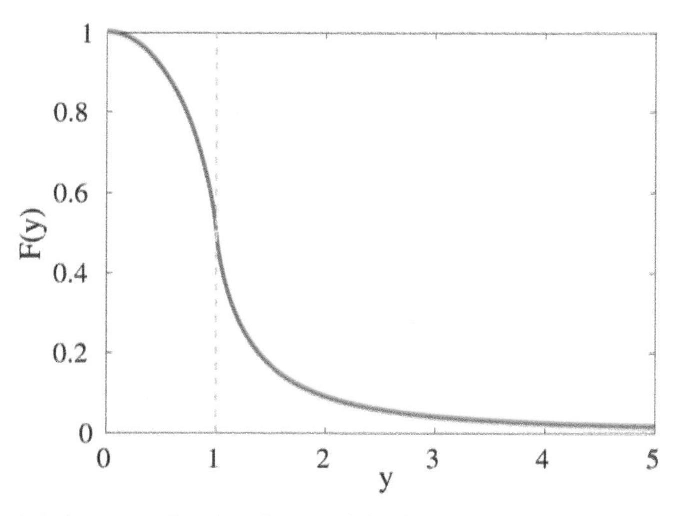

Figure 1.5. $F(y)$ is shown as a function of y. $F(y)$ falls off at large values of y, that is, for $q \to 2k_F$.

[10] This divergence controls the long-range behaviour of the screened Coulomb potential discussed below.

$y \to \infty$, $F(y) \to 0$ as shown in figure 1.5. This yields the dielectric constant to have a form,

$$\epsilon(\mathbf{q}, 0) = 1 + \frac{\kappa^2}{q^2}F(y). \tag{1.81}$$

Hence the calculation of the potential $\phi^{\text{total}}(\mathbf{r})$ can be done using $\phi^{\text{total}}(\mathbf{q}) = \frac{\phi^{\text{ext}}}{\epsilon(\mathbf{q})}$. It is interesting to look at the real space variation of $\phi^{\text{total}}(\mathbf{r})$,

$$\begin{aligned}\phi^{\text{total}}(\mathbf{r}) &= \sum_{\mathbf{q}} e^{i\mathbf{q}.\mathbf{r}} \frac{\phi^{\text{ext}}(\mathbf{q})}{1 - \frac{\kappa^2}{q^2}F(y)} \\ &= \frac{y^2}{(2 + y^2)^2}\frac{\cos(2k_{\text{F}}r)}{r^3}.\end{aligned} \tag{1.82}$$

Now one can see that the potential no longer has an exponential fall off, but acquires a convolution of $\frac{1}{r^3}$ and an oscillatory behaviour due to $\cos(2k_{\text{F}}r)$. This is known as Friedel oscillation, which governs the behaviour of the potential at a (large) distance r from the location of the impurity in a medium.

Often the response of the electron liquid is expressed in terms of the conductivity, σ of the medium which is an experimentally accessible quantity. The conductivity is related to the dielectric constant in the limit $\mathbf{q} \to 0, \quad \omega \to 0$ via,

$$\epsilon(\mathbf{q} \to 0, \omega \to 0) = 1 + \frac{4\pi i}{\omega}\sigma(\mathbf{q} \to 0, \omega \to 0). \tag{1.83}$$

Replacing $\epsilon(\mathbf{q} \to 0, \omega \to 0)$ from equation (1.75),

$$\sigma(\mathbf{q} \to 0, \omega \to 0) = -\frac{i\kappa^2\omega}{4\pi e^2}. \tag{1.84}$$

Thus the conductivity is imaginary which implies that it is non-dissipative and vanishes in the limit $\omega \to 0$.

Finally to complete the story, we look at the opposite limit, that is, $\omega \gg v_{\text{F}}q$. Writing down the non-interacting susceptibility once more,

$$\chi_0(\mathbf{q}, \omega) = 2\sum_{k<k_{\text{F}}} \frac{\xi_{\mathbf{k}+\mathbf{q}} - \xi_{\mathbf{k}}}{(\omega + i\eta)^2 - (\xi_{\mathbf{k}+\mathbf{q}} - \xi_{\mathbf{k}})^2}. \tag{1.85}$$

Neglecting the second term in the denominator for large ω, and also taking $q \to 0$ limit,

$$\chi_0(\mathbf{q} \to 0, \omega) = \frac{1}{m\omega^2}\sum_{k<k_{\text{F}}}(q^2 + 2\mathbf{k}.\,\mathbf{q}) \;=\; \frac{q^2}{m\omega}\sum_{k<k_{\text{F}}}\left(1 + \frac{2\mathbf{k}.\,\mathbf{q}}{q^2}\right) \tag{1.86}$$

$\mathbf{k}.\,\mathbf{q}$ involves the angle between the two vectors $\mathbf{q}$ and $\mathbf{k}$ and the sum over $\mathbf{k}$ implies average over the direction of $\mathbf{k}$. Thus $\sum_{\mathbf{k}} 1$ counts the number of $\mathbf{k}$-states which gives the total number of states, say, n. Finally one gets,

$$\chi(\mathbf{q} \to 0, \omega) = \frac{nq^2}{m\omega^2}. \tag{1.87}$$

This yields the dielectric constant to have the form,

$$\epsilon(0, \omega) = 1 - \frac{4\pi ne^2}{m\omega^2} = 1 - \frac{\omega_p^2}{\omega^2} \tag{1.88}$$

where ω_p $(=\frac{4\pi ne^2}{m})$ is the plasmon frequency. Thus $\epsilon = 0$ at $\omega = \omega_p$, which says that the dielectric response vanishes when the frequency of oscillation of the system matches that of the plasmon mode. As opposed to the particle–hole excitations, which involves promoting a particle from inside the Fermi sea to outside, the plasma oscillation involves motion of all the particles. Thus it is a collective phenomenon demonstrated by the system.

1.3.4 Conductivity

If we now calculate the conductivity, we get both the real (σ') and the imaginary (σ'') parts such that,

$$\sigma(\mathbf{q}, \omega) = \sigma'(\mathbf{q}, \omega) + i\sigma''(\mathbf{q}, \omega)$$

where the real part denotes dissipation and the imaginary component denotes the non-dissipative part. The non-dissipative part is given by,

$$\sigma''(0, \omega) = \frac{ne^2}{m\omega}. \tag{1.89}$$

The real part, σ' is related to σ'' via the Kramers–Kronig relation,

$$\sigma''(0, \omega) = \frac{1}{\pi}\mathcal{P}\int_{\infty}^{\infty} \frac{\sigma'(0, \omega')d\omega'}{\omega - \omega'}.$$

Substituting the imaginary part from equation (1.89), the dc conductivity is,

$$\sigma'(0, \omega) = \frac{ne^2\pi}{m}\delta(\omega). \tag{1.90}$$

The $\delta(\omega)$ part in $\sigma'(0, \omega)$ denotes an infinite conductivity, that is, it behaves like a perfect conductor at $\omega = 0$. This is clearly an unphysical result and is an artefact of the RPA, where at $T = 0$, there is no interaction between the electrons. Of course at finite temperatures, there is a finite dc conductivity. The same effect can be brought about if electronic interaction is included, that is, going beyond the RPA. We stop here, and the effects of considering interaction effects are beyond the scope of the book.

However, there is one way to include the interaction effects in a phenomenological fashion which is incorporated by a relaxation time, τ in the expression for conductivity, namely,

$$\sigma(0, \omega) = \frac{ne^2}{m}\frac{1}{1 - i\omega\tau} \tag{1.91}$$

which yields the correct non-interacting limit ($\tau \to \infty$). For the dielectric response, it yields,

$$\sigma(0, \omega) = 1 - \frac{\omega_p^2}{\omega^2 + \frac{i\omega}{\tau}} = 1 - \frac{\omega_p^2}{\omega\left(\omega + \frac{i}{\tau}\right)}. \tag{1.92}$$

This implies that the dielectric response acquires an imaginary part, or equivalently the conductivity acquires a real part at the poles of $\chi_0(\mathbf{q}, \omega)$, that is,

$$\xi_{\mathbf{k+q}} - \xi_{\mathbf{k}} = \pm\omega. \tag{1.93}$$

The susceptibility diverges under this condition and hence a particle–hole pair is created. The frequency regime for such excitation to occur is:

$(i)\ \omega < \frac{q^2 + 2qk_\text{F}}{m}$

$$(ii)\ \omega > \begin{Bmatrix} 0 & \text{for } q < 2k_\text{F} \\ \left(\frac{q^2 - 2qk_\text{F}}{m}\right) & \text{for } q > 2k_\text{F} \end{Bmatrix}$$

The dispersion is shown in figure 1.6.

It is also interesting to look at the rate of production of the particle–hole pairs in the system. Let us denote this by τ_p^{-1} which assumes a form,

$$\frac{1}{\tau_p} \simeq 4\pi \int \frac{d^3\mathbf{k}}{(2\pi)^3}\delta(\xi_\mathbf{k} - \xi_{\mathbf{k+q}} + \omega)\left[n(\xi_\mathbf{k})(1 - n(\xi_{\mathbf{k+q}})) - n(\xi_{\mathbf{k+q}})(1 - n(\xi_\mathbf{k}))\right] \tag{1.94}$$

where $n(\xi_\mathbf{k})(1 - n(\xi_{\mathbf{k+q}}))$ is the rate of creation of pairs, and $n(\xi_{\mathbf{k+q}})(1 - n(\xi_\mathbf{k}))$ is the rate for their dissociation. A rough estimate of the integral can be done via changing over to the energy variable, ξ using $kdk = md\xi$ ($\epsilon_k = \frac{k^2}{2m}$, $\hbar = 1$),

$$\begin{aligned} \frac{1}{\tau_p} &\simeq \frac{m^2}{\pi q^2} \int d\xi \int d\xi' \delta(\xi - \xi' + \omega)[n(\xi) - n(\xi')] \\ &\simeq \frac{m^2}{\pi q} d\xi[n(\xi) - n(\xi + \omega)]. \end{aligned} \tag{1.95}$$

Using Taylor expansion for small ω,

$$n(\xi + \omega) = n(\xi) + \omega\frac{dn}{d\xi}$$

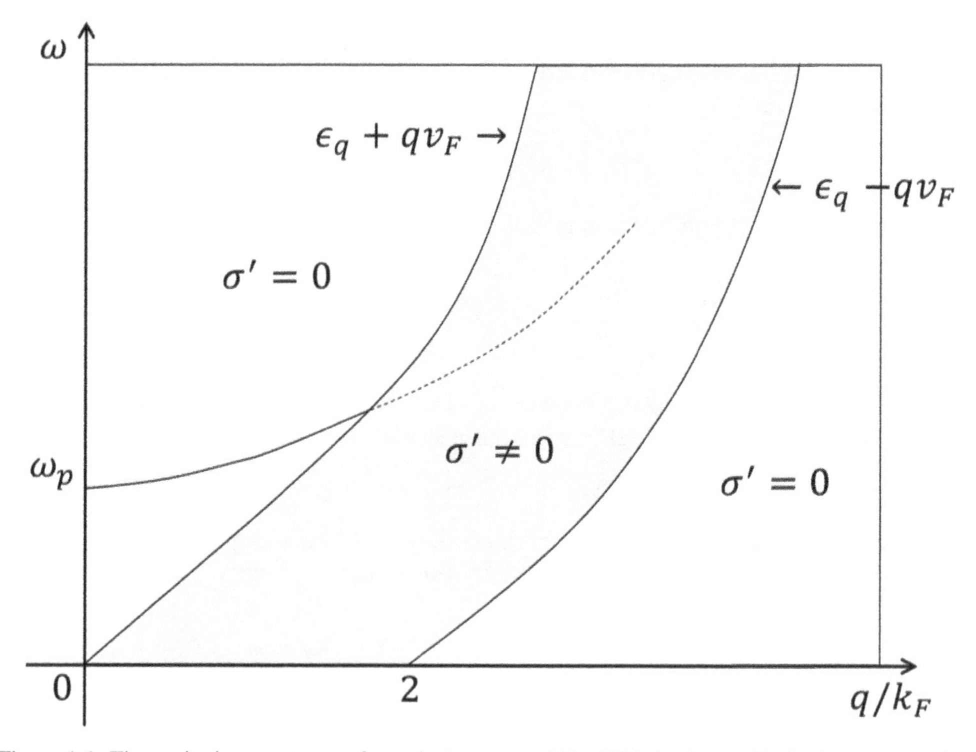

Figure 1.6. The excitation spectrum of an electron gas within RPA is shown. Excitations occur where $\epsilon(\mathbf{q}, \omega) = 0$ and are denoted by $\omega = \frac{q^2 \pm 2qk_\mathrm{F}}{2m}$ which is shown by the coloured region.

$$\frac{1}{\tau_p} \simeq \frac{\omega m^2}{\pi q} \int d\xi \left(-\frac{dn}{d\xi} \right) = \frac{\omega m^2}{\pi q}. \tag{1.96}$$

Thus τ_p is inversely proportional to ω (restricted to small ω only). The coloured area in figure 1.6 which denotes the real part of the conductivity, or the imaginary part of the dielectric constant to be non-zero is bounded by the lines defined by $\omega = \epsilon_q \pm v_\mathrm{F} q$. Thus the figure describes the excitation spectrum of the electron gas to density fluctuations at all values of q/k_F. Between $0 < q/k_\mathrm{F} < 2$, the plasmon modes are highly damped.

1.4 Determination of the Fermi surface: the de Haas–Van Alphen effect

The properties of electronic systems, for example, those of metals depend primarily on the electrons residing near the Fermi surface. Thus a knowledge of the shape of the Fermi surface is central to the understanding of their properties. The Fermi surface is a surface in the reciprocal space at zero temperature that demarcates the occupied electronic states from the unoccupied ones. The shape of the Fermi surface, which is often intricate and complex can provide a useful guide to the properties of metals.

The shape of the Fermi surface of metals is determined by de Haas–Van Alphen (dHVA) experiment. It involves oscillation of the diamagnetic susceptibility as a function of the applied magnetic field strength, B. The result yields detailed features of the extremal areas of the Fermi surface. The experimental observations were done on the magnetization of the semi-metal bismuth (Bi) at a temperature of about 14.2 K and high magnetic fields by de Haas and Van Alphen in 1930 [10]. The susceptibility scaled by the magnetic field $\left(\frac{M}{B}\right)$ shows oscillations as a function of $1/B$. Similar oscillations are also shown in the magnetoresistance, a phenomenon known as the Shubnikov–de Haas effect.

The dHVA effect is a direct consequence of the quantization of the closed electron orbits in the presence of an external magnetic field and thus manifests purely quantum phenomenon. For the realization of the dHVA effect, there are a few pre-requisites, namely:

1. the orbits of the electrons should remain well-defined, even if they scatter.
2. The quantization of the orbits is robust and is not smeared by the thermal effects (it may be noted that the temperature at which the original experiment was performed is very low).
3. High magnetic fields are required since the energy difference between the quantized orbits is $\hbar\omega_{\rm B}$ ($\omega_{\rm B} = \frac{eB}{m}$), such that $\hbar\omega_{\rm B} \gg k_{\rm B}T$.

In fact the first condition is stated as, $\omega_{\rm B}\tau > 2\pi$ where τ denotes the relaxation time. Equivalently, a high purity sample is required for an efficient observation of the effect. For an analysis of the phenomenon, the classical equation of motion in a magnetic field B is a good starting point, which is,

$$\mathbf{F} = \hbar\dot{\mathbf{k}} = -e(\mathbf{v} \times \mathbf{B}). \tag{1.97}$$

Considering the component of the velocity (v_t) perpendicular to the magnetic field, one may write,

$$dk = \frac{eB}{\hbar} v_t dt. \tag{1.98}$$

The time taken to complete a full cycle of motion (which should be same in real and momentum spaces) is given by,

$$T = \frac{2\pi}{\omega_{\rm B}} = \frac{\hbar}{eB}\oint \frac{dk}{v_t}. \tag{1.99}$$

Since the magnetic force is transverse to the velocity of the electron, a constant magnetic field cannot change the energy of the electron, and hence it should move on the constant energy surface. The velocity of such electrons is given by,

$$\mathbf{v} = \frac{1}{\hbar}\delta_k E_k \tag{1.100}$$

which says that the motion is perpendicular to the constant energy surface. Thus the orbit is confined to the intersection of the constant energy surface and a plane perpendicular to the magnetic field. Since the electrons near the Fermi surface are important for observing such effects, the constant energy surfaces are indeed the Fermi surfaces.

The dHVA effect relates the energy of the electron to the area of the orbit in k-space. This yields the shape of the Fermi surface. To understand how this is done, one may consider two close by orbits in the k-space that differ in energy by a small amount ΔE. Let us assume the velocity perpendicular to the orbit is $\vec{v}$ (see figure 1.7), which is given by,

$$|\mathbf{v}| = \frac{1}{\hbar}\frac{\Delta E}{\Delta k}. \tag{1.101}$$

From figure 1.7,

$$v_t = u \sin\theta = \frac{1}{\hbar}\frac{\Delta E}{\Delta k}\sin\theta = \frac{1}{\hbar}\frac{\Delta E}{\frac{\Delta k}{\sin\theta}} = \frac{1}{\hbar}\frac{\Delta E}{\Delta k_t}. \tag{1.102}$$

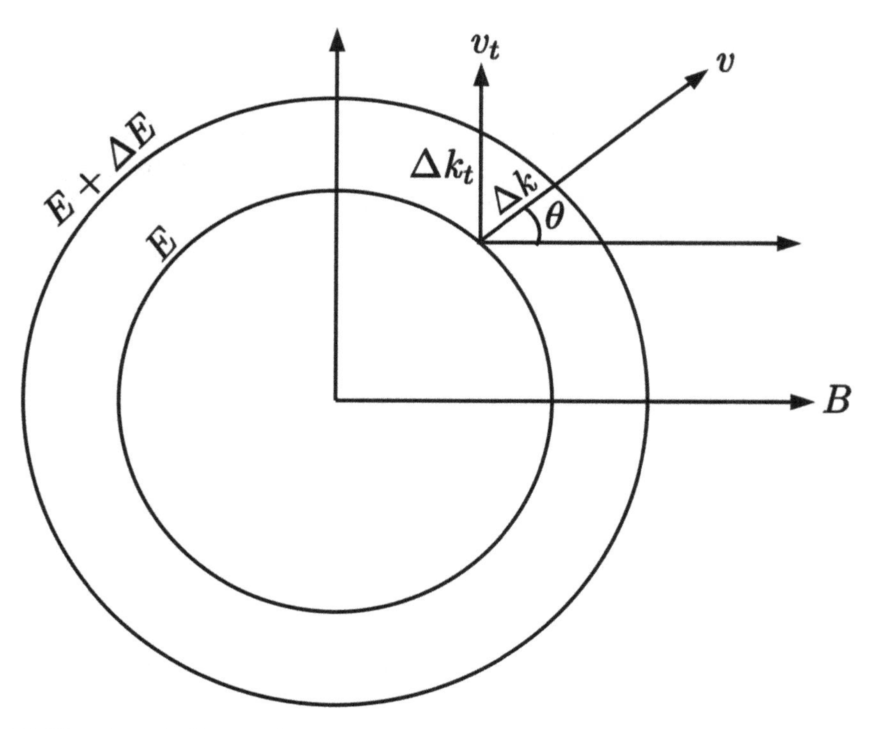

Figure 1.7. The schematic diagram of two close energy contours with energy E and $E + dE$ in the momentum space.

Hence the time-period can now be written as,

$$T = \frac{2\pi}{\omega_B} = \frac{\hbar}{eB} \oint \frac{dk}{\frac{\Delta E}{\Delta k_t}} = \frac{\hbar}{eB\Delta E} \oint \Delta k_t dk = \frac{\hbar}{eB} \frac{\Delta A}{\Delta E} \tag{1.103}$$

where ΔA denotes the area between two constant energy surfaces.

We know that the energy levels of the electrons in a magnetic field applied along the z-direction are given by,

$$E_{n,k_z} = \left(n + \frac{1}{2}\right)\hbar\omega_B + \frac{\hbar^2 k_z^2}{2m}. \tag{1.104}$$

Thus the difference between different orbits with the same value of k_z is $\hbar\omega_B$. One now has to relate ΔE (in the preceding discussion) to $\hbar\omega_B$. To do this, the area of the space between the orbits, ΔA is written as (replacing ΔE by $\hbar\omega_B$ in equation (1.101)),

$$\Delta A = \frac{eB}{\hbar^2} \frac{2\pi}{\omega_B} \hbar\omega_B = \frac{2\pi eB}{\hbar}. \tag{1.105}$$

Thus ΔA depends only on the magnetic field B.

Now consider an extension of the problem in three dimensions. The effect of the magnetic field along the z-axis is to create a quantization in the momentum space so that the circular energy contours because of the $(n + \frac{1}{2})\hbar\omega_B$ term transform into cylindrical energy surface (with a small extension[11]) form whose axis lies along the z-axis and perpendicular to the cross-section area in figure 1.7.

Each such cylinder possess different quantum number with corresponding energy given by,

$$E_n = \left(n + \frac{1}{2}\right)\hbar\omega_B + \frac{\hbar^2 k^2}{2m} \tag{1.106}$$

with a finite width along the z-axis. Now if we increase the magnetic field, the cross-sectional area (or the radius) of the cylinder increases. For some particular value of the field, the cross-sectional area grows so much that the cylindrical pillbox, for a fixed value of n, breaks away from the Fermi surface. Why it breaks away can be understood in the following manner. The energy E is considered to be in the vicinity of the Fermi surface. Thus as the cross-sectional area grows, the energy tends to become larger than the Fermi energy, which is not permissible and the breakaway occurs. When that happens, the states within the Fermi surface (the so-called allowed or occupied states) are pulled out. In this situation, the electrons with energy just greater than the Fermi energy hop on to the next cylinder[12]).

This results in lowering of the energy and eventually induces an oscillation of energy as a function of (increasing) magnetic field. These oscillations are detected as

[11] The structure is like a pillbox that is used in the calculation of the electric field due to a sheet of charge using Gauss's law in electrostatics.

[12] The next cylinder can accommodate a larger number of electrons as the density of states scale with the magnetic field B.

oscillations of the magnetic susceptibility and called the dHVA effect. An oscillating profile from real experiments is presented in figure 1.8. Thus the area difference between the cylinder corresponding to an index 0 and the one corresponding to an index n is,

$$|A_n - A_0| = \Delta A = \frac{2\pi e B}{\hbar}n. \tag{1.107}$$

Thus the area of the nth tube (up to a constant) is,

$$A_n = \frac{2\pi e B}{\hbar}n \tag{1.108}$$

where ($n = 0, 1, 2, \ldots$). Now if A_0 denotes the cross-sections of the Fermi surface perpendicular to B, and B_1, B_2 denotes two values of the magnetic field that create two adjacent cylinders whose area is A_0, in which case,

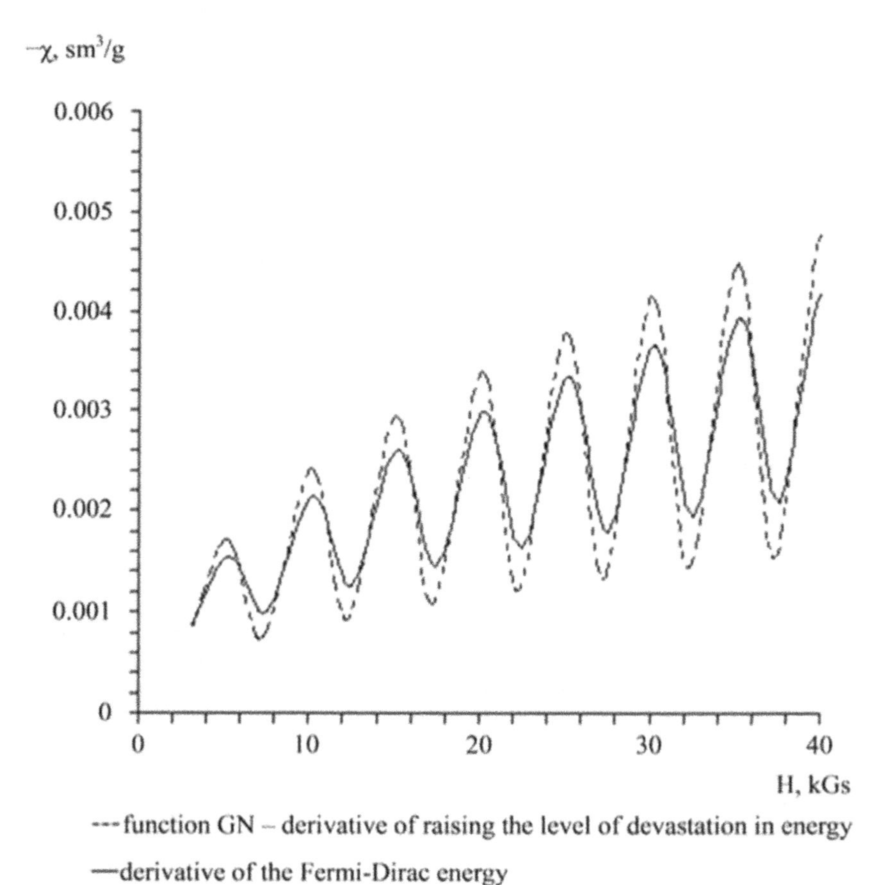

Figure 1.8. Schematic sketch of (dHvA) oscillations of the magnetic susceptibility as a function of the magnetic field, B. Taken from reference [11], copyright 2014 by authors and Scientific Research Publishing Inc.

$$\frac{1}{B_2} = \frac{2\pi e}{\hbar A_0}(n+1) \tag{1.109}$$

and

$$\frac{1}{B_1} = \frac{2\pi e}{\hbar A_0}n \tag{1.110}$$

$$\frac{1}{B_2} - \frac{1}{B_1} = \Delta\left(\frac{1}{B}\right) = \frac{2\pi e}{\hbar A_0}. \tag{1.111}$$

The last expression denotes the reciprocal of the field necessary to create one oscillation of the magnetic susceptibility. The parameters under experimental control are B_1 and B_2 and A_0 can be calculated from the above expression [14]. A_0 yields sufficient information on the Fermi surface.

Onsager pointed out that the change in $\frac{1}{B}$, that is $\Delta(\frac{1}{B})$, through a single oscillation period is related to the frequency of oscillation via,

$$\frac{1}{f} = \Delta\left(\frac{1}{B}\right) = \frac{2\pi e}{\hbar A_0} \tag{1.112}$$

where $1/f$ is the period of the dHVA oscillation in $\frac{1}{B}$, that is, f is the frequency (whose unit is either Gauss or Tesla), and A_0 is an external cross-sectional area of the Fermi surface in a plane normal to the magnetic field such that,

$$\frac{dA_0}{dk_z} = 0. \tag{1.113}$$

The maximum and minimum cross-sections are the extremal ones. One can also alter the directions of the magnetic field to map out the Fermi surface in various directions.

1.5 Fermi liquid theory

One last thing we wish to pursue very briefly is the essence of Fermi liquid theory (FLT). The subject demands more attention than what it gets here. Since the motivation is to provide readers with a bird's eye view of the topic, we introduce the basic concepts here. The mathematical derivations are beyond the purview of the discussions presented here.

The success of the single-electron picture of metals rests on the FLT theory by Landau (1956–1958) [12]. In the FLT, the weakly interacting electrons in a metal form quasiparticles that follow usual fermionic properties. The FL theory is remarkably successful, despite a few failures, especially where the interactions between electrons are quite strong and hence cannot be ignored.

In order to understand the conditions of applicability of FLT, we can define the mean electronic separation, r_s and density ρ_e via,

$$\rho_e \times \frac{4}{3}\pi r_s^2 = 1.$$

The average Coulomb interaction per electron is given by,

$$\langle PE \rangle \sim \frac{1}{2}\frac{e^2}{4\pi\epsilon_0 r_s}.$$

Further, the mean kinetic energy is,

$$\langle KE \rangle \sim \frac{\hbar^2}{2m}\frac{1}{r_s^2}$$

$$\frac{\langle PE \rangle}{\langle KE \rangle} \sim \frac{e^2}{8\pi\epsilon_0 r_s} \times \frac{2mr_s^2}{\hbar^2} = \frac{me^2}{4\pi\epsilon_0\hbar^2} r_s = \frac{r_s}{a_0}. \tag{1.114}$$

Thus the ratio of the mean potential to the mean kinetic energies is of the order of electron–electron separation, which further can be measured in units of the Bohr radius, a_0. Some typical values are,

$$\begin{aligned} r_s/a_0 &= 1.9 \quad \text{for Be} \\ &= 5.6 \quad \text{for Cs.} \end{aligned}$$

Thus the ratio r_s/a_0 is large, and for Wigner crystals, $r_s/a_0 > 20$. This means that the mean electronic separation is larger than the Bohr radius, which in turn, advocates that FLT will provide an appropriate description of electronic systems.

The need for an FLT in the first place comes from the fact that the specific heat of electrons, according to the classical (and non-interacting) theory, should be $3k_B/2$, however, one usually gets a value much lower than this. Also, the susceptibility, χ, of the free moments deviates significantly from the classical behaviour, namely, $\chi \sim \frac{1}{T}$. Thus the non-interacting theory is found to be insufficient in a variety of cases.

These puzzles are solved by the FLT which says that only a small fraction of the electrons near the Fermi surface take part in contributing to the physical observable, such as the specific heat or the susceptibility, etc. These electrons are promoted from just inside the Fermi surface to just outside, and such phenomena are known as particle–hole excitations. The same formalism as that of the non-interacting systems will continue to be valid, except for the bare electronic mass, m is to be replaced by the effective mass, m^*. Only the electrons within k_BT of the Fermi energy contribute to the specific heat so that the specific heat is proportional to T and is small. Also, the electrons within an energy slice $\mu_B B$ of the Fermi surface can be magnetized with a moment, μ_B leading to a temperature independent Pauli susceptibility. These dependencies were matched with experiments on metals and are found to have good support.

However, although FLT solved these riddles, there remain questions that are yet to be answered, such as how does a non-interacting theory explain behaviour of systems where the interaction affects are fairly important. The answer provided by Landau rests on the concept of '*adiabatic continuity*'. The adiabatic continuity says that the labels associated with eigenstates are more robust against perturbation than the eigenstates themselves. To have a clarity in understanding, consider a one-dimensional box whose eigenstates are given by sinusoidal function, that is, how many zeros it has (see left panel of figure 1.9). The larger the energy, the greater the number of nodes. Now if we perturb the system by a small harmonic oscillation potential, $V(x) = \frac{1}{2}\epsilon x^2$ (ϵ is small), the new eigenstates of the system will no longer remain as simple sine waves, but involve a mixing of all the eigenstates of the original unperturbed problem. However, the number of nodes still remains a good indicator to describe the eigenstates of this more complicated (interacting) problem. The correspondence is shown schematically in figure 1.9.

Landau applied this idea to the interacting gas of electrons. He imagined turning on the interaction effects slowly, and observed how the eigenstates of the system behave. He postulated that there will be a one-to-one mapping of the low energy states of that of the interacting system with that of the non-interacting Fermi gas. The assumption is that the good quantum numbers associated with the excitations of the non-interacting systems will continue to remain valid, even after the interactions are turned on. Just as Pauli's exclusion principle holds for the non-interacting electrons, this would remain so even in the presence of interactions. We can therefore retain the picture of excitation of particles and holes, with them carrying the same quantum numbers as their electronic counterparts of the free Fermi gas. These excitations are called quasiparticles, whose wavefunctions and the energies are different from those of the corresponding electrons in the non-interacting problem. These quasiparticles are core to the understanding of FLT, and account for the

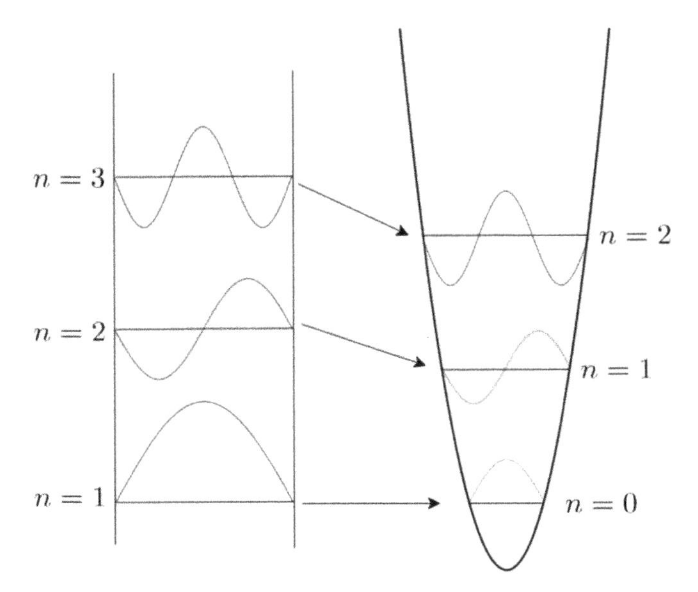

Figure 1.9. The one-to-one correspondence between the states of a particle in an infinite potential and a particle subjected to a harmonic potential is shown. The arrows denote the corresponding energy states for which the number of nodes remains fixed.

observed temperature dependencies of specific heat and susceptibility, with the only requirement of having a Fermi surface.

The energies of the quasiparticles are not the same as that of the non-interacting electrons. In Landau's theory, the modified energy appears through two terms. Their origin can be understood as follows.

1. First, when a quasiparticle moves there will be a backflow in the filled Fermi sea (due to momentum conservation), as the quasiparticles '*push*' the ground state out of the way. This modifies inertial mass of the quasiparticles, m is replaced by m^*.
2. Second, the quasiparticle energy must depend on the distribution of other quasiparticles, which Landau had included via some 'f' function, where the total energy of the interacting system can be expanded as a functional of variation in density, namely,

$$E = \sum_{\vec{k},\sigma} \frac{p_{\rm F}}{m^*}(\hbar k - p_{\rm F})\delta n_{\vec{k},\sigma} + \frac{1}{2}\sum_{\vec{k},\, \vec{k}',\, \sigma,\, \sigma'} f_{\vec{k}\sigma,\vec{k}'\sigma'}\delta n_{\vec{k},\sigma}\delta n_{\vec{k}',\sigma'}$$

where $p_{\rm F}$ denotes the Fermi momentum, and $\delta n_{\mathbf{k},\sigma}$ ($\delta n_{\mathbf{k}',\sigma'}$) denotes the change in the Fermi-distribution function. We skip a detailed discussion of the above and encourage readers to look at more specialized articles [15]. Also, for a short review see reference [16].

Let us look at the additional constraints on the validity of Landau FLT. The quasiparticles must be long lived in the vicinity of the Fermi sphere. In fact the inverse of the QP lifetime, τ_{el} can be evaluated by using Fermi's golden rule,

$$\frac{1}{\tau_{el}} = \frac{2\pi}{\hbar}\sum_f |V_{fi}|^2\delta(\epsilon - \epsilon_p). \tag{1.115}$$

Assuming V_{fi} to be a constant and independent of the energy transfer ϵ, one can convert the above sum into an integral. Hence integrate over the final (continuum) states. This goes as,

$$\frac{1}{\tau_{el}} \sim \epsilon^2.$$

Also, τ_{el} is found to have the temperature dependence of the form,

$$\frac{1}{\tau_{el}} \sim T^2.$$

This decay rate is important in transport properties of metals, as it yields a T^2 resistivity.

Here we present a few examples, which from a theoretical perspective generate non-Fermi liquid (NFL) characteristics in interacting electronic systems. These are:

1. metals close to the quantum critical point, where a phase transition occurs close to $T = 0$, the quasiparticles scatter so strongly that they cease to follow usual results of FLT.

2. Metals in one dimension are called Luttinger liquid. In 1D, electrons are unstable and decay into two separate particles (spinons and holons) that carry the spin and charge, respectively.
3. Disordered Kondo metals. Here the scattering from the magnetic impurities are too strong to allow for stable quasiparticles to form.

1.6 Summary and outlook

We began with a brief introduction of quantum mechanics that we believe a condensed matter physicist should be acquainted with. The simplest variety of the solid phase are metals, however, a completely free electron theory is found to be inadequate to explain a host of features found in them. In a bid to understand the general properties of metals, we have introduced the jellium model, where in addition to the kinetic energy of the non-interacting electrons, inter-particle interaction terms between the electrons, the interaction between the electrons and the background positive charges, and the interaction among the background charges are included. The resultant Hamiltonian is solved within a Hartree–Fock approximation, and obtains the ground state energy as a function of electron density. Further, the magnetic and the dielectric properties are discussed, and physical observables, such as, magnetization, dielectric response and conductivity, etc are obtained.

Electronic systems, such as, metals are usually characterized by their Fermi surfaces, which have often complex structures in the Brillouin zone. The experimental determination of the Fermi surface is done via dHvA effect which involves oscillation of the magnetic susceptibility as a function of the applied magnetic field. This aids in mapping out of the extremal regions of the Fermi surface, and thus provides hints on its shape, which eventually contribute to the understanding of the transport properties of the metals.

Finally, we have introduced the FLT that is central to the understanding of the behaviour of fermionic systems. The need to go beyond the free electron theory and establish a connection between the non-interacting and the interacting systems is emphasized in understanding the properties of real materials. The main postulates of the FLT are laid down, with a brief note on their inadequacies in explaining the properties of more complicated and strongly interacting systems.

References

[1] Schiff L I 1949 *Quantum Mechanics* 1st edn (New York: McGraw-Hill)
[2] Merzbacher E 1998 *Quantum Mechanics* 3rd edn (New York: Wiley)
[3] Chen-Tannoudji C, Diu B and Laloë F 1977 *Quantum Mechanics, vol I and vol II* 1st edn (New York: Wiley)
[4] Sakurai J J and Napolitano J 2020 *Modern Quantum Mechanics* 3rd edn (Cambridge: Cambridge University Press)
[5] Mahan G D 2000 *Many Particle Physics* 3rd edn (Berlin: Springer)
[6] Fetter A L and Walecka J D 1971 *Quantum Theory of Many Particle Systems* (New York: McGraw-Hill)

[7] Ziman J M 1972 *Principals of the Theory of Solids* 2nd edn (Cambridge: Cambridge University Press)
[8] Kramers H A 1926 Wellenmechanik und halbzahlige quantisierung *Z. Phys.* **39** 828 Also see Griffiths D 2016 *Introduction to Quantum Mechanics* 4th edn (Cambridge: Cambridge University Press)
[9] Gell-Mann M and Brueckner A 1957 *Phys. Rev.* **106** 364
[10] De Haas W J and Van Alphen P M 1930 *Proc. Acad. Sci. Amst.* **33** 1106
[11] Gulyamov G, Erkaboev U I and Sharibaev N Y 2014 *J. Mod. Phys.J. Mod. Phys.* **05** 51908
[12] Ter H D (ed) 1965 The theory of a Fermi liquid *Collected Papers of L D Landau* (Oxford: Pergamon)
[13] Zhong F H, Lin C and Ceperley D M 2002 *Phys. Rev.* B **66** 036703
[14] Onsager L 1952 *Phil. Mag.* **43** 1006
[15] Kinza M 2018 *Theory of Fermi Liquids in Metals* (Berlin: Springer)
[16] Neilson D 1996 *Aust. J. Phys.* **49** 79–102

IOP Publishing

Condensed Matter Physics: A Modern Perspective

Saurabh Basu

Chapter 2

Magnetic phenomena in solids

2.1 Introduction

A large variety of metals, insulators, and even superconductors demonstrate magnetic properties. The importance of the magnetic phenomena gets enhanced due to the vast technological applications, which include equipment in our surroundings on a daily basis. Some of them are transformers, storage in computers, permanent magnets in motors, etc.

In the last couple of decades or so, magnetic materials have made inroads into several intriguing discoveries, such as dilute magnetic semiconductors, iron based superconductors with very large resistance in presence of magnetic fields (colossal magnetoresistance (CMR), and giant magnetoresistance (GMR)), etc, and more interestingly, applications in the field of spintronics, where the spin degrees of freedom (as opposed to the charge) are employed to carry information.

In the first course on solid state physics, we have seen that independent electron approximation is suitable for understanding certain properties of materials, such as, metallicity, etc. Even the insulating properties are well explained by the free electrons subjected to an effective potential due to the presence of ionic cores (remember the Kronig–Penny model learned in the first course on solid state physics) via computation of the band structure within certain approximations, most familiar of them being the tight binding model. Thus within an independent electron approximation, it is possible to conceptually understand the energy spectrum resulting from an effective potential, phonons, etc.

However there are more interesting phenomena, such as, magnetic phenomena in solids, specifically ferromagnetism and antiferromagnetism, etc where the many-electron aspects show up in a way that necessitates going beyond the single-particle picture. For example, consider spin waves, where the spin of one electron is flipped, while all the (valance) electrons take part, resulting in a collective behaviour. The collective and the local aspects of the electron correlations are intertwined in a complicated way in giving rise to magnetic correlations. The most familiar type of

doi:10.1088/978-0-7503-3031-2ch2

magnetism that is widely discussed in literature is the ferromagnetism in 3d metals, such as, Fe, Co, Ni, etc where the exchange interaction between the largely delocalized 3d electrons facilitates ferromagnetism, whereas, in the $4f$ transition metals and their compounds (the Lanthanides) require a localized description. Even explanation of antiferromagnetic arrangement of spins requires exchange interaction to be invoked among the localized electrons.

Before we start discussing magnetic phenomena in solids, let us list just a few known magnetic materials, and their molar susceptibilities, χ_m defined by,

$$\chi_m = \kappa \mathcal{V}_m = \kappa \frac{M}{\rho} \tag{2.1}$$

($\mathcal{V}_m$: molar volume, M: Molar mass, ρ: mass density) where κ is the volume susceptibility which appears in the proportionality of $\mathbf{M}$ (magnetization) and $\mathbf{B}$ (magnetic field) as, $\mathbf{M} = \kappa \mathbf{H}$. The magnetic susceptibilities of a few common magnetic elements are listed in table 2.1.

2.2 Magnetic ordering: diamagnetism and paramagnetism

To begin with, we shall review atomic magnetism with a view to understanding magnetic phenomena in insulators. It may be noted that these properties can quantitatively be understood by the independent electron approximation. In the following, we discuss how the atomic susceptibilities are computed. In the presence of an external magnetic field $\mathbf{B}$, the kinetic energy operator assumes the form [1],

$$\mathcal{K} = \frac{1}{2m} \sum_i (\mathbf{p}_i - e\mathbf{A}_i)^2 \tag{2.2}$$

where $\mathbf{A}_i$ is the local vector potential derivable from the magnetic field $\mathbf{B}$. Choosing a symmetric gauge $\mathbf{A}_i = \frac{1}{2}(\mathbf{r}_i \times \mathbf{B})$, one gets,

$$\begin{aligned} \mathcal{K} &= \frac{1}{2m} \sum_i \left(\mathbf{p}_i - \frac{e}{2}\mathbf{r}_i \times \mathbf{B}\right)^2 \\ &= \frac{1}{2m} \sum_i \left[\mathbf{p}_i^2 + \frac{e^2}{4}(\mathbf{r}_i \times \mathbf{B})^2 - \frac{e}{2}\{\mathbf{p}_i \cdot (\mathbf{r}_i \times \mathbf{B}) + (\mathbf{r}_i \times \mathbf{B}) \cdot \mathbf{p}_i\} \right]. \end{aligned} \tag{2.3}$$

One can split the above expression as,

$$\mathcal{K} = \mathcal{K}_0 + \frac{e^2}{8m} B^2 \sum_i \left[r_i^2 - 2e\mathbf{p}_i \cdot (\mathbf{r}_i \times \mathbf{B}) \right] \tag{2.4}$$

where $\mathcal{K}_0$ is the kinetic energy without an external field. Using the vector identity,

$$\mathbf{a} \cdot (\mathbf{b} \times \mathbf{c}) = \mathbf{b} \cdot (\mathbf{c} \times \mathbf{a}) = \mathbf{c} \cdot (\mathbf{a} \times \mathbf{b}).$$

Table 2.1. Magnetic susceptibility of some common magnetic elements.

Name	χ_m ($\times 10^{-6}$cm^{3}mol^{-1})
Aluminum (Al)	+16.5
Antimony (Sb)	−37.0
Bismuth (Bi)	−280.1
Boron (B)	−6.7
Calcium (Ca)	+40.0
Carbon (C)	−6.0
Cesium (Cs)	+29.0
Chromium (Cr)	+16.7
Copper (Cu)	−5.46
Cobalt (Co)	Ferro
Gallium (Ga)	−21.6
Gold (Au)	−28
Indium (In)	−64.0
Iridium (Tr)	+32.1
Iron (Fe)	Ferro
Lanthanum (La)	+118.0
Lead (Pb)	−23.0
Lithium (Li)	+14.2
Manganese-α (Mn)	+529.0
Molybdenum (Mo)	−96.5
Niobium (Nb)	+195.0
Nickel (Ni)	Ferro
Molybdenum (Mo)	+72.0
Phosphorus, Black (P)	−26.6
Platinum (Pt)	+201.9
Potassium (K)	+20.8
Rhodium (Rh)	+117.0
Rubidium (Ru)	+17.0
Selenium (Se)	−25.0
Silicon (Si)	−3.9
Silver (Ag)	−19.5
Strontium (Sr)	+92.0
Sulfur α (S)	−15.5
Sodium (Na)	+16.0
Tantalum (Ta)	+154.0
Thorium (Th)	+132.0
Thallium (Tl)	−50.0
Tin Gray (Sn)	−37.0
Titanium (Ti)	+151.5
Tungsten (W)	+59.0
Vanadium (V)	+255.0
Zinc (Zn)	−11.4

That is, writing,

$$\sum_i \mathbf{p}_i \cdot (\mathbf{r}_i \times \mathbf{B}) = \sum_i \mathbf{B} \cdot (\mathbf{r}_i \times \mathbf{p}_i) = \sum_i \mathbf{B} \cdot \mathbf{L}_i = \mathbf{B} \cdot \mathbf{L} \tag{2.5}$$

where $\mathbf{L}$ $(=\sum_i \mathbf{L}_i)$ is the total electronic orbital angular momentum. Thus the kinetic energy (and also the Hamiltonian $\mathcal{H}$) becomes,

$$\mathcal{H} = \mathcal{K} = \mathcal{K}_0 + \mu_{\mathrm{B}} \mathbf{L} \cdot \mathbf{B} + \frac{e^2}{8m} B^2 \sum_i (x_i^2 + y_i^2). \tag{2.6}$$

Treating the last two terms on the right-hand side (RHS) as perturbation, that is, treating $\mathcal{H} = \mathcal{H}_0 + \mathcal{H}'$, we have for $\mathcal{H}'$ (adding the spin of the electrons),

$$\mathcal{H}' = \mu_{\mathrm{B}}(\mathbf{L} + g\mathbf{S}) \cdot \mathbf{B} + \frac{e^2}{8m} B^2 \sum_i (x_i^2 + y_i^2) \tag{2.7}$$

where $\mu_{\mathrm{B}} = \frac{eh}{2m}$ is the Bohr magneton having a value 0.579×10^{-8} eV G^{-1} and the Landé g factor is given by, $g \approx 2$. The energy correction can be computed via a perturbation theory using the unperturbed states of the Hamiltonian $\mathcal{H}_0$.

It is instructive to note that the magnetization, M is defined as the first derivative of the free energy, F with respect to the fixed B. That is,

$$M = -\frac{1}{\mathcal{V}} \frac{\partial F}{\partial B} \tag{2.8}$$

where $\mathcal{V}$ is the volume. Thus, the magnetic susceptibility, χ is obtained from the magnetization as,

$$\chi = \frac{\partial M}{\partial B} \tag{2.9}$$

thereby implying that the susceptibility is the second derivative of the free energy with respect to B. Hence, we need to evaluate both the terms in $\mathcal{H}'$ up to second order in B, that is, B^2. Hence, performing a second order perturbation theory one gets [1],

$$\Delta E_n = \sum_{n \neq m} \frac{|\langle \phi_n | \mu_{\mathrm{B}} \mathbf{B} \cdot (\mathbf{L} + g\mathbf{S}) | \phi_m \rangle|^2}{E_n - E_m} + \frac{e^2}{8m} B^2 \langle \phi_n | \sum_i (x_i^2 + y_i^2) | \phi_n \rangle. \tag{2.10}$$

2.3 Magnetic properties of filled and partially filled shell materials

Here we shall discuss the magnetic properties of insulating materials, that is, whose valence electronic shells are all filled. Hence the total orbital and spin angular momentum is zero, that is, the only surviving term is given by,

$$\Delta E_n = \frac{e^2}{8m} B^2 \langle \phi_n | \sum_i (x_i^2 + y_i^2) | \phi_n \rangle. \tag{2.11}$$

Let us evaluate the change in the ground state energy (we call the ground state wavefunction as $|\phi_0\rangle$),

$$
\begin{aligned}
\Delta E_0 &= \frac{e^2}{8m}B^2\langle\phi_0|\sum_i(x_i^2 + y_i^2)|\phi_0\rangle \\
&= \left(\frac{2}{3}\right)\frac{e^2}{8m}B^2\langle\phi_0|\sum_i(r_i^2)|\phi_0\rangle \\
&= \frac{e^2}{12m}B^2\langle\phi_0|\sum_i(r_i^2)|\phi_0\rangle.
\end{aligned}
\tag{2.12}
$$

This should suffice to calculate the susceptibility of a material using,

$$
\begin{aligned}
\chi &= -\frac{\mathcal{N}}{\mathcal{V}}\frac{\partial^2(\Delta E_0)}{\partial B^2} \\
&= -\frac{e^2}{6m}\frac{\mathcal{N}}{\mathcal{V}}\langle\phi_0|\sum_i(r_i^2)|\phi_0\rangle \\
&= -\frac{e^2}{6m}n\langle\phi_0|\sum_i(r_i^2)|\phi_0\rangle
\end{aligned}
\tag{2.13}
$$

where $\mathcal{N}$ denotes the number of ions and $n\,(=\frac{\mathcal{N}}{\mathcal{V}})$ is the density. The negative sign in front of the above expression indicates diamagnetic properties, where the moment is induced opposite to the applied field. χ in the above equation is known as the Larmor diamagnetic susceptibility. One may note that in the sum, electrons in the outermost shells contribute maximally owing to their large mean square distance from the nucleus. Consider Z_{out} to be the number of electrons in the outermost shell and r_{out} be the corresponding distances, then the largest term in the sum yields the susceptibility to have a form,

$$
\chi = -\frac{e^2}{6m}nZ_{\text{out}}r_{\text{out}}^2. \tag{2.14}
$$

The above formula (equation (2.14)) correctly explains the magnetic behaviour of the alkali halides, such as, He, Ne, Ar, Kr, Xe and their ionic configurations.

Exploration of the magnetic properties of materials with partially filled shells is easier and often taught in the first course on statistical mechanics [2]. A brief recap is presented in the following.

Consider $\mathcal{N}$ identical non-interacting spin-S particles in the presence of an external magnetic field B. The corresponding Zeeman Hamiltonian can be written as,

$$
\mathcal{H} = -\mu_{\text{B}}\mathbf{S}\cdot\mathbf{B}. \tag{2.15}
$$

It may be noted that electronic degrees of freedom are not important for this discussion. The canonical partition (orbital) function is written as,

$$
\mathcal{Z} = \sum_{\{S\}} e^{-\beta\mathcal{H}\{S\}} = \sum_{|\mathbf{S}|=-S}^{+S} e^{-\beta\mu_{\text{B}}SB} \tag{2.16}
$$

where it is assumed that lowest $(2S + 1)$ states ($-S$ to $+S$) are thermally excited at a temperature, T with appreciable probability. The above sum in equation (2.16) is computed in the form of a geometric progression (GP) series,

$$e^{-\beta\mathcal{H}} = \frac{e^{\beta\mu_B B\left(S+\frac{1}{2}\right)} - e^{-\beta\mu_B B\left(S+\frac{1}{2}\right)}}{e^{\beta\mu_B B/2} - e^{-\beta\mu_B B/2}}. \tag{2.17}$$

As earlier, the magnetization, M can be calculated using,

$$M = -\frac{\partial F}{\partial B} = \mu_B S B_S(\beta\mu_B S B) \tag{2.18}$$

$B_S(x)$ is called the Brillouin function, and is defined by,

$$B_S(x) = \frac{2S+1}{2S}\coth\left(\frac{2S+1}{2S}x\right) - \frac{1}{2S}\coth\left(\frac{1}{2S}x\right) \tag{2.19}$$

where $x = \mu_B B/k_B T$. The Brillouin function $B_S(x)$ as a function of x for a number of S values is shown in figure 2.1. As can be seen from the plot, for $x \gg 1$, that is, $\mu_B B \gg k_B T$, $B_S(x) \to 1$. Thus, at lower temperatures and large values of external fields, the magnetization reaches its saturation value. While in the other limit, that is, for $\mu_B B \ll k_B T$, one may do a small x-expansion which yields,

$$\coth(x) \approx \frac{1}{x} + \frac{1}{3}x + \mathcal{O}(x^3). \tag{2.20}$$

Thus,

$$B_S(x) \approx \frac{S+1}{3S}x + \mathcal{O}(x^3). \tag{2.21}$$

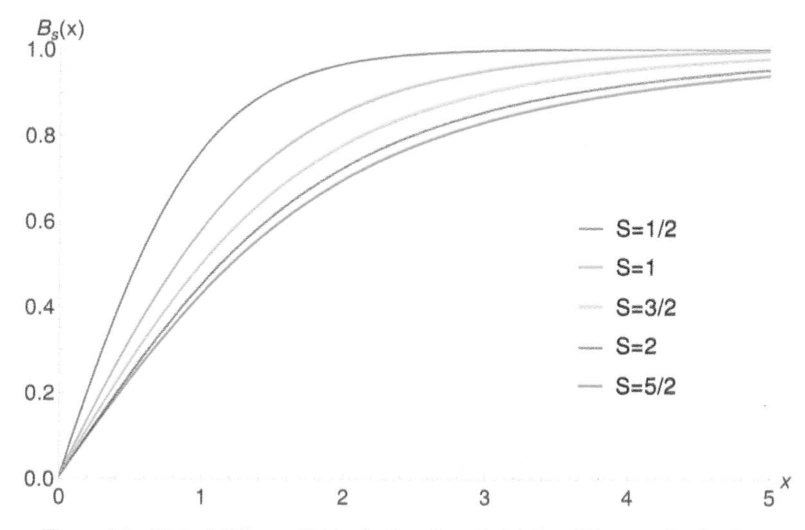

Figure 2.1. Plot of different Brillouin functions $B_S(x)$ for different spin S values.

Hence the susceptibility is computed as,

$$\chi = \frac{\mu_B^2}{3}\frac{S(S+1)}{k_B T} = \frac{C_0}{T}. \tag{2.22}$$

This is called Curie's law and is schematically shown in figure 2.2. We may denote the above expression by χ_{Curie} (that is, replace χ by χ_{Curie} in equation (2.22)) and C_0 is called as the Curie constant. Thus, the susceptibility behaving inversely with temperature ($1/\chi$ versus T is shown via the blue line in figure 2.2) is a feature of 'paramagnetic solids' whose ordering of the magnetic moments is facilitated by a magnetic field, which, however, may be hindered by thermal effects. The law is found to be valid at large temperatures even when considerable magnetic interactions exist among the magnetic moments. The magnetic properties of the rare earth materials, such as, La, Lu, Nd, Ce, Dy, etc are adequately described by Curie's law. However, in transition metals, such as, Fe, Zn, Cr, etc, there are additional phenomena, such as, crystal field splitting, etc which are dominant for partially filled *d*-shells. In order to understand the crystal field splitting in partially filled shells (say *d* shells) one needs to understand the Russell–Saunders (RS) coupling and the Hund's rules (three of them). We postpone their discussion here and include them in appendix.[1]

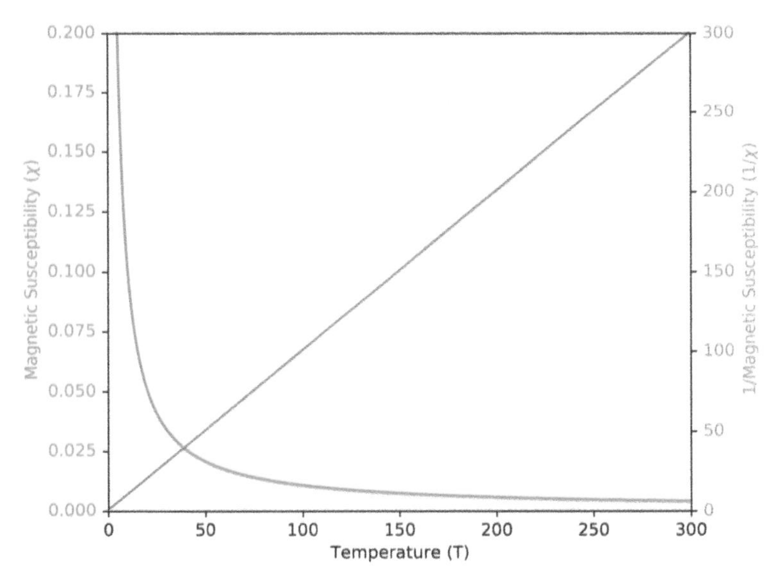

Figure 2.2. χ_{Curie} (in red) and its inverse (in blue) are plotted as a function of temperature (T) and $1/T$ respectively. The slope of the linear plot yields C_0.

[1] Classic texts such as Ashcroft–Mermin [1], Kittel [2], etc exist which treat these topics quite elaborately, and second, the focus of this text is to understand the magnetic properties from an electronic perspective.

Having discussed Curie's law, one should be aware that there is another type of paramagnetic phenomenon applicable for metals, namely, Pauli paramagnetism which refers to the magnetic moments of conduction electrons. Standard techniques of statistical mechanics can be applied (for example, see Pathria [3]) to obtain the magnetic susceptibility, χ_{Pauli} which is a constant (as opposed to having an inverse temperature dependence like that in χ_{Curie}). In fact the expression is,

$$\chi_{\text{Pauli}} = \mu_B^2 N(\varepsilon_F) \tag{2.23}$$

where $N(\varepsilon_F)$ refers to the density of states at the Fermi level. Thus the susceptibility, χ of free electrons (Pauli) is a constant (independent of temperature), while for electrons bound to atoms, χ depends inversely on temperature (Curie).

A deeper introspection reveals that the Pauli susceptibility can be cast into Curie's law with the temperature, T replaced by T_F, the Fermi temperature. Since $T_F \approx 50\,000$ K for typical metals, χ_{Pauli} is typically three orders of magnitude lower than χ_{Curie} even at room temperature.

Further, there is a finite magnetization of the conduction electrons that is diamagnetic in nature, giving rise to a phenomenon called Landau diamagnetism. The susceptibility, let us call it χ_{Landau}, which can be shown to be [3],

$$\chi_{\text{Landau}} = -\frac{1}{3}\chi_{\text{Pauli}}. \tag{2.24}$$

The total susceptibility thus is a combination of the Pauli term, Landau susceptibility, Larmor susceptibility, etc. It is prudent to say that it is quite complicated to extricate a particular contribution from a combination of these effects.

2.4 Ferromagnetism and antiferromagnetism

From the discussion that we had so far, it is clear that electronic interactions which cannot be neglected in order to explain certain types of magnetic ordering, such as ferromagnetism and antiferromagnetism. Without such interactions (which are called magnetic interactions, although the origin is electronic in nature), thermal effects would have randomized the magnetic moments (or the spins of the electrons) resulting in the absence of any sort of magnetic order. Even in antiferromagnetic materials where there is no net magnetization, there is still antiparallel order among the neighbouring spins.

Before we bring about the electronic interactions to explain magnetism, let us discuss how one may get the magnetic Hamiltonian involving only the spins (or the magnetic moments) of the charge carriers. From a classical perspective, it is natural to expect that the dipolar force between moments, $\mathbf{m}_1$ and $\mathbf{m}_2$ separated by a distance $|\mathbf{r}|$ described by a potential of the form,

$$V(\mathbf{r}) = \frac{\mathbf{m}_1 \cdot \mathbf{m}_2 - 3(\mathbf{m}_1 . \hat{\mathbf{r}})(\mathbf{m}_2 . \hat{\mathbf{r}})}{r^3} \tag{2.25}$$

will be operative. However, putting $\mathbf{m}_i$ to be of the order of atomic moments, that is Bohr magneton (μ_B), and $|\mathbf{r}|$ to be of the order of Bohr radius[2], a_0, V comes out in the range 10^{-4} eV, which is far smaller than the typical atomic energies and hence cannot account for magnetic ordering, that typically involves energy scale of the order of an eV. Thus, quantum mechanical exchange is really the dominant player, and dipolar interactions can safely be ignored.

An immediate motive is to arrive at a Heisenberg-like model[3], although here we restrict ourselves to an Ising-like Hamiltonian. To keep the matter even simpler, let us talk about two Ising-like spins (spins only have discrete degrees of freedom, for example, two degrees of freedom, pointing up (↑) and down (↓)). Consider S_1 and S_2 both having values $s = 1/2$, which means that the spins will have two possible orientations where they can either be pointing up or down. Now consider addition of these two $s = \frac{1}{2}$ particles,

$$S = S_1 + S_2. \tag{2.26}$$

Writing the problem in the (s, m_s) basis (S^2 has the eigenvalue $s(s+1)$ and S_z has the eigenvalue m_s, with $\hbar = 1$), the direct product space $(s_1, m_{s_1}) \otimes (s_2, m_{s_2})$ becomes four-dimensional. Now $m_s = \pm\frac{1}{2}\hbar$ denote the $|\uparrow\rangle$-spin and the $|\downarrow\rangle$-spin states with eigenvalues $+\frac{\hbar}{2}$ and $-\frac{\hbar}{2}$, respectively. Thus the basis states are spanned by $|\uparrow\uparrow\rangle$, $|\uparrow\downarrow\rangle$, $|\downarrow\uparrow\rangle$ and $|\downarrow\downarrow\rangle$. These are eigenstates of S_1^2, S_2^2, S_{1z} and S_{2z}. The eigenvalues for the total spin, S, that is, s $(= s_1 + s_2)$ are 0 and 1. For the total z-component of the spin, S_z, that is, m_s $(= m_{s_1} + m_{s_2})$ are 1, 0, 0, −1. One can check that,

$$\begin{aligned} S_z|\uparrow\uparrow\rangle &= \hbar|\uparrow\uparrow\rangle; \qquad S_z|\uparrow\downarrow\rangle = 0 \\ S_z|\downarrow\downarrow\rangle &= -\hbar|\downarrow\downarrow\rangle; \qquad S_z|\uparrow\downarrow\rangle = 0. \end{aligned} \tag{2.27}$$

Thus we can take linear combination of the above basis states to write the wavefunctions as,

$$\chi_{00} = \frac{1}{\sqrt{2}}[|\uparrow\downarrow\rangle - |\downarrow\uparrow\rangle] \quad : \quad s = 0 \quad \text{singlet} \tag{2.28}$$

$$\left.\begin{aligned} \chi_{11} &= \frac{1}{\sqrt{2}}[|\uparrow\uparrow\rangle] \\ \chi_{10} &= \frac{1}{\sqrt{2}}[|\uparrow\downarrow\rangle + |\downarrow\uparrow\rangle] \\ \chi_{1-1} &= \frac{1}{\sqrt{2}}[|\downarrow\downarrow\rangle] \end{aligned}\right\} \quad : \quad \text{s = 1 triplets.} \tag{2.29}$$

[2] In quantum mechanical systems, a length scale is usually taken as the Bohr radius.
[3] The Heisenberg model is discussed later.

The first one denotes a singlet wavefunction, χ_s, while the last three denote triplet states, χ_t. The singlet state is odd (changes sign) under the exchange of spins, while the triplet states are even (no change in sign). These are the eigenstates of the total spin operators, namely, S^2 and S_z. It can be checked that the S^2 operator has an eigenvalue 0 for the singlet state, and 1 for the triplet states. Also the total spin, S satisfies,

$$\begin{aligned} S^2 = (S_1 + S_2)^2 &= S_1^2 + S_2^2 + 2S_1 \cdot S_2 \\ &= \frac{3}{4}\hbar^2 + \frac{3}{4}\hbar^2 + 2S_1 \cdot S_2 \\ &= \frac{3}{2}\hbar^2 + 2S_1 \cdot S_2. \end{aligned} \tag{2.30}$$

Thus, for the singlet state, the operator $S_1 \cdot S_2$ has an eigenvalue $-\frac{3}{4}\hbar^2$, and $\frac{1}{4}\hbar^2$ for the triplet states. Denoting these eigenvalues by E_s and E_t, respectively, we can write down a spin-only Hamiltonian, namely, (with $\hbar = 1$),

$$\mathcal{H} = \frac{1}{4}(E_s + 3E_t) - (E_s - E_t)S_1 \cdot S_2. \tag{2.31}$$

Readers can check that the Hamiltonian has energies E_s for the singlet state and E_t for the triplet states by operating the Hamiltonian in equation (2.31) on the states in equation (2.28) and equation (2.29). We may ignore the constant term, $\left(\frac{E_s + 3E_t}{4}\right)$ (or re-define the zero energy which is common to all states), and re-write the Hamiltonian as,

$$\mathcal{H} = -JS_1 \cdot S_2 \tag{2.32}$$

where $J = E_s - E_t$. If J is positive, the system favours parallel alignment of spins and if J is negative, it favours antiparallel alignment.

One may wish to extend the above scenario to an array of spins with full spin rotational symmetry, and interacting via nearest neighbour exchange coupling to arrive at the Heisenberg model,

$$\mathcal{H} = -J\sum_{i,\delta} \mathbf{S}_i \cdot \mathbf{S}_{i+\delta} \tag{2.33}$$

where δ refers to the neighbours of site i. This model was solved exactly by Bethe [4] and later on by the others in one dimension.

Now if one includes the orbital wavefunctions, in addition to the spin states, one can write the total eigenfunction as,

$$\begin{aligned} \psi_{1,2}(\mathbf{r}_1, \mathbf{r}_2) &= \psi_{\text{sym}}(\mathbf{r}_1, \mathbf{r}_2)\chi_s(1, 2) \\ &= \psi_{\text{antisym}}(\mathbf{r}_1, \mathbf{r}_2)\chi_t(1, 2) \end{aligned} \tag{2.34}$$

where ψ_{sym} and ψ_{antisym} denote the symmetric and antisymmetric states which are, respectively, even and odd under exchange of $\mathbf{r}_1$ and $\mathbf{r}_2$. This takes care of the total fermionic wavefunction being antisymmetric. Further, the total Hamiltonian can be written as,

$$\mathcal{H} = \mathcal{H}_1 + \mathcal{H}_2 + \mathcal{H}_{12} \tag{2.35}$$

where $\mathcal{H}_1$ and $\mathcal{H}_2$ are the single-particle Hamiltonians, and $\mathcal{H}_{12}$ denotes the potential energy term due to the exchange interaction obtained above. These can be written as the stationary state wavefunctions, $\phi(\mathbf{r}_i)$ centred at the lattice points, $\mathbf{r}_i$ where the particles are located, via,

$$J = \int \int \phi_1^*(\mathbf{r}_1)\phi_2^*(\mathbf{r}_2)\mathcal{H}_{12}(\mathbf{r}_1, \mathbf{r}_2)\phi_2(\mathbf{r}_1)\phi_1(\mathbf{r}_2)d^3\mathbf{r}_1 d^3\mathbf{r}_2. \tag{2.36}$$

The exchange interaction is very strong and can be of the order of a fraction of an eV, which is equivalent to several hundreds of Kelvin. Thus the exchange interaction is sufficient to align the spins even at room temperature. However, it decays exponentially with distance.

Let us discuss the exchange interaction, $\mathcal{H}_{12}$ in some more detail. It is the Coulomb interaction between the particles and for the simplest case, corresponding to two one-electron atoms (such as two hydrogen (H) atoms),

$$\mathcal{H}_{12}(\mathbf{r}_1, \mathbf{r}_2) = \frac{e^2}{|\mathbf{r}_1 - \mathbf{r}_2|} + \frac{e^2}{|\mathbf{R}_1 - \mathbf{R}_2|} - \frac{e^2}{|\mathbf{r}_1 - \mathbf{R}_1|} - \frac{e^2}{|\mathbf{r}_2 - \mathbf{R}_2|} \tag{2.37}$$

where $\mathbf{r}_{1,2}$ refer to the coordinates of the electrons and $\mathbf{R}_{1,2}$ denote the nuclei of the two atoms. Here the true ground state is the Heitler–London (HL) state, namely,

$$\psi_{HL} = \psi_s = \frac{1}{\sqrt{2}}[\psi_1(\mathbf{r}_1)\psi_2(\mathbf{r}_2) + \psi_1(\mathbf{r}_2)\psi_2(\mathbf{r}_1)] \tag{2.38}$$

ψ_s is the singlet state associated with χ_{00}. Of course, the HL state applies well to the atoms that are physically separated, however, for atoms in a real solid, the magnetic interaction is severely complex, and may not be restricted to a 4×4 Hilbert space (that is a simple two-body term) which we have discussed earlier. Still in cases, where the magnetic atoms (or ions) are well separated, one can extend the two-spin interaction picture for the entire system.

The prospects of ordering also depend on the dimensionality of the lattice. For example, one-dimensional spin systems really do not order at any finite temperature. The reason can be stated through the following illustration with Ising spins.

Assuming a nearest neighbour spin-only Hamiltonian as in equation (2.33), left and right in figure 2.3 denote two different phases, where the left denotes a perfectly ordered phase with an energy $-NJ$ (N being the number of bonds), while the right denotes a perfectly disordered phase with an energy $-(N - 1)J$ where only one bond

Figure 2.3. A perfectly ordered (left) and a completely disordered (right) phase in a one-dimensional spin chain.

is broken. Thus the relative energy difference between the phases in them, that is, $\frac{\Delta E}{E} \sim \frac{1}{N}$, vanishes in the limit of large N. Such a vanishingly small energy difference cannot stabilize an ordered state. Thus dimensions higher than 'one' are crucially required for magnetic ordering to exist.

In the following let us describe a couple of different methods of solving the Heisenberg model using certain approximations. The methods are:

(i) mean field theory (MFT);

(ii) Holstein–Primakoff transformation and linear spin wave theory.

2.5 Mean field theory

In the mean field approximation, each spin '*feels*' an average field due to all the other spins of the system. The approximation is valid in any dimension, however, it is more accurate as the dimensionality grows larger (the fluctuations of the mean field state compared to the exact one diminishes with increase in the number of nearest neighbours or the dimensionality). In order to implement the MFT, one decouples $\mathbf{S}_i$ from $\mathcal{H}$ in equation (2.33) which makes it a single-site Hamiltonian (that is, at a given site i) of the form,

$$\mathcal{H}_{MF}(i) = -\ \mathbf{S}_i.\ \langle \mathbf{S}_{i+\delta} \rangle\ -\ g\mu_{\mathrm{B}}\ \mathbf{B}\cdot \sum_i \mathbf{S}_i \tag{2.39}$$

where one can notice that we have included an additional applied magnetic field $\mathbf{B}_{\mathrm{ext}}$, and the second term denotes coupling with $\mathbf{B}_{\mathrm{ext}}$. Since the bracketed expression in the first term above is just a constant (being summed over nearest neighbours), the above Hamiltonian in equation (2.39) becomes a single-spin Hamiltonian in an effective field, $\mathbf{B}_{\mathrm{eff}}$, namely,

$$\mathcal{H}_{MF}(i) = -\mathbf{S}_i \cdot \mathbf{B}_{\mathrm{eff}} \tag{2.40}$$

where $\mathbf{B}_{\mathrm{eff}} = \mathbf{B}_{\mathrm{ext}} - \frac{J}{g\mu_{\mathrm{B}}} \sum_\delta \mathbf{S}_{i+\delta}$. Here g is the Landé g factor, and μ_{B} denotes the Bohr magneton. In fact $\mathbf{S}_{i+\delta}$ can be replaced by its thermal average $\langle \mathbf{S}_{i+\delta} \rangle$, so that

$$\begin{aligned} \mathbf{B}_{\mathrm{eff}} &= \mathbf{B}_{\mathrm{ext}} - \frac{J}{g\mu_{\mathrm{B}}} \sum_\delta \langle \mathbf{S}_{i+\delta} \rangle \\ &= \mathbf{B}_{\mathrm{ext}} - z\frac{J}{g\mu_{\mathrm{B}}} M \end{aligned} \tag{2.41}$$

where $M = \langle \mathbf{S}_{i+\delta} \rangle$ is the magnetization and z denotes the coordination number.

Thus essentially we have a non-interacting system, and the thermal average can be calculated as if one particular spin is in a bath of an effective magnetic field. In this scenario, the standard approach to calculate the magnetization is via computing the free energy, $F = -\ k_{\mathrm{B}}T \ln \mathcal{Z}$ where $\mathcal{Z}$ is the canonical partition function. The method is detailed in section 3.1. For convenience the results are quoted again. The partition function is written as,

$$\mathcal{Z} = \sum e^{-\beta\mathcal{H}}_{\mathcal{H}_{MF}} = \sum e^{-\beta_i \mathcal{H}_{\text{eff}}} = \sum e^{-\beta\gamma S B_{\text{eff}}} \\ = \frac{e^{\beta\gamma B_{\text{eff}}\left(S+\frac{1}{2}\right)} - e^{-\beta\gamma B_{\text{eff}}\left(S+\frac{1}{2}\right)}}{e^{\beta\gamma B_{\text{eff}}/2} - e^{-\beta\gamma B_{\text{eff}}/2}} \tag{2.42}$$

where $\gamma = g\mu_{\text{B}}$. Hence the magnetization is computed using $M = \frac{\partial F}{\partial B_{\text{eff}}}$. Thus one arrives at,

$$M = \gamma S B_S(S\beta\gamma B_{\text{eff}}) = \gamma S B_S\left(S\beta\gamma(B - \frac{zJ^2}{\gamma}M)\right) \tag{2.43}$$

$B_S(x)$ being the Brillouin function discussed earlier, and

$$x = \left(S\beta\gamma\left(B - \frac{zJ^2}{\gamma}\right)M\right).$$

The above equation is non-linear, and can be solved for the magnetization, M for given values of B and T (remember $\beta = \frac{1}{k_{\text{B}}T}$). Assuming positive values of J, the transition from a paramagnet to a ferromagnetic state is indicated by the appearance of spontaneous magnetization, (finite value of M) in the zero magnetic field limit ($B \to 0$) below a certain critical temperature, T_{c}. Thus in the limit $H \to 0$, $B_S(x)$ can be expanded for small x,

$$M \simeq -\gamma S\left(\frac{S+1}{3S}\right)\frac{zJ}{\gamma k_{\text{B}}T_{\text{c}}}M. \tag{2.44}$$

Solving for the transition temperature, T_{c},

$$T_{\text{c}} = \frac{S(S+1)}{3}zJ. \tag{2.45}$$

Thus the critical temperatures scales with the exchange interaction, J, implying that T_{c} is large for large J. Further, the magnitude of the spin S and the coordination number z (which means dimensionality) play roles in deciding the value of T_{c}.

In order to see how M varies with T below T_{c}, equation (2.43) has to be solved numerically. The solution is shown schematically in figure 2.4. Analytic results exist at the extreme limits, that is for (i) $T \ll T_{\text{c}}$ and (ii) $T \to T_{\text{c}}$ below. Let us demonstrate how to get these.

(i) At $T \ll T_{\text{c}}$, β is large, so the Brillouin function, $B_S(x)$ has to be examined for large x, which yields,

$$B_S(x) \simeq = 1 + \frac{2S+1}{S}\exp\left(-\frac{2S+1}{S}x\right) - \frac{1}{S}\exp\left(-\frac{x}{S}\right) \\ = 1 + \frac{1}{S}((2S+1)\exp(-2x) - 1)\exp\left(\frac{x}{S}\right) \\ = 1 - \frac{1}{S}\exp\left(-\frac{x}{S}\right). \tag{2.46}$$

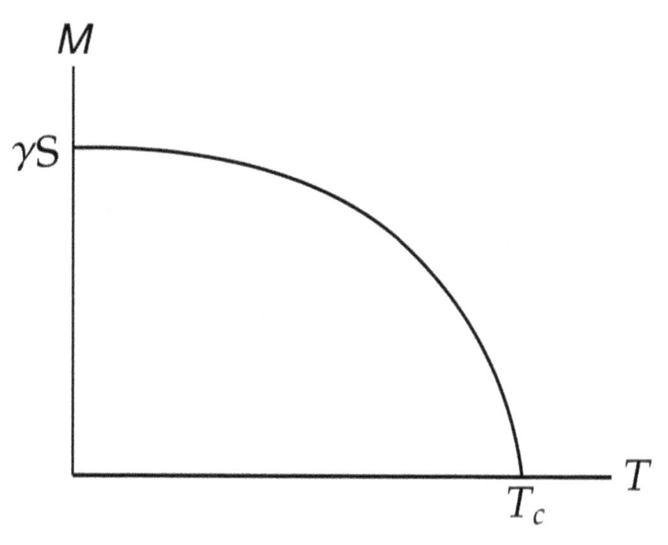

Figure 2.4. Plot of magnetization as a function of temperature. The magnetization vanishes as $(T - T_c)^{1/2}$.

Thus,

$$
\begin{aligned}
M = \gamma S B_S(x) &= \gamma S - \exp\left(-\frac{JzM}{\gamma k_B T}\right) \\
&= \gamma S - \exp\left(-\frac{3}{S+1}\frac{T_c}{\gamma k_B T}\right).
\end{aligned} \tag{2.47}
$$

In the limit of $T \to 0$ (that is, $T \ll T_c$), the second term (the exponential term) is small. Thus one gets $M \simeq \gamma S$.

(ii) For T close to T_c (from below), we need to expand $B_S(x)$ for small x, where,

$$
B_S(x) = \frac{S+1}{3S}x - \frac{(S+1)(2S^2+2S+1)}{90S^3}x^3. \tag{2.48}
$$

This yields,

$$
\begin{aligned}
M &\simeq \frac{S+1}{3}\frac{SM}{k_B T} - \frac{(S+1)(2S^2+2S+1)}{90S^3}\frac{(zJ)^3 S^3 M^{2\,3}}{\gamma k_B T} \\
&= \frac{T_c}{T}M - \frac{3(2S^2+2S+1)}{10\gamma^3 S^2(S+1)^2}\left(\frac{T_c}{T}\right)^3 M^3.
\end{aligned} \tag{2.49}
$$

Since $M \neq 0$, one can solve for M^2,

$$
M^2 \simeq \frac{10S^2(S+1)^2}{3(2S^2+2S+1)\gamma^3}\left(\frac{T}{T_c} - 1\right). \tag{2.50}
$$

Thus $M(T) \sim (T - T_c)^{1/2}$ for T approaching T_c from below, that is, from the ordered regime. The exponent 1/2 is a signature of the MFT. The temperature dependence of the magnetization is schematically shown in figure 2.4. It falls off slowly from a value γS at $T = 0$, however, there is a rapid decline in the vicinity of $T = T_c$.

2.6 Linear spin wave theory

To continue our discussion on ferromagnets, let us again consider the Heisenberg model given by equation (2.33) in a magnetic field B applied in the z-direction. For positive J, the system minimizes its energy by having all the spins aligned in the z-direction at zero temperature. At small, but finite temperatures, investigating elementary excitations of a spin system is hard to determine owing to the non-commutativity of the components of the spin operator. However, such an endeavour would still be possible if we can transform the spin operators to bosonic operators via a canonical transformation[4], known as the Holstein–Primakoff (HP) transformation [5]. The spin operators at a site i are denoted as,

$$S_i^+ = S_{ix} + iS_{iy} = \sqrt{2S}\left(1 - \frac{b_i^\dagger b_i}{2S}\right)^{1/2} b_i \tag{2.51}$$

$$S_i^- = S_{ix} - iS_{iy} = \sqrt{2S}\, b_i^\dagger\left(1 - \frac{b_i^\dagger b_i}{2S}\right)^{1/2} \tag{2.52}$$

where $S_i^\pm$ are the spin raising and the lowering operators and $b_i^\dagger (b_i)$ are the bosonic creation (annihilation) operators. It is easy to verify that the components of the spin operator obey the commutation relation,

$$\left[S_x, S_y\right] = iS_z, \qquad (\hbar = 1) \tag{2.53}$$

where the bosonic operators obey $[b, b^\dagger] = 1$ at each lattice site i. Further, S^2 and S_z commute with the Hamiltonian $\mathcal{H}$ in equation (2.33). Some essential mathematical steps include writing down,

$$\mathbf{S}^2 = S_x^2 + S_y^2 + S_z^2.$$

Thus $S_z^2 = \mathbf{S}^2 - S_x^2 - S_y^2$. Using $S^\pm = S_x \pm iS_y$

$$S_z^2 = \mathbf{S}^2 - \frac{1}{2}(S^+S^- + S^-S^+) \qquad \forall\, i \tag{2.54}$$

$\mathbf{S}^2$ and S_z acting on the common eigenfunction of $\mathbf{S}^2$ and S_z yield, $S(S+1)$ and S_z. Using the spin-boson transformation relations (equations (2.51) and (2.52)), one gets,

$$S_z = S - b^\dagger b. \tag{2.55}$$

[4] Such transformations always preserve the commutation relations of both the spin and the boson operators.

Assuming translational invariance of the system, we can do a Fourier transform,

$$b_i = \frac{1}{\sqrt{N}} \sum_k e^{-ik.\mathbf{r}_i} b_k. \tag{2.56}$$

Note that the Fourier transformed operators obey,

$$[b_k, b_{k'}^\dagger] = \delta_{k,k'}.$$

A priori, b_k and $b_k^\dagger$, which denote the annihilation and the creation of a quasiparticle, respectively, are called magnons or spin-wave excitations. The spin-wave excitations can be denoted by a small rotation of one spin with respect to its preceding neighbour, starting with a perfectly aligned configuration for the first spin. Thus it is a long wavelength excitation and requires several lattice sites (where the atoms and the ions are localized) for the spins to come back to their original orientation (see figure 2.5). In view of this slow variation (refer to figure 2.5), the expansion of the square root in the HP transformation may converge quickly. The reason being $b_i^\dagger b_i$, which denotes a spin deviation at site i has a slow variation. Thus expanding the square roots,

$$S_i^+ \simeq \sqrt{2S}\left[b_i - \frac{b_i^\dagger b_i b_i}{4S} + \cdots\right]. \tag{2.57}$$

Fourier transforming the RHS,

$$S_i^+ = \left(\frac{2S}{N}\right)^{1/2}\left[\sum_k e^{-ik.R_i} b_k - \frac{1}{4SN} \sum_{k,k',k''} e^{i(k-k'-k'').R_i} b_k^\dagger b_{k'} b_{k''} + \cdots\right]. \tag{2.58}$$

Similarly,

$$S_i^- = \left(\frac{2S}{N}\right)^{1/2}\left[\sum_k e^{i\vec{k}.R_i} b_k^\dagger - \frac{1}{4SN} \sum_{k,k',k''} e^{i(k+k'-k'').R_i} b_k^\dagger b_{k'}^\dagger b_{k''} + \cdots\right] \tag{2.59}$$

S^+ and S^- are called magnon operators. Also without employing any approximation, the z-component of the spin can be written as,

$$S_{iz} = S - b_i^\dagger b_i = S - \frac{1}{N}\sum_{k,k'} e^{i(k-k').R_i} b_k^\dagger b_{k'}. \tag{2.60}$$

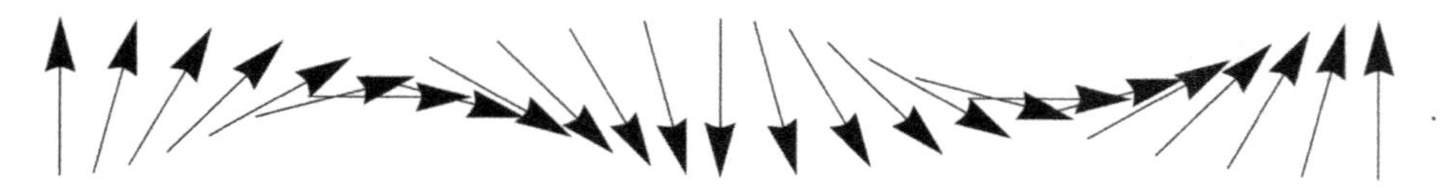

Figure 2.5. Plot showing spin wave excitation. Each spin is slightly rotated with respect to its neighbour. A complete cycle of rotation is shown. These spin wave excitations are called '*magnons*' (analogous to phonons denoting lattice excitations).

Summing over all the spins along the z-axis that is along the direction of the magnetic field,

$$S^{\text{tot}} = \sum_i S_{iz} = NS - \sum_k b_k^{\dagger} b_k \tag{2.61}$$

where, $\sum_i e^{i(k-k').R_i} = \delta_{k,k'}$ is used. Neglecting the terms cubic in magnon operators, namely, S_i^+ and S_i^-, the Hamiltonian (equation (2.33)) in k-space can be written as,

$$\begin{aligned}\mathcal{H} = \frac{JS}{N} \sum_{k,k',\vec{\delta}_i} & [e^{i(k-k').R_i} e^{ik'.\delta_i} b_k b_{k'}^{\dagger} + e^{i(k-k').R_i} e^{ik'.\delta_i} b_k^{\dagger} b_{k'} \\ & - e^{i(k-k').R_i} b_k^{\dagger} b_{k'} - e^{i(k'-k').R_i} e^{i(k-k').\delta_i} b_k^{\dagger} b_{k'}] \\ & - \frac{\mu_{\text{B}} B}{N} \sum_{i,k,k'} e^{i(k-k').R_i} b_k^{\dagger} b_{k'} + \frac{1}{2} JNZS^2\end{aligned} \tag{2.62}$$

δ connects to the neighbours, and z is the coordination number. Using definitions of the δ-function one can write,

$$\mathcal{H} = JzS \sum_k [\gamma_k b_k^{\dagger} b_k + \gamma_{-k} b_k^{\dagger} b_k - 2 b_k^{\dagger} b_k] + \frac{\mu_{\text{B}} B}{N} \sum_k b_k^{\dagger} b_k + \frac{1}{2} JNzS^2 \tag{2.63}$$

where $\gamma_k = \frac{1}{z} \sum_{\delta} e^{ik.\delta}$, and thus $\gamma_k = \gamma_{-k}$[5]. The Hamiltonian can further be simplified to arrive at,

$$\mathcal{H} = \sum_k \left[-2JzS(1 - \gamma_k) + \mu_{\text{B}} B \right] b_k^{\dagger} b_k + \frac{1}{2} JNzS^2 \tag{2.64}$$

where N denotes the total number of spins. Importantly, $\mathcal{H}$ is diagonalized with the first term being quadratic in the basis of magnon operators ($b_k^{\dagger} b_k$ denotes number operator for magnons) and the second term is merely a constant. Thus the energy eigenvalue for the magnon Hamiltonian is,

$$\omega_k = -2JSz(1 - \gamma_k) + \mu_{\text{B}} B. \tag{2.65}$$

Hence the magnons (or spin waves) disperse according to this relation. The long wavelength behaviour, say, in a square lattice is parabolic, that is, goes as $\sim k^2$ (k being the inverse of the wavelength) and owing to the $1 - \gamma_k$ factor, it vanishes as $k \to 0$. This vanishing of ω_k is in accordance with the Goldstone theorem [6], and the zero energy mode is called Goldstone mode which arises due to the spontaneous breaking of the symmetry.

[5] For example, $\gamma_k = \frac{1}{4}(\cos k_x a + \cos k_y a)$ for a two-dimensional square lattice.

2.6.1 Quantum XY model

The Heisenberg Hamiltonian in one dimension was exactly solved by H Bethe in 1931 [4]. In spite of the solution being quite elegant, it still does not enlighten us much about the basic properties, such as long-range order, etc. Rather a simple and more intuitive picture of interacting $s = \frac{1}{2}$ particles emerges from the similarity between the spin and the fermion operators. This similarity was originally exploited by Jordan and Wigner [7] who had converted an interacting problem of $s = \frac{1}{2}$ particles to that of spinless fermions via a canonical transformation, which, for obvious reasons is known as Jordan–Wigner transformation. It is applicable to a simpler variant of the Heisenberg Hamiltonian, where the coupling between the z-component of the spins is switched off. However, quite fortunately, it captures the low energy properties of the antiferromagnetic Heisenberg model.

The model is described by the Hamiltonian,

$$\mathcal{H} = J\sum_i (S_{i,x}S_{i+1,x} + S_{i,y}S_{i+1,y}). \tag{2.66}$$

The interaction Hamiltonian is restricted to the x and y components of spins among the nearest neighbours in a one-dimensional chain. The components of the spins obey usual commutation relations,

$$\left[S_i^\alpha, S_j^\beta\right] = i\epsilon_{\alpha\beta\gamma}\delta_{ij}S_i^\gamma \tag{2.67}$$

where $\hbar = 1$ and ϵ_{ijk} is the Levi-Civita tensor. The interaction does not include the z-component of spin, and hence the name XY model. Defining raising and lowering operators,

$$S_i^\pm = S_{i,x} \pm iS_{i,y}. \tag{2.68}$$

In terms of $S^\pm$, the Hamiltonian takes a form,

$$\mathcal{H} = \frac{J}{2}\sum_i (S_i^+S_{i+1}^- + S_i^-S_{i+1}^+). \tag{2.69}$$

Now, since $\mathbf{S}_i = \frac{1}{2}\boldsymbol{\sigma}_i$ and $S_i^\pm = \sigma_i^\pm$, σ being the Pauli matrices, the Hamiltonian can be written as,

$$\mathcal{H} = \frac{J}{2}\sum_i (\sigma_i^+\sigma_{i+1}^- + \text{h. c.}) \tag{2.70}$$

where h.c. denotes the Hermitian conjugate. Owing to the 2×2 structure of the Pauli matrices, the vector space is two-dimensional. Now we can derive a fermionic Hamiltonian in terms of c, $c^\dagger$ by defining,

$$c_i^\dagger = \left(\Pi_{j<1}\sigma_j^z\right)\sigma_i^+, \qquad c_i = \left(\Pi_{j<1}\sigma_j^z\right)\sigma_i^-. \tag{2.71}$$

Again these transformations preserve the commutation relations for the fermions and the spin-1/2 particles. From the commutation relation of the Pauli matrices, one can check for the fermionic anticommutation relations for c, $c^{\dagger}$. The reader is advised to go through a few steps of algebra to derive,

$$\sigma_i^z = 1 - 2c_i^{\dagger}c_i \tag{2.72}$$

or,

$$\sigma_i^z = (-1)^i c_i^{\dagger}c_i \tag{2.73}$$

such that σ_i^z can take values ±1 for $n = 0, 1$. Also the $\sigma_i^{\pm}$ are defined by,

$$\begin{aligned}\sigma_i^+ &= c_i^{\dagger} \exp\left[-\pi i \sum_{j=1}^{i-1} c_j^{\dagger}c_j\right]\\ \sigma_i^- &= \exp\left[-\pi i \sum_{j=1}^{i-1} c_j^{\dagger}c_j\right] c_i.\end{aligned} \tag{2.74}$$

One can check that,

$$\{c_i, c_i^{\dagger}\} = \frac{\{\sigma_i^-, \sigma_i^+\}}{\Pi_{j=i}^{i-1}(-1)^j c_j^{\dagger}c_j \Pi_{j'=1}^{i-1}(-1)^{j'} c_{j'}^{\dagger}c_{j'}} = 1. \tag{2.75}$$

Also, one can show that, $c_i^{\dagger}c_i = \sigma_i^-\sigma_i^+$.

A special mention is required for the commutation relations. c operators at the same site obey anticommutation relations, while at different sites obey commutation relations (similar to bosons), such that the unitary rotations obey neither bosonic nor fermionic relations.

Going back to equation (2.70), for $1 \leqslant i \leqslant (N-1)$,

$$\begin{aligned}\sigma_i^{-1}\sigma_{i+1}^{+} &= \exp\left[-\pi i \sum_{j=1}^{i-1} c_j^{\dagger}c_j\right] c_i c_{i+1}^{\dagger} \exp\left[-\pi i \sum_{j=1}^{i-1} c_j^{\dagger}c_j\right]\\ &= c_i \exp\left[-\pi i \sum_{j=1}^{i-1} c_j^{\dagger}c_j\right] \exp\left[-\pi i \sum_{j=1}^{i-1} c_j^{\dagger}c_j\right] c_{i+1}^{\dagger}\\ &= c_i \exp\left[-\pi i \sum_{j=1}^{i-1} c_j^{\dagger}c_j\right] c_{i+1}^{\dagger} = c_i(1 - 2c_i^{\dagger}c_i)c_{i+1}^{\dagger}\\ &= c_i^{\dagger}c_i.\end{aligned} \tag{2.76}$$

Similarly the conjugate is written as,

$$\sigma_{i+1}^{-}\sigma_i^{+} = c_i^{\dagger}c_{i+1}. \tag{2.77}$$

Thus after expressing the '*boundary particle*' operators $\sigma_N^-\sigma_1^+ + \sigma_1^-\sigma_N^+$ in terms of the fermion operators c_i and $c_i^\dagger$ one gets,

$$\begin{aligned}\mathcal{H} =& \frac{J}{2}\sum_i (c_{i+1}^\dagger c_i + c_i^\dagger c_{i+1}) \\ &- \frac{J}{2}\sum_i (c_1^\dagger c_N + c_N^\dagger c_1)\left(\exp\left[-\pi i \sum_{j=1}^{i-1} c_j^\dagger c_j + 1\right]\right).\end{aligned} \tag{2.78}$$

The first term on the RHS is quadratic in c operators, and describes free fermions in a closed chain. The effect of the boundary enters through the second term and can be neglected for large N, as it denotes merely a $1/N$ correction to the first term. Thus the Hamiltonian becomes,

$$\mathcal{H} \simeq \frac{J}{2}\sum_i (c_i c_{i+1}^\dagger + c_{i+1} c_i^\dagger). \tag{2.79}$$

This finally yields a quadratic term, and hence a non-interacting fermionic Hamiltonian in one dimension which can be solved exactly. The Hamiltonian in equation (2.79) commutes with the number operator, $\mathcal{N} = c_i^\dagger c_i$, that is, $[\mathcal{H}, \mathcal{N}] = 0$. Further, the z-component of the spin operator can be written as,

$$S_i^z = \frac{1}{2}[\sigma_i^\dagger, \sigma_i^-] = c_i^\dagger c_i - \frac{1}{2} \tag{2.80}$$

and $S^z = \sum_i S_i^z$. Thus each spinless fermion created by $c^\dagger$ carries $S^z = 1$.

Further, Fourier transforming the fermionic operators using,

$$c_i = \frac{1}{\sqrt{N}}\sum_k e^{ik.R_i} c_k \tag{2.81}$$

one arrives at a tight binding form,

$$H = \sum_k \epsilon_k c_k^\dagger c_k = J\sum_k \cos ka \;\; c_k^\dagger c_k \tag{2.82}$$

where $k \in [-\pi, \pi]$, and a is the lattice constant which, without any loss of generality, can be taken to be unity. Thus, from an interacting spin problem, we arrived at a non-interacting fermionic problem. The spectrum is gapless, that is, there may be gapless excitations, which implies that one extra fermion can be added to the system without any additional cost of energy at the Fermi level. However, the gapless situation will vanish if a nearest neighbour interaction term among the z-components, that is, $J_z S_i^z S_{i+1}^z$ is included.

Finally, the z-component of the spin, S^z yields,

$$S^z = \sum_i c_i^\dagger c_i = \sum_k c_k^\dagger c_k - \frac{N}{2}. \tag{2.83}$$

If we split the above sum in $k < 0$ and $k > 0$ (including $k = 0$)

$$S^z = \sum_{k>0} c_k^\dagger c_k + \sum_{k<0} (1 - c_k^\dagger c_k) - \frac{N}{2} \tag{2.84}$$

which can also be written as,

$$S^z = \sum_k sgn(\epsilon_k) c_k^\dagger c_k + \sum_{k<0} 1 - \frac{N}{2}. \tag{2.85}$$

Thus excitations with $|k| < \frac{\pi}{2}$ carry $S^z = +1$ while those with $|k| > \frac{\pi}{2}$ carry $S^z = -1$. Thus the z-component of the total spin of the ground state equals zero and hence it is non-degenerate. The same result holds for the unrestricted (that includes interaction between the z-component of the spins) antiferromagnetic Heisenberg model. In fact the ground state energy and the excitation spectrum are identical to the antiferromagnetic case where the ground state is non-degenerate. However, this is very unlike the ferromagnetic Heisenberg model, where the ground state is hugely degenerate and it carries a value for the z-component of the spin, namely, $S_z = \frac{N}{2}$.

Just to put things into perspective, here we have discussed a magnetic Hamiltonian in one dimension which has an exact solution. The solution yields a magnetic metal with gapless excitations. In the event one additionally includes a z component of the spin interaction, The spectrum becomes gapped, and hence corresponds to an insulating scenario.

2.7 Ising model of ferromagnetism: transfer matrix

Consider a spin-only model interacting via nearest neighbour exchange interaction in the presence of an external magnetic field, B. In order to solve the problem we shall use a transfer matrix technique. Again, we consider $s = \frac{1}{2}$ particles which can assume two different orientations, namely, ↑ and ↓. The Hamiltonian of such a system is written as,

$$\mathcal{H} = g\mu_B \mathbf{B} \cdot \sum_i \mathbf{S}_i - \sum_{\langle ij \rangle} J_{ij} \mathbf{S}_i \cdot \mathbf{S}_j \tag{2.86}$$

where g and μ_B are the Landé g factor and Bohr magneton, respectively. Because of the discrete possibilities of the spin orientation, the Hamiltonian can be written in a scalar form as,

$$\mathcal{H} = g\mu_B B \sum_i S_i^z - \sum_{\langle ij \rangle} J_{ij} S_i^z S_j^z. \tag{2.87}$$

Writing $m_i = S_i^z$,

$$\mathcal{H} = g\mu_B B \sum_i m_i - \sum_{\langle ij \rangle} J_{ij} m_i m_j. \tag{2.88}$$

A further variable transform, $\alpha_i = 2m_i$ and assuming a homogeneous J, that is $J_{ij} = JA\langle ij \rangle$ yields,

$$\mathcal{H} = \frac{g\mu_B B}{2}\sum_i \alpha_i - \frac{J}{2}\sum_{i=1}^{N}\alpha_i\alpha_{i+1}. \tag{2.89}$$

We can assume a periodic boundary condition denoted by, $\alpha_{N+1} = \alpha_1$ (N being the number of spins in the chain). Thus the last spin is connected to the first one, and the system is in the shape of a closed loop with no free edge. The canonical partition function for the above Hamiltonian is written as,

$$\mathcal{Z} = \sum_{\{\alpha_i\}} e^{-\beta E_n\{\alpha_i\}} = \sum_{\{n_i\}} e^{-\beta\mathcal{H}\{\alpha_i\}}, \quad \text{where} \quad \beta = \frac{1}{k_B T}. \tag{2.90}$$

Expanding the partition function, one gets [8],

$$\mathcal{Z} = \sum_{\alpha_1=\pm 1}\sum_{\alpha_2=\pm 1}\cdots\sum_{\alpha_N=\pm 1} K(\alpha_1, \alpha_2)K(\alpha_2, \alpha_3)\ldots K(\alpha_N, \alpha_1) \tag{2.91}$$

where

$$K(\alpha_1, \alpha_2) = \exp\left[-\frac{\beta g\mu_B B}{2}(\alpha_1 + \alpha_2) + \frac{\beta J}{2}\alpha_1\alpha_2\right]. \tag{2.92}$$

Writing K for $\alpha_i = \pm 1$,

$$K = \begin{bmatrix} e^{(-x+a)} & e^{-a} \\ e^{-a} & e^{(x+a)} \end{bmatrix} \tag{2.93}$$

where $x = \frac{g\mu_B B}{2k_B T}$ and $a = \frac{J}{2k_B T}$. Thus the partition function becomes,

$$Z = \mathrm{Tr}(K^N). \tag{2.94}$$

In order to obtain the partition function in the closed form, please note that, K being a 2×2 matrix, has two eigenvalues. Let us call them λ_1 and λ_2 which yields,

$$Z = \mathrm{Tr}(K^N) = \lambda_1^N + \lambda_2^N = \lambda_1^N\left[1 + \left(\frac{\lambda_2}{\lambda_1}\right)^N\right]. \tag{2.95}$$

Assuming one of them to be greater than the other, that is, $\lambda_1 > \lambda_2$. Since the partition function involve terms raised to the power N and with N being large, one can write (neglecting the second term inside the bracket),

$$Z = \lambda_1^N \tag{2.96}$$

where $\lambda_{1,2}$ are given by,

$$\lambda_{1,2} = e^a[\cosh x \pm (\sinh^2 x + e^{-4a})^{1/2}]. \tag{2.97}$$

Keeping the relevant one[6] for computing the partition function,

$$\lambda_1 = e^{a}[\cosh x + (\sinh^2 x + e^{-4a})^{1/2}]. \tag{2.98}$$

The free energy is written as,

$$F = -k_B T \ln Z = -Nk_B T \ln \lambda_1. \tag{2.99}$$

One can hence compute the magnetization, M using,

$$M = \lim_{B\to 0} -\frac{\partial F}{\partial B} = \lim_{B\to 0}\frac{Ng\mu B}{2}\left\{\frac{\sinh(g\mu_B B/2k_B T)}{\left[\sinh^2(\frac{g\mu_B B}{2k_B T}) + e^{-2J/k_B T}\right]^{1/2}}\right\} \tag{2.100}$$

which yields,

$$M = \left[\frac{Ng^2\mu_B^2}{4k_B}\frac{e^{J/k_B T}}{T}\right]B. \tag{2.101}$$

The magnetic susceptibility is given by,

$$\chi = \lim_{B\to 0}\frac{\partial M}{\partial B} = \frac{e^{J/k_B T}}{T}. \tag{2.102}$$

Finally one can derive Curie's law,

$$\begin{aligned}\frac{1}{\chi} &= Te^{-J/k_B T}\\ &= T(1 - J/k_B T + \cdots)\\ &= T - J/k_B.\end{aligned} \tag{2.103}$$

Rewriting the above relation in a more familiar form,

$$\chi = \frac{C}{T - \theta} \tag{2.104}$$

where θ is a characteristic temperature called the Curie temperature ($\theta = J/k_B$) at which ferromagnetic ordering takes place. Thus the system is paramagnetic above a temperature $T = \theta$ where the spins are randomly oriented owing thermal effects, while they align below θ. A schematic $1/\chi$ versus T plot in figure 2.6 identifies the Curie temperature, θ. This is the simplest discussion of ferromagnetism in a model Hamiltonian which arises out of exchange interaction among the neighbouring spins in the presence of a magnetic field. The known ferromagnets, such as, Fe and Ni have Curie temperatures 1093 K and 650 K, respectively.

[6] The other one will vanish in the limit of large N.

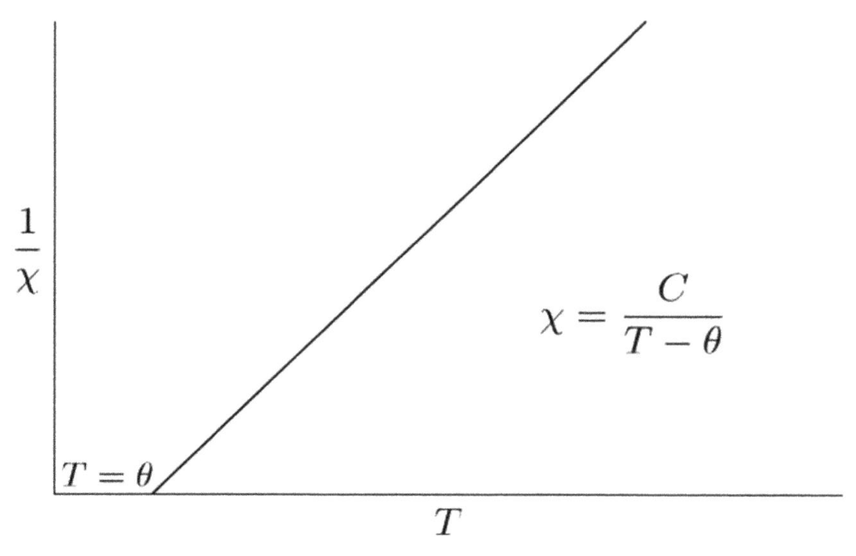

Figure 2.6. $1/\chi$ is schematically shown as a function of temperature, T. It vanishes at $T = \theta$.

It is worthwhile to point out that the above model has no spontaneous magnetization, which means that as the external magnetic field is switched off, the magnetization vanishes.

2.8 Critical exponent and the universality class

Phase transition is associated with some complex interplay between two or more competing factors. Often we can move from one phase to another by changing any control parameter, such as temperature, pressure or the magnetic field. For example, consider a ferromagnet to a paramagnet phase transition. One can go from one phase to the other only by changing the temperature (T) of a magnetic system. The spin–spin interaction (or the exchange interaction), and the thermal energy are the two competing factors here. The temperature is the controlling parameter.

A few thermodynamic observables become singular, that is, they diverge as $T \to T_c$. We can describe such a singularity of a thermodynamic quantity in terms of a power series in the vicinity of the critical temperature (T_c). The leading singularity of the power series in the asymptotic limit of the controlling parameter can be described as a critical exponent. In other words, critical exponents describe the behaviour of a physical quantity near continuous phase transitions.

Let us consider a physical quantity $F(t)$ in terms of a power series around the critical temperature as

$$F(t) = At^{k}(1 + at^{k_1} + \cdots) \tag{2.105}$$

where $t = \frac{T - T_c}{T_c}$[7] is the reduced temperature, as $T \to T_c$, $t \to 0$. Then, the leading order singularity around $t \approx 0$ will look like,

$$F(t) \sim t^k. \tag{2.106}$$

The critical exponent k can be defined as,

$$k = \lim_{t \to 0} \frac{\ln|F(t)|}{\ln|t|}. \tag{2.107}$$

The value of the critical exponent also depends on the space dimension. It may be possible that completely different systems in the same dimension may have the same values for the critical exponents. For example, the magnetic phase transition in the 2D Ising model, and the geometrical phase transition of the random percolation problem [9] have the same critical exponents. Also the critical exponents of 3D Ising model are the same as that of liquid–vapour transition. A universality class is defined by a collection of different models which have the same set of critical exponents. So same critical exponents corresponding to different phase transitions, place them in the same universality class.

The magnetization (m) with a critical exponent β around critical regime behaves as,

$$m \sim (-t)^{\beta} \tag{2.108}$$

remember, for $T < T_c$, the reduced temperature t is negative. The mean field exponent β has a value 0.5 for the Ising universality class.

2.9 Quantum antiferromagnet

Antiferromagnetism requires at least two sublattices, say A and B, inter-penetrating with each other to explain their structures and properties. Considering the Heisenberg model,

$$\mathcal{H} = -J \sum_{\langle ij \rangle} \mathbf{S}_i . \mathbf{S}_j \tag{2.109}$$

now with $J < 0$, such energy is minimized when spins in one sublattice point in the $+z$-direction (up spin), with those in the other sublattice pointing in the $-z$-direction (down spin). This state with alternate up and down spins is referred to as the classical Néel state. The lattice is called the bipartite lattice. In such a bipartite lattice, one can define a staggered magnetization, $M_S = \sum_i (-1)^i M_i$, where $M_i = \langle S_i \rangle$. Because of the pre-factor $(-1)^i$, it takes a value $+1$ for the sites on the A sublattice, while it takes -1 for those in the B sublattice or vice versa. Thus M_S assume a value NS (N being the total number of spins).

One can also define a sublattice magnetization pertaining to each of the two sublattices which takes a value $NS/2$ for the Néel state. However, the true ground state is far from classical, which is especially true at low dimensions owing to the presence of quantum fluctuations. These fluctuations lower the saturation value (Néel) of the magnetization. Let us illustrate this in the following by employing a

[7] t is not to be confused with time.

Holstein–Primakoff transformation to an antiferromagnetic Heisenberg model where one has to separately deal with the sublattices A and B in a bipartite lattice.

The transformations for the A sublattice can now be written as,

$$S_i^{A+} = S_{ix}^A + iS_{iy}^A = \sqrt{2S}\left(1 - \frac{a_i^\dagger a_i}{2S}\right)^{1/2} a_i \tag{2.110}$$

and,

$$S_i^{A-} = S_{ix}^A - iS_{iy}^A = \sqrt{2S}\, a_i^\dagger\left(1 - \frac{a_i^\dagger a_i}{2S}\right)^{1/2} \tag{2.111}$$

where, $a_i^\dagger$ (a_i) denote bosonic creation (annihilation) operators for the A sublattice. Repeating the same for the B sublattice,

$$S_i^{B+} = S_{ix}^B + iS_{iy}^B = \sqrt{2S}\left(1 - \frac{b_i^\dagger b_i}{2S}\right)^{1/2} b_i \tag{2.112}$$

and,

$$S_i^{B-} = S_{ix}^B - iS_{iy}^B = \sqrt{2S}\, b_i^\dagger\left(1 - \frac{b_i^\dagger b_i}{2S}\right)^{1/2} \tag{2.113}$$

where $b_i^\dagger$ (b_i) again denote bosonic creation (annihilation) operators for the B sublattice. The z-component of the spin operators are written as,

$$S_{iz}^A = (S - a_i^\dagger a_i), \qquad S_{iz}^B = -(S - b_i^\dagger b_i). \tag{2.114}$$

Now, as was done earlier for a ferromagnet, we introduce the Fourier transformed operators,

$$\begin{aligned} \alpha_k &= \frac{1}{\sqrt{N}}\sum_{i\in A} a_i e^{ik.R_i}, \qquad & \beta_k &= \frac{1}{\sqrt{N}}\sum_{i\in B} b_i e^{ik.R_i} \\ \alpha_k^\dagger &= \frac{1}{\sqrt{N}}\sum_{i\in A} a_i^\dagger e^{-ikR_i}, \qquad & \beta_k^\dagger &= \frac{1}{\sqrt{N}}\sum_{i\in B} b_i^\dagger e^{-ikR_i}. \end{aligned} \tag{2.115}$$

Recognizing that the periodicity of the sublattice is twice that of the crystal lattice, the effective Brillouin zone (over which the momentum index k runs) is half. Also, α_k and β_k correspond to excitation of magnons in A and B sublattices. Expanding the spin operators up to linear order for the A sublattice in α_k and β_k,

$$\begin{aligned} S_i^{A+} &\simeq \left(\frac{2S}{N}\right)^{1/2}\left[\sum_k e^{-ikR_i}\alpha_k + \cdots\right] \\ S_i^{A+} &\simeq \left(\frac{2S}{N}\right)^{1/2}\left[\sum_k e^{ikR_i}\alpha_k + \cdots\right]. \end{aligned} \tag{2.116}$$

Similarly for the B sublattice,

$$S_i^{B+} \simeq \left(\frac{2S}{N}\right)^{1/2}\left[\sum_{\vec{k}} e^{-ikR_i}\beta_k + \cdots\right]$$
$$S_i^{A+} \simeq \left(\frac{2S}{N}\right)^{1/2}\left[\sum_{k} e^{ikR_i}\beta_k + \cdots\right]. \tag{2.117}$$

The z-components yield exact expressions,

$$S_{iz}^{A} = S - \frac{1}{N}\sum_{k,k'} e_i^{i(k-k')R}\alpha_{\vec{k}}^{\dagger}\alpha_{k'}$$
$$S_{iz}^{B} = -\left(S - \frac{1}{N}\sum_{k,k'} e_i^{i(k-k')R}\beta_k^{\dagger}\beta_{k'}\right). \tag{2.118}$$

Inserting these into the Heisenberg Hamiltonian, the Hamiltonian becomes quadratic and it reads as,

$$\mathcal{H} \simeq -NzJS^2 + JSz\sum_k\left[\gamma_k(\alpha_k^{\dagger}\beta_k^{\dagger} + \alpha_k\beta_k) + (\alpha_k^{\dagger}\alpha_k + \beta_k^{\dagger}\beta_k)\right] \tag{2.119}$$

where,

$$\gamma_{\vec{k}} = \frac{1}{z}\sum_{\vec{\delta}\in \mathrm{nn}} e^{ik\delta} \tag{2.120}$$

where nn denotes the number of nearest neighbours. Although the Hamiltonian is bilinear in $\alpha_{\vec{k}}$ and β_k, unlike the ferromagnetic case, it is not readily diagonalizable. Another canonical transformation[8] involving linear combinations of α_k and β_k as,

$$\eta_{\vec{k}} = u_k\alpha_k - v_k\beta_k^{\dagger}; \qquad \zeta_k = u_k\beta_k - v_k\alpha_k^{\dagger}$$
$$\eta_k^{\dagger} = u_k\alpha_k^{\dagger} - v_{\vec{k}}\beta_k; \qquad \zeta_{\vec{k}}^{\dagger} = u_k\beta_k^{\dagger} - v_k\alpha_k. \tag{2.121}$$

To have the bosonic commutation relations intact for η_k and ζ_k, we demand that the coefficients u_k and v_k are related by,

$$u_{\vec{k}}^2 - v_{\vec{k}}^2 = 1. \tag{2.122}$$

This allows us to choose (though not uniquely), $u_k = \cosh\theta_k$ and $v_k = \sinh\theta_k$. For each k, one can put the anomalous terms (the ones which diagonalizing the Hamiltonian) to zero which yields the following condition,

$$\tanh 2\theta_k = -\gamma_k. \tag{2.123}$$

[8] These are called Bogoliubov or Bogoliubov–Valatin transformation and will be used in the chapter on superconductivity.

Thus the Heisenberg Hamiltonian becomes,

$$\mathcal{H} \simeq -NzJS^2 + NJSz\sum_k E_k\left(\eta_k^\dagger\eta_k + \zeta_k^\dagger\zeta_k + 1\right) \tag{2.124}$$

where the dispersion, E_k is given by,

$$E_k = NJSz\sqrt{1 - \gamma_k^2}. \tag{2.125}$$

A comparison with the ferromagnetic case (which yields $E_k \sim 1 - \gamma_k$), the antiferromagnetic dispersion is $\sqrt{1 - \gamma_k^2}$. Thus, here we get a linear behaviour, that is, $E_k \sim k$ as opposed to parabolic ($E_k \sim k^2$) for ferromagnets. Additionally, there are two degenerate modes for each k in the Brillouin zone owing to two sublattice structure of the lattice.

At zero temperature, there are no magnons (or excitations of the spin wave). Thus $\langle \eta_k^\dagger \eta_k \rangle$ or $\langle \zeta_k^\dagger \zeta_k \rangle$ vanish and one gets the ground state energy as,

$$E_{GS} = -NJzS(S + 1) + \sum_k E_k. \tag{2.126}$$

Introducing a constant κ, the expression for E_{GS} can be re-written as,

$$E_{GS} = -NJzS^2\left(1 + \frac{\kappa}{S}\right) \tag{2.127}$$

where κ is given by,

$$\kappa = \left(\frac{2}{N}\right)\sum_k\left[1 - (1 - \gamma_k^2)^{1/2}\right]. \tag{2.128}$$

Thus E_{GS} can be obtained for a crystal lattice.

The spin-wave excitations lower the value of the sublattice magnetization compared to its saturation value. It is possible to have a quantitative estimate for a given lattice. For either of the sublattices (A sublattice is considered here for concreteness), the average value of the z-component of the spin is written as,

$$\langle S_z^A \rangle = S - \frac{1}{N}\sum_k \langle \alpha_k^\dagger \alpha_k \rangle. \tag{2.129}$$

Using the Bogoliubov transformation,

$$\begin{aligned}\langle S_z^A \rangle &= S - \frac{1}{N}\sum_k \langle (u_k\eta_k^\dagger + v_k\zeta_k)(u_k\eta_k + v_k\zeta_k^\dagger) \rangle \\ &= S - \frac{1}{N}\sum_k (u_k^2\langle \eta_k^\dagger\eta_k \rangle + v_k^2\langle \zeta_k^\dagger\zeta_k \rangle + v_k^2) \\ &= S - \delta S_A \qquad \text{(say)}.\end{aligned} \tag{2.130}$$

Clearly the second term denotes the departure from the saturation values and hence is denoted as fluctuations. The averages $\langle \eta_k^\dagger \eta_k \rangle$ and $\langle \zeta_k^\dagger \zeta_k \rangle$ yield the Bose distribution function at finite temperatures, $f_B(\omega_k)$ $(= \frac{1}{e^{\beta\omega_k} - 1})$ such that they yield the number of bosons (spin-waves) when summed over all k in the Brillouin zone.

$$n_k = \langle \eta_k^\dagger \eta_k \rangle = \langle \zeta_k^\dagger \zeta_k \rangle = f_B(\omega_k). \tag{2.131}$$

The coefficients u_k and v_k satisfy,

$$u_k^2 + v_k^2 = \cosh 2\theta_k = \frac{1}{\sqrt{1 - \gamma_k^2}}. \tag{2.132}$$

Thus the magnitude of the fluctuations, δS_A can be written as,

$$\delta S_A = \frac{1}{N} \sum_k \left(n_k + \frac{1}{2} \right) \frac{1}{\sqrt{1 - \gamma_k^2}}. \tag{2.133}$$

Thus δS_A can be computed at a given temperature and a crystal lattice.

2.10 Itinerant electron magnetism

It is by and large true that identifying a magnetic order present in the system is far easier than understanding the origin of the order. There are competing views of studying magnetic order, which are exchange interaction between the localized electrons, and those between the itinerant electrons. Even though we feel that conceptualizing magnetic order arising either from a local or an itinerant view should not be viewed as a regimented criterion, it is often perceived as a convenience of description. We have seen so far how model Hamiltonians arise out of exchange interaction between the localized moments, so it is instructive to discuss the role of itinerant electrons in producing magnetic effects in materials.

One simple way to consider the role of itinerant electrons resulting in magnetic effects is via a magnetic impurity in a free-electron model or in the presence of bands. Such a scenario can indeed be realized in dilute magnetic semiconductors. We have shown in chapter 1 that the localized magnetic impurity causes an oscillation in the susceptibility at large distances away from the scatterer (the magnetic impurity). In a slightly different context, if a spin polarized electron gas is either ferromagnetically or antiferromagnetically coupled to an impurity, a particular kind of exchange interaction ensues, and is known as the RKKY interaction [13].

An alternate route to magnetic phenomena in itinerant systems arises from the competition between the kinetic and the potential energies. While we describe this scenario later, *a priori* it can be understood via a population imbalance of one kind of spin with respect to the other, thereby causing a decrease in the potential energy and increase in kinetic energy. If such an interplay can lower the total energy, such that it becomes favourable for the bands to split, and give rise to a net magnetization. The necessary condition for such a scenario to occur is determined by the density of states at the Fermi level and the energy scale associated with the Coulomb interaction[9].

In a physical situation, it is impossible to extricate the contributions arising out of the localized electrons from the itinerant ones. For example, magnetism in light actinides (where partially extended $5f$ orbitals are involved), is potentially different from the heavy actinides (or the lanthanides) where the $5f$ orbitals are strongly hybridized with the $6d$ and the $7s$ bands, thereby making the itinerant description to be more suited in the former, while a localized description is preferred for the latter. The same is true for the heavy fermion compounds[10], such as, $CeAl_3$, UPt_3, etc, where the localized f electrons strongly hybridize with the conduction band electrons, creating a significantly complicated scenario.

Kübler [14] first attempted to explain magnetic correlations in materials via local density functional approximation (LDA). The method incorporates electronic correlations and indicates the importance of the electronic structure in understanding magnetic properties. Specifically, the itinerant electron picture makes concrete contributions to the exploration of half-metallic ferromagnets, giant magnetoresistance (GMR) observed in multilayered systems, etc.

In the previous discussion we have seen ferromagnetism in the presence of a spin exchange interaction. It is worthwhile to investigate the ordering scenario in detail in an itinerant electron model, such as the Hubbard model. Since this discussion may be new to readers, we include an introduction of the model and emphasize its properties. We apprise readers that an elementary knowledge of band theory and second quantization are essential to understand the subsequent discussions.

A brief introduction to an interacting electronic model in terms of creation and annihilation operators is as follows,

$$\mathcal{H} = \sum_{i,i',\sigma,\sigma'} c_{i\sigma}^{\dagger} t_{ii'}^{\sigma,\sigma'} c_{i'\sigma'} + \sum_{i,i',j,j',\sigma,\sigma'} U_{ii'jj'} c_{i\sigma}^{\dagger} c_{j\sigma} c_{i'\sigma'}^{\dagger} c_{j'\sigma'} \tag{2.134}$$

where i, i', j, j' refer to site indices and σ, σ' denote the spins. The Hamiltonian includes a single-particle term (the kinetic energy), and a two-particle interaction term. There could be the presence of interaction effects involving a larger number of particles, however, they are mostly weaker (other than being unsolvable), as compared to the two-particle term, thereby making two-particle interactions good enough for describing most of the interacting systems. It may be noted that we have considered most general forms for both the terms in the Hamiltonian, however, either of them, or both may not depend upon the spin indices as shown in equation (2.134).

2.11 Magnetic susceptibility: Kubo formula

Here we shall show the calculation of the magnetic susceptibility using linear response theory. We shall be deriving Kubo formula which is essential in a variety of systems. For example, calculation of resistivity for an electron gas in the presence of

[9] It will be introduced as the Stoner criterion.

[10] The effective mass of the fermions is several times larger than the corresponding bare mass.

an external magnetic field (for example, the quantum Hall effect), or the polarizability of a dielectric in the presence of an electric field, etc. We give a thorough derivation of the Kubo formula in chapter 2.

Consider an applied magnetic field $\vec{B}(\vec{r}, t)$. It is useful to assume it to be function of both $\vec{r}$ and t for reasons that will be clear later. Assume $\mathcal{H}'$ denotes the coupling between the spin and the magnetic field and is given by,

$$\mathcal{H}' = -\int \mathbf{B}(\mathbf{r}, t) \cdot \mathbf{S}(\mathbf{r})\, d\mathbf{r} \tag{2.135}$$

where $\mathbf{S}(\mathbf{r}) = \sum_i \delta(\mathbf{r} - \mathbf{r}_i)\mathbf{S}_i$. $\mathbf{S}_i$ denotes the spin vectors at a given site i written in the Heisenberg representation. Also, here we have dropped the Bohr magneton, $\mu_B = -\frac{e\hbar}{2m}$, by taking $\mu_B = 1$. We shall compute the magnetization defined by $\langle \mathbf{S}(\mathbf{r}, t)\rangle$ induced by the magnetic field $\mathbf{B}$ via,

$$\langle \mathbf{S}(\mathbf{r}, t)\rangle = \langle \psi_0(t)|\mathbf{S}(\mathbf{r})|\psi_0(t)\rangle \tag{2.136}$$

where $|\psi_0(t)\rangle$ denotes the ground state at time t. Using the results of linear response theory, at the first order in B in a translationally invariant system,

$$\langle \mathbf{S}_i(\mathbf{r}, t)\rangle_B = \langle \mathbf{S}_i(\mathbf{r}, t)\rangle_{B=0} + \sum_j \int dt' \int d\mathbf{r}' \chi_{ij}(\mathbf{r} - \mathbf{r}', t - t')\mathbf{B}_j(\mathbf{r}, t') \tag{2.137}$$

where $\chi_{ij}(\vec{r}, t)$ is a retarded two-particle propagator, and denotes the susceptibility tensor. The susceptibility tensor is the same as the spin–spin correlation function defined by,

$$\chi_{ij}(\mathbf{r} - \mathbf{r}', t - t') = i\theta(t - t')\left\langle \left[\mathbf{S}_i(\mathbf{r}, t), \mathbf{S}_j(\mathbf{r}', t')\right]\right\rangle \tag{2.138}$$

where,

$$\begin{aligned} \theta(t - t') &= 1 \quad \text{for} \quad t > t' \\ &= 0 \quad \text{for} \quad t < t'. \end{aligned} \tag{2.139}$$

In order to proceed further, we shall Fourier transform the spin operators,

$$\mathbf{S}(\mathbf{r}) = \sum_q e^{iq\cdot\mathbf{r}}\mathbf{S}(q) \tag{2.140}$$

where the time variable (t) is withheld for brevity. $\mathbf{S}(q)$ can be written in terms of the fermion operators as,

$$\mathbf{S}(q) = \sum_{k,\alpha,\beta} c_{k+q}^{\alpha\dagger}\sigma^{\alpha\beta}c_k^{\beta}. \tag{2.141}$$

The components of $\mathbf{S}$, or the linear combinations thereof allow us to write the spin raising and lowering operators,

$$S^{\pm} = \frac{1}{2}(S_x \pm iS_y) \tag{2.142}$$

where,

$$S^{+}(\mathbf{r}) = \sum_{q} e^{iq\cdot\mathbf{r}} \sum_{k} a^{\dagger}_{k+q\uparrow} a_{k\downarrow}$$

and,

$$S^{-}(\mathbf{r}) = \sum_{q} e^{-iq\cdot\mathbf{r}} \sum_{k} a^{\dagger}_{k+q\downarrow} a_{k\uparrow}$$

Now, we can write down the transverse and the longitudinal susceptibilities χ^{-+} and χ_{zz} using the raising, lowering and z-component of the spin operators,

$$\chi^{-+}(\mathbf{r} - \mathbf{r}', t - t') = i\theta(t - t')\langle[\mathbf{S}(\mathbf{r}, t), \mathbf{S}^{\dagger}(\mathbf{r}', t)]\rangle \tag{2.143}$$

and,

$$\chi_{zz}(\mathbf{r} - \mathbf{r}', t - t') = i\theta(t - t')\langle[S_z(\mathbf{r}, t), S_z(\mathbf{r}', t)]\rangle. \tag{2.144}$$

The equation of motion for χ^{-+} can now be written as (with $\hbar = 1$),[11]

$$\begin{aligned} i\frac{\partial}{\partial t}\chi^{-+}(k, q, t) = & -\delta(t)\langle[c^{\dagger}_{k+q\downarrow}c_{\mathbf{k}\uparrow}, S^{+}(0, 0)]\rangle \\ & + i\theta(t)\langle[[c^{\dagger}_{k+q\downarrow}c_{k\uparrow}, \mathcal{H}], S^{+}(0, 0)]\rangle. \end{aligned} \tag{2.145}$$

Here the initial time t' has been set to zero and the derivative of $\theta(t)$ is $\delta(t)$. Substituting the Hubbard Hamiltonian, $\mathcal{H}$ above,

$$\begin{aligned} \left[c^{\dagger}_{k+q\downarrow}c_{k\uparrow}, \mathcal{H}\right] = & -(\epsilon_{k+q} - \epsilon_k)c^{\dagger}_{k+q\downarrow}c_{k\uparrow} \\ & + \frac{U}{N}\sum_{k',q'}(c^{\dagger}_{k+q\downarrow}c_{k-q'\uparrow}c^{\dagger}_{k'-q'\downarrow}c_{k'\uparrow} \\ & - c^{\dagger}_{k'+q'\uparrow}c_{k'\downarrow}c^{\dagger}_{k+q-q'\downarrow}c_{k\uparrow}). \end{aligned} \tag{2.146}$$

Now, the quartic operators in the second term of RHS can be dealt with in a Hartree–Fock approximation, and the combinations, such as, $\langle c^{\dagger}c\rangle$ can be retained[12].

At finite temperatures, the average values are written as,

$$\langle c^{\dagger}_{k\sigma}c_{k'\sigma'}\rangle = \delta_{kk'}\delta_{\sigma\sigma'}f_{k\sigma} \tag{2.147}$$

where $f_{k\sigma}$ denotes the distribution function for fermions with spin σ and momentum k and has a form,

$$f_{k\sigma} = \frac{1}{e^{\beta(\epsilon_{k\sigma}-\mu)} + 1} \tag{2.148}$$

[11] A time dependent external magnetic field was taken earlier precisely for this reason so that a time derivative can be taken.

[12] A $\langle c^{\dagger}c^{\dagger}\rangle$ or $\langle cc\rangle$ is relevant for studying superconducting correlations.

where $\epsilon_{k\sigma}$ denotes the band energies. Thus the second term in the above commutator is written as,

$$\frac{U}{N}\sum_{k'}(f_{k\uparrow} - f_{k+q\downarrow})c^{\dagger}_{k+k'+q\downarrow}c_{k+k'\uparrow} + (f_{k'\downarrow} - f_{k'\uparrow})c_{k+q\downarrow}c_{k\uparrow}.$$

Going back to the equation of motion (EOM), the term,

$$\langle[c^{\dagger}_{k+q\downarrow}c_{k\uparrow}, S^{+}(0)]\rangle = \sum_{k,q}\langle[c^{\dagger}_{k+q\downarrow}c_{k\uparrow}, c^{\dagger}_{k+q\uparrow}c_{k\downarrow}]\rangle$$

can be simplified to yield $(f_{k+q\downarrow} - f_{k\uparrow})$. Thus, the EOM for the transverse susceptibility, χ^{-+} can be written as,

$$\begin{aligned}\left[i\frac{d}{dt} + (\tilde{\epsilon}_{k+q\uparrow} - \tilde{\epsilon}_{k\downarrow})\right]\chi^{-+}(k, q, t) = &- \delta(t)(f_{k+q\downarrow} - f_{k\uparrow}) \\ &- (f_{k+q\downarrow} - f_{k\uparrow})\frac{U}{N}\sum_{k'}\chi^{-+}(k', q, t)\end{aligned} \tag{2.149}$$

where the renormalized band energies $\tilde{\epsilon}_{k\sigma} = \epsilon_k - \frac{U}{N}\sum f_{k\sigma}$. Thus the one-particle energies are modified by the interaction term as shown here.

The dynamical susceptibility, $\chi(\omega)$, is calculated by doing a Fourier transform,

$$\chi(\omega) = \int_{-\infty}^{\infty} dt\chi(t)e^{i\omega t}. \tag{2.150}$$

Thus,

$$\chi(k, q, \omega) = \frac{(f_{k\uparrow} - f_{k+q\downarrow})\left(1 + \frac{U}{N}\chi^{-+}(q, \omega)\right)}{\omega + \tilde{\epsilon}_{k+q\uparrow} - \tilde{\epsilon}_{k\downarrow}} \tag{2.151}$$

where $\chi^{-+}(q, \omega) = \sum_k \chi^{-+}(k, q, \omega)$, one can write the expression for susceptibility as,

$$\chi^{-+}(q, \omega) = \frac{\chi_0^{-+}(q, \omega)}{1 - U\chi_0^{-+}(q, \omega)} \tag{2.152}$$

where

$$\chi_0^{-+}(q, \omega) = \frac{1}{N}\sum_k \frac{f_{k\uparrow} - f_{k+q\downarrow}}{\omega - (\tilde{\epsilon}_{k\downarrow} - \tilde{\epsilon}_{k+q\uparrow}) + i\eta} \tag{2.153}$$

is the free susceptibility tensor. The $+i\eta$ is added in the denominator as is usually done for a propagator [15], with the positive sign referring to the retarded propagator. The divergence of χ^{-+} denotes an instability in the system, and hence signals a phase transition. This instability is indicated by the divergence of $\chi^{-+}(q, \omega)$, which happens when,

$$1 - U\chi_0^{-+}(q, \omega) = 0 \tag{2.154}$$

is satisfied. In particular, at $\omega = 0$, that is, when the external magnetic field is time independent, the system is lossless. Thus, $\Im\chi^{-+}(q, \omega = 0) = 0$. In such a static situation, any instability in the system is a signature of the ground state instability. Further, the energies of the low-lying excited states begin to merge with that of the ground state. Hence, we can rephrase the instability criterion as,

$$U\chi_0^{-+}(q, 0) = 1 \tag{2.155}$$

where,

$$\chi_0^{-+}(q, 0) = \frac{1}{N}\sum_k \frac{f_k - f_{k+q}}{\epsilon_{k+q} - \epsilon_k}. \tag{2.156}$$

Thus, at any arbitrary value of q, the instability condition is associated with a critical value of the interaction U,

$$U_c = \frac{1}{\chi_0^{-+}(q, 0)}. \tag{2.157}$$

In order to relate $\chi_0^{-+}(q, 0)$ to something physical, we may note that the terms function f_{k+q} and ϵ_{k+q} can be expanded in a Taylor series in the large wavelength limit, namely, $q \to 0$,

$$f_{k+q} \approx f_k + q \cdot \frac{\partial \epsilon_k}{\partial k}\frac{\partial f_k}{\partial \epsilon_k} \tag{2.158}$$

$$\epsilon_{k+q} \approx \epsilon_k + q \cdot \frac{\partial \epsilon_k}{\partial k} \tag{2.159}$$

$$\lim_{q\to 0}\chi^{-+}(q, 0) = \frac{1}{N}\sum_k\left(-\frac{\partial f_k}{\partial \epsilon_k}\right) = \frac{1}{N}\sum_k \delta(\epsilon - \epsilon_k) \tag{2.160}$$

where we have replaced the derivative of the Fermi function by a δ-function which is valid at zero temperature. Moreover,

$$\frac{1}{N}\sum_k \delta(\epsilon - \epsilon_k) = N(\epsilon)$$

$N(\epsilon)$ being the density of states (DOS). This finally yields us the Stoner criterion that we are familiar with, namely,

$$UN(\epsilon_{\mathrm{F}}) = 1. \tag{2.161}$$

Thus, an instability corresponding to $q = 0$ leads to a tendency for the system to acquire a finite magnetization.

However, instability may occur even at finite value of q. To remind ourselves, the spin arrangement in antiferromagnets has an ordering wave vector $q = (\frac{\pi}{a}, \frac{\pi}{a}, \frac{\pi}{a})$ such that $e^{iq \cdot r}$ changes sign as one goes from one lattice site to its neighbour. In this case, the Fermi surface for the half-filled band (one electron per lattice site) coincides with the Brillouin zone. Thus, as opposed to a ferromagnet, an infinitesimal interaction strength, U may cause an instability in a system with one conduction electron per atom, inducing a transition from a metal to an insulating antiferromagnet.

To make the ongoing discussion more concrete, let us consider a crystal lattice, and more importantly, a specific form of the interaction in this case, the interaction is between the electrons at the same lattice site, and obeying Pauli exclusion principle. In particular, we wish to talk about the Hubbard model, study its properties, and explore its utility for studying magnetic properties of solids.

2.12 Hubbard model: an introduction

In a real solid there are atoms or ions which are periodically placed and the electrons are usually free to move through such an array. The atoms or ions have very complex energy levels (or orbitals). The Hubbard model [10, 11] simplifies the description of the constituent atoms with a periodic array of sites with a single energy level. This serves as an approximate description for materials where one energy band is in the vicinity of the Fermi surface, and hence only one orbital is important. With such a postulate, the Hilbert space of the model is restricted to four choices, which are, $|0\rangle$, $|\uparrow\rangle$, $|\downarrow\rangle$, $|\uparrow\downarrow\rangle$. When the electron density is sufficiently large in such a scenario, the electrons can interact pairwise via a screened Coulomb potential with the largest interaction being between the two electrons residing on the same site, that is, for a '*doublet*' state, $|\uparrow\downarrow\rangle$. Thus all the other three states are approximated in a manner as if they are experiencing no potential. The Hubbard model enunciates an interaction energy, U, if a particular site is doubly occupied (needless to say, by opposite spins to conform with the Pauli exclusion principle). At a site i, this simplified model is represented by an interaction energy of the form $Un_{i\uparrow}n_{j\downarrow}$. Any term such as $Vn_{i\sigma}n_{j\sigma}$, where $i \neq j$ are excluded here and are included in the extended version of the Hubbard model.

Thus, putting these ideas together, the Hubbard Hamiltonian in the grand canonical ensemble is written as,

$$\mathcal{H} = -t\sum_{\langle ij\rangle,\sigma}(c_{i\sigma}^{\dagger}c_{j\sigma} + \text{h. c.}) - \mu\sum_{i,\sigma}n_{i,\sigma} + U\sum_{i}n_{i\uparrow}n_{i\downarrow}. \tag{2.162}$$

The first term being the kinetic energy, which illustrates hopping of electrons from a lattice site j to a site i with an energy scale t that could be determined by the overlap of orbitals for the neighbouring atoms (c, $c^{\dagger}$ being the fermion annihilation and creation operators). The second term is the chemical potential which fixes the electron density and the last term is the most 'simplified' Hubbard (onsite) interaction term that we have discussed earlier. It says that a doubly occupied site, such as, $|\uparrow\downarrow\rangle$ will have an energy U. Thus if U is large (see discussion below), it will cost large energy to form a doubly occupied site, and would correspond to an insulator, known as a Mott insulator.

2.13 Symmetries of the Hubbard model

Exploring symmetries in a model Hamiltonian can be quite helpful in understanding concepts that govern various physical phenomena. If the onsite interaction, U is uniform at all lattice sites (most commonly made assumption), then the U-term is invariant under all symmetry operations of the lattice. Additionally, the spin rotational invariance and the particle–hole symmetry are important ingredients for understanding magnetism and the electronic properties.

2.13.1 Spin-rotational invariance

The kinetic energy of the model describe hopping of electrons for both spins from one lattice site to another. Since this term has got nothing to do with spins, it is invariant under the rotation of spins. Now, let us consider the interaction term, $n_{i\uparrow}n_{i\downarrow}$.

$$\begin{aligned} n_{i\uparrow}n_{i\downarrow} &= c_{i\uparrow}^{\dagger}c_{i\uparrow}c_{i\downarrow}^{\dagger}c_{i\downarrow} = c_{i\uparrow}^{\dagger}c_{i\uparrow}(1 - c_{i\downarrow}c_{i\downarrow}^{\dagger}) \\ &= n_{i\uparrow} - c_{i\uparrow}^{\dagger}c_{i\uparrow}c_{i\downarrow}c_{i\downarrow}^{\dagger} = n_{i\uparrow} - c_{i\uparrow}^{\dagger}c_{i\downarrow}c_{i\uparrow}c_{i\downarrow}^{\dagger} \\ &= n_{i\uparrow} - S_i^{+}S_i^{-}. \end{aligned} \tag{2.163}$$

Here we have introduced the relation between the spin and the electron operators as,

$$S_i^{\gamma} = \sum_{\alpha,\beta} c_{i\alpha}^{\dagger}\sigma_{\alpha\beta}^{\gamma}c_{i\beta}. \tag{2.164}$$

Here, for once, we use α, β to denote $\uparrow$ and $\downarrow$ spins. Since σ^{γ} denote $(\sigma_x, \sigma_y, \sigma_z)$ are the components of the Pauli matrix. We can also write, following the above, for the LHS,

$$n_{i\uparrow}n_{i\downarrow} = n_{i\downarrow} - S_i^{-}S_i^{+} \tag{2.165}$$

owing to the exclusion principle, $n_{i\sigma}^2 = n_{i\sigma}$ (which is the property of an idempotent matrix) and squaring $S_i^z = \frac{1}{2}(n_{i\uparrow} - n_{i\downarrow})$, one gets,

$$(S_i^z)^2 = \frac{1}{4}(n_{i\uparrow} + n_{i\downarrow} - 2n_{i\uparrow}n_{i\downarrow}). \tag{2.166}$$

Thus the interaction term takes the form,

$$\mathcal{H}_{\text{int}} = \frac{UN}{2} - \frac{2U}{3}\sum_i \vec{S}_i^{\,2}. \tag{2.167}$$

It is clearly seen that this term has spin rotational invariance as $\vec{S_i}^2$ has eigenvalues given by,

$$\begin{aligned} \vec{S_i}^2 &= \frac{3}{4} \qquad \text{for} \quad |\uparrow\rangle, |\downarrow\rangle \\ &= 0 \qquad \text{for} \quad |0\rangle, |\uparrow\uparrow\rangle, |\downarrow\downarrow\rangle |\uparrow\downarrow\rangle. \end{aligned} \tag{2.168}$$

If one notices the second term in equation (2.167) for a moment, it is large for uncompensated configurations, such as $|\uparrow\rangle$ and $|\downarrow\rangle$. Thus the Hubbard term seeks for such configurations, compared to the other two.

2.13.2 Particle–hole symmetry

We can write down the interaction term in a symmetrized form as,

$$\mathcal{H}_{\text{int}} = U\sum_i \left(n_{i\uparrow} - \frac{1}{2}\right)\left(n_{i\downarrow} - \frac{1}{2}\right). \tag{2.169}$$

We are going to show particle–hole symmetry of the Hamiltonian which is important, since it provides useful mappings between the repulsive (positive U) and the attractive (negative U) Hubbard Hamiltonians.

It is important to introduce the concept of bipartite lattice in this context. The entire lattice here splits into two sublattices of the types A and B (which may or may not mean same type of atoms or ions), where A-atoms have neighbours as B-atoms, and vice versa. A square lattice and a honeycomb lattice are examples of bipartite lattices.

Under a particle–hole transformation, the $c_{i\sigma}$ and $c_{i\sigma}^\dagger$ operators transform into a different set of operators via,

$$d_{i\sigma}^\dagger = (-1)^i c_{i\sigma}. \tag{2.170}$$

The $(-1)^i$ takes a value +1 in one sublattice and −1 in the other sublattice. The nomenclature of particle–hole transformation is aptly justified since,

$$d_{i\sigma}^\dagger d_{i\sigma} = 1 - c_{i\sigma}^\dagger c_{i\sigma}. \tag{2.171}$$

It can be easily checked that the particle and hole occupations $n = c_{i\sigma}^\dagger c_{i\sigma} = 0, 1$ are interchanged under this transformation. The kinetic energy of course is an invariant under particle–hole transformation, that is,

$$c_{i\sigma}^\dagger c_{j\sigma} \rightarrow (-1)^{i+j} d_{i\sigma} d_{j\sigma}^\dagger = d_{j\sigma}^\dagger d_{i\sigma} \tag{2.172}$$

where we have used fermionic anticommutation relation, and in a bipartite lattice $(-1)^{i+j} = -1$.

Now let us look at the interaction term, namely $U\left(n_{i\uparrow} - \frac{1}{2}\right)\left(n_{i\downarrow} - \frac{1}{2}\right)$. It is easy to check that,

$$U\left(n_{i\uparrow}-\frac{1}{2}\right)\left(n_{i\downarrow}-\frac{1}{2}\right)=Un_{i\uparrow}n_{i\downarrow}-\frac{U}{2}(n_{i\uparrow}+n_{i\downarrow})+\frac{U}{4} \tag{2.173}$$

with the last two terms being constants, the first term on the RHS represents the familiar Hubbard interaction. One can trivially show that the form is preserved under the particle–hole transformation, and one gets the same form for the Hubbard Hamiltonian in terms of the d (hole)-operators. Readers are encouraged to complete a few steps of algebra to convince themselves.

2.13.3 Extreme limits of the Hubbard model

It is instructive to explore two opposite limits of the Hubbard model. They are non-interacting limit, ($U = 0$) and the opposite of that, which is, $t = 0$, when the system splits into collection of individual atoms. The two limits are more familiarly categorized as $U/t \to 0$ and $U/t \to \infty$, respectively. In the non-interacting or the band limit, one gets the tight binding Hamiltonian,

$$\mathcal{H}=\sum_{k,\sigma}(\varepsilon_k-\mu)n_{k\sigma}=\sum_{k}\xi_k n_{k\sigma} \tag{2.174}$$

with $\xi_k = (\varepsilon_k - \mu)$. Now let us make the case more concrete by taking a two-dimensional square lattice, for which the band energies are given by,

$$\varepsilon_k = -2t(\cos k_x a + \cos k_y a) \tag{2.175}$$

a being the lattice constant. Changing μ smoothly from $-4t$ to $+4t$ ($\pm 4t$ denote the band edges, resulting in a bandwidth of $8t$), n changes from 0 to 2. Of course, in the absence of the interaction, the Hamiltonian in equation (2.174) endorses a metallic state, which can further be confirmed from the calculation of the compressibility, $\kappa = (\frac{\partial^2 E}{\partial n^2})^{-1}$ at $T = 0$ which is proportional to the density of states. $\kappa \neq 0$ points towards a metallic behaviour. Further, the band energies given in equation (2.175) are shown in the surface plots in figure 2.7 where we have taken $t = 1$ and $a = 1$.

On the other hand, in the extreme strong coupling limit or the atomic limit ($t = 0$), the Hubbard Hamiltonian is,

$$\mathcal{H}=\sum_{i}Un_{i\uparrow}n_{i\downarrow} \tag{2.176}$$

where the chemical potential or the atomic energy has been set to zero. The ground state of this model is tremendously degenerate. For N sites at half-filling (number of electrons is also N) the spin degeneracy alone is 2^N. Thus the degeneracy corresponding to $U \to \infty$ is exponentially large. However, an effective spin Hamiltonian can be obtained which is also useful for the study of magnetism. At half-filling, in the limit $U/t \to \infty$, one can derive a Heisenberg model from the Hubbard model. Here we sketch the derivation pictorially for two sites i and j (see figure 2.8). Thus the energy cost is $E_{\uparrow\downarrow} = -\frac{2t^2}{U}$. The factor 2 in the right arises because hopping can occur from left to right or right to left. Thus a state with two spins in a

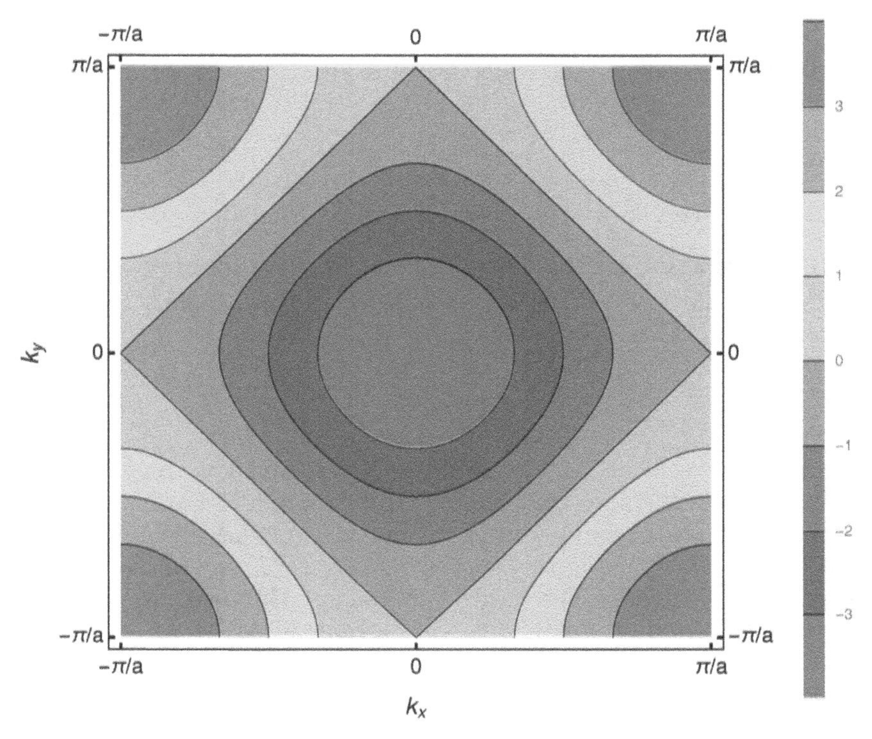

Figure 2.7. Contour plot of the dispersion surface $\varepsilon_k = -2t(\cos k_x a + \cos k_y a)$. The shapes shown here, for example, blue circle at the center (corresponding to low filling), green rhombus (half-filling), etc, demonstrate constant energy surfaces.

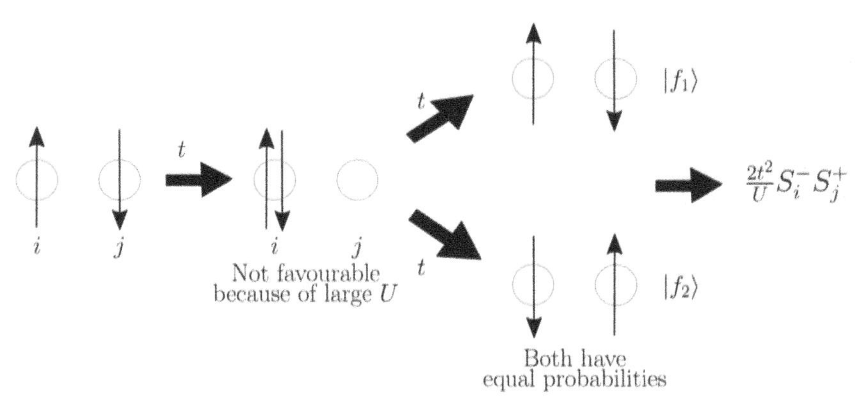

Figure 2.8. Second order in t (think of perturbation theory with an energy denominator).

singlet configuration can gain kinetic energy by tunnelling to doubly occupied states. However, there will be no hopping corresponding to,

$$\uparrow\uparrow \cdot\ E_{\uparrow\uparrow} = 0$$

owing to the Pauli principle. This leads to an effective Heisenberg Hamiltonian valid for two spins,

$$\mathcal{H}_{\text{Heisenberg}} = JS_1 \cdot S_2 \tag{2.177}$$

where $J = \frac{2t^2}{U}$. In fact in a d-dimensional hypercubic lattice with z neighbours, $J = \frac{zt^2}{U}$. If we extend the case for the two-site problem to an array of sites with full spin rotational symmetry (over a unit sphere) for the spin angular momentum, we obtain the familiar Heisenberg model,

$$\mathcal{H} = J\sum_{\langle ij \rangle} \mathbf{S}_i \cdot \mathbf{S}_j. \tag{2.178}$$

We wish to make it clear that we shall study the magnetic phenomena via an electronic model, for example, the fermionic Hubbard model. Except in $d = 1$, where the model can be solved exactly using Bethe ansatz, in higher dimensions, there are no exact solutions. We shall mostly concentrate on $d = 2$, and hence use controlled approximations to solve the Hubbard Hamiltonian. The reader should be careful and clear in mind about the inadequacies of the approximation used, and the extent of their validity.

Consider a generic Hamiltonian of the form,

$$\mathcal{H} = \mathcal{H}_0 + \mathcal{H}'. \tag{2.179}$$

Usually the eigensolutions of $\mathcal{H}_0$ are completely known. Further, it may be assumed that the eigenvalues of $\mathcal{H}_0$ are bunched into different groups. The energy eigenvalues in each group are located in close vicinity to each other, however, the levels belonging to different groups are separated far apart. Let us label the groups by α, β, etc, while the energy levels in each group are denoted by, p, q, etc. For example, $E_{p\alpha}$ denotes the pth energy level in the group α.

Without any further conditions imposed, $\mathcal{H}'$ should have matrix elements involving eigenstates from different groups. Mathematically stated, this is equivalent to $\langle \alpha, p|\mathcal{H}'|\beta, q\rangle \neq 0$ for $\alpha \neq \beta$. However, if $\mathcal{H}'$ is weak, that is, it can be used as a perturbation term, then we can do a unitary transformation on $\mathcal{H}'$, and hence obtain an effective Hamiltonian, $\mathcal{H}_{\text{eff}}$. Thus $\mathcal{H}_{\text{eff}}$ does not have matrix elements between different groups and only the elements involving states within a group. The matrix elements of $\mathcal{H}_{\text{eff}}$ up to second order in $\mathcal{H}'$ are written as [16],

$$\begin{aligned}\langle \alpha, p|\mathcal{H}_{\text{eff}}|\beta, q\rangle = {}& E_\alpha \delta_{p,q} + \langle p, \alpha|\mathcal{H}'\alpha, q\rangle \\ & + \frac{1}{2}\sum_{\beta,k\neq\alpha} \langle \alpha, p|\mathcal{H}'|\beta, k\rangle\langle \beta, k|\mathcal{H}'|\alpha, q\rangle \\ & \left[\frac{1}{E_{\alpha,p} - E_{\beta,k}} + \frac{1}{E_{\alpha,q} - E_{\beta,k}}\right]\end{aligned} \tag{2.180}$$

$\mathcal{H}_{\text{eff}}$ thus yields matrix elements involving a particular group, say α, and hence is diagonal.

We shall apply the above calculations to the Hubbard Hamiltonian (equation (2.162)) and aim to formally arrive at a superexchange Hamiltonian (as depicted earlier pictorially) in the following. Being one of the simplest cases, let us consider two sites, namely, 1 and 2, and two electrons with spins $\uparrow$ and $\downarrow$. The kinetic energy, t can be taken as the perturbation, that is, $t \ll U$. There are six states corresponding to the extreme strong coupling limit, that is, $t = 0$. They are,

(i) both are $\uparrow$-spin particles and hence belong to different sites (Pauli's exclusion principle), that is,

$$|1\rangle = c_{1\uparrow}^{\dagger}c_{2\uparrow}^{\dagger}|0\rangle, \qquad \text{with energy,} \qquad E = 0$$

(ii) both are $\downarrow$-spin particles and hence belong to different sites (Pauli's exclusion principle), that is,

$$|2\rangle = c_{2\downarrow}^{\dagger}c_{2\downarrow}^{\dagger}|0\rangle, \qquad \text{with energy,} \qquad E = 0$$

(iii) $\uparrow$-spin particle in site 1, $\downarrow$-spin particle in site 2, that is,

$$|3\rangle = c_{1\uparrow}^{\dagger}c_{2\downarrow}^{\dagger}|0\rangle, \qquad \text{with energy,} \qquad E = 0$$

(iv) $\downarrow$-spin particle in site 1, $\uparrow$-spin particle in site 2, that is,

$$|4\rangle = c_{1\downarrow}^{\dagger}c_{2\uparrow}^{\dagger}|0\rangle, \qquad \text{with energy,} \qquad E = 0$$

(v) $\uparrow$-spin particle in site 1, $\downarrow$-spin particle in site 1, that is,

$$|5\rangle = c_{1\downarrow}^{\dagger}c_{1\uparrow}^{\dagger}|0\rangle, \qquad \text{with energy,} \qquad E = U$$

(vi) $\uparrow$-spin particle in site 2, $\downarrow$-spin particle in site 2, that is,

$$|6\rangle = c_{2\downarrow}^{\dagger}c_{2\uparrow}^{\dagger}|0\rangle, \qquad \text{with energy,} \qquad E = U.$$

The purpose of writing down the effective Hamiltonian is to eliminate the last two states, that is, (v) and (vi), with large energy (of magnitude U).

We can now switch on a small hopping term, t, so that the Hamiltonian can be written as, $\mathcal{H} = \mathcal{H}_0 + \mathcal{H}'$, where,

$$\begin{aligned} \mathcal{H}_0 &= U(n_{1\uparrow}n_{1\downarrow} + n_{2\uparrow}n_{2\downarrow}) \\ \mathcal{H}' &= -\,t(c_{1\uparrow}^{\dagger}c_{2\uparrow}^{\dagger} + c_{1\downarrow}^{\dagger}c_{2\downarrow}^{\dagger} + c_{2\uparrow}^{\dagger}c_{1\uparrow}^{\dagger} + c_{2\downarrow}^{\dagger}c_{1\downarrow}^{\dagger}). \end{aligned} \tag{2.181}$$

Let us evaluate a non-zero matrix element, for example,

$$\langle 6|\mathcal{H}'|3\rangle = -\,t\langle 0|c_{2\downarrow}c_{2\uparrow}\left(c_{1\uparrow}^{\dagger}c_{2\uparrow} + c_{1\downarrow}^{\dagger}c_{2\downarrow} + c_{2\uparrow}^{\dagger}c_{1\uparrow} + c_{2\downarrow}^{\dagger}c_{1\downarrow}\right)\left(c_{2\downarrow}^{\dagger}c_{1\uparrow}^{\dagger}\right)|0\rangle.$$

There are four terms here, and they can be shown to have the following values, namely,

(a) $-t\langle 0|\left(c_{2\downarrow}c_{2\uparrow}c_{1\uparrow}^{\dagger}c_{2\uparrow}c_{2\downarrow}^{\dagger}c_{1\uparrow}^{\dagger}\right)|0\rangle|0\rangle = 0.$

(b) $-t\langle 0|\left(c_{2\downarrow}c_{2\uparrow}c_{1\downarrow}^{\dagger}c_{2\downarrow}c_{2\downarrow}^{\dagger}c_{1\uparrow}^{\dagger}\right)|0\rangle|0\rangle = 0.$
(c) $-t\langle 0|\left(c_{2\downarrow}c_{2\uparrow}c_{2\uparrow}^{\dagger}c_{1\uparrow}c_{2\downarrow}^{\dagger}c_{1\uparrow}^{\dagger}\right)|0\rangle|0\rangle = -t.$
(d) $-t\langle 0|\left(c_{2\downarrow}c_{2\uparrow}c_{2\downarrow}^{\dagger}c_{1\downarrow}c_{2\downarrow}^{\dagger}c_{1\uparrow}^{\dagger}\right)|0\rangle|0\rangle = 0.$
(e) $\langle 5|\mathcal{H}'|3\rangle = -t.$
(f) $\langle 6|\mathcal{H}'|4\rangle = -t.$
(g) $\langle 5|\mathcal{H}'|4\rangle = t.$

Now, up to second order in t, using equation (2.180), one gets,

$$\begin{aligned}
\langle 3|\mathcal{H}_{\text{eff}}|3\rangle &= -\frac{2t^2}{U} = \langle 4|\mathcal{H}_{\text{eff}}|4\rangle \\
\langle 3|\mathcal{H}_{\text{eff}}|4\rangle &= \frac{2t^2}{U} = \langle 4|\mathcal{H}_{\text{eff}}|3\rangle \\
\langle 1|\mathcal{H}_{\text{eff}}|1\rangle &= \langle 1|\mathcal{H}_{\text{eff}}|2\rangle = \langle 2|\mathcal{H}_{\text{eff}}|1\rangle = \langle 2|\mathcal{H}_{\text{eff}}|2\rangle = 0.
\end{aligned} \tag{2.182}$$

Thus $\mathcal{H}_{\text{eff}}$ connects the states $|3\rangle$ and $|4\rangle$ only up to order $\frac{t^2}{U}$. Thus in the basis of $|3\rangle$ and $|4\rangle$, one gets the following form for $\mathcal{H}_{\text{eff}}$, namely,

$$\mathcal{H}_{\text{eff}} = \begin{bmatrix} -\dfrac{2t^2}{U} & \dfrac{2t^2}{U} \\ \dfrac{2t^2}{U} & -\dfrac{2t^2}{U} \end{bmatrix}. \tag{2.183}$$

The eigenvalues are $E = 0$ and $E = -\frac{4t^2}{U}$. We show the energies and the corresponding states schematically in figure 2.9. Thus we get an effective superexchange Hamiltonian of the form,

$$\mathcal{H}_{\text{eff}} = J\mathbf{S}_1 \cdot \mathbf{S}_2, \quad \text{with} \quad J = \frac{4t^2}{U}. \tag{2.184}$$

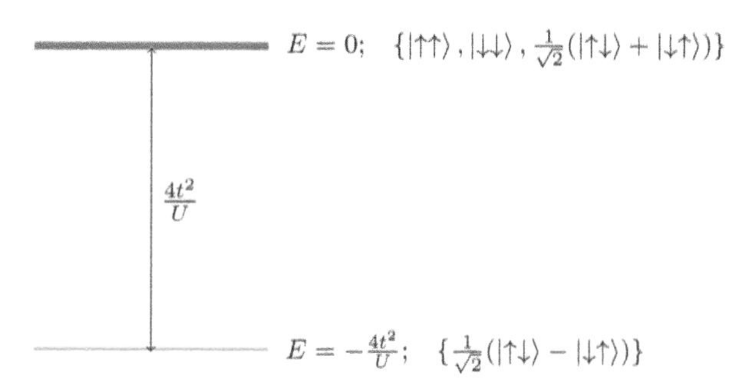

Figure 2.9. The energy levels and the corresponding states are shown up to order t^2/U.

2.14 Ferromagnetism in Hubbard model: Stoner criterion

To explore ferromagnetism in Hubbard model, we write the Hamiltonian once again on a lattice,

$$\mathcal{H} = -t\sum_{\langle ij\rangle,\sigma}(c_{i\sigma}^{\dagger}c_{j\sigma} + \text{h.c.}) + U\sum_{i} n_{i\uparrow}n_{i\downarrow}. \tag{2.185}$$

The model can be solved within a mean field approximation, which decouples the quartic term (in terms of the fermionic operator) into quadratic ones. This is known as the Hartree–Fock approximation which allows,

$$U\sum_{i} n_{i\uparrow}n_{i\downarrow} \rightarrow U\sum_{i}\langle n_{i\uparrow}\rangle n_{i\downarrow} + U\sum_{i}\langle n_{i\downarrow}\rangle n_{i\uparrow}. \tag{2.186}$$

Let us provide a physical feel for the mean field approximation and its validity. Suppose a student is sitting in the class and, he (she) has friends who are sitting right beside him (her) and there are some friends who are sitting very far away, so that no communication (or interaction) is possible while the class is going on. However, the students who are sitting in the vicinity, can interact with that particular student. The situation is similar to a many-body system, where the charge carriers interact strongly with other carriers that are in the vicinity, and less interaction ensues with the ones that are far off, and even less interaction with those that are farther off. However, one can reduce the complexity of the problem by making an approximation that an average field is acting on the charge carrier (that is, one particular electron) due to the presence of all other carriers (just like an average effect on a particular student due to all other students in the class). This reduces the complicated many-body phenomenon to a single-particle problem (and hence solvable) at the expense of ignoring fluctuations which arise out of a differential nature of the interaction between charge carriers at the vicinity with those which are further away. This allows us to replace the operators n_i by their expectation values, thereby making the quartic operators at each lattice site into quadratic ones.

Coming back to the context, in a ferromagnet, since one species of spin, that is, say $\uparrow$-spin outnumbers the other one, that is, $\downarrow$-spin (see figure 2.10), we can assume,

$$\langle n_{i\uparrow}\rangle \gg \langle n_{i\downarrow}\rangle. \tag{2.187}$$

Thus magnetization, m at a given lattice site defined by, $m_i = \langle n_{i\uparrow}\rangle - \langle n_{i\downarrow}\rangle = m$ can be computed via,

$$m = \int_{\varepsilon_A}^{\varepsilon_F} d\varepsilon N_{\uparrow}(\varepsilon) - \int_{\varepsilon_B}^{\varepsilon_F} d\varepsilon N_{\downarrow}(\varepsilon). \tag{2.188}$$

where $N_\sigma(\varepsilon)$ denotes the density of states for spin-σ and ϵ_A and ϵ_B denote the bottom of the bands for $\uparrow$ and $\downarrow$-spins, respectively. We may choose a level symmetrically between the bottom of the $\uparrow$ and $\downarrow$-spin bands denoted by the dashed horizontal line in the figure. Let us assume that for the energy gap between this line and $\varepsilon_{A,B}$ be denoted by Δ one may write the integral as,

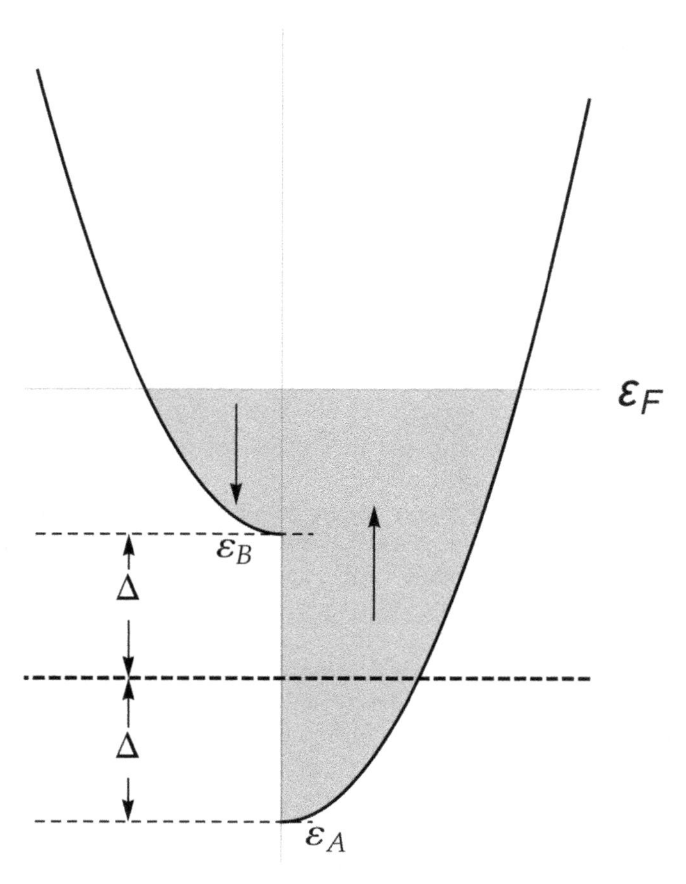

Figure 2.10. The bands for ↑ and ↓-spins are shown. The Fermi energy (ε_{F}) is marked by a horizontal line, while the dashed horizontal line denotes a reference which is separated by an energy Δ from the bottom of the ↑ and ↓-spin bands. The shaded area denotes filled energy levels.

$$m = \int_{-\Delta}^{\varepsilon_{\mathrm{F}}} d\varepsilon N(\varepsilon + \Delta) - \int_{\Delta}^{\varepsilon_{\mathrm{F}}} d\varepsilon N(\varepsilon - \Delta). \tag{2.189}$$

Here the ↑ and ↓-spin density of states are denoted by $N(\varepsilon \pm \Delta)$. Assuming Δ to be small, one can do a Taylor expansion of the density of states,

$$N(\varepsilon + \Delta) = \frac{dN}{d\varepsilon}\Delta \tag{2.190}$$

$$N(\varepsilon - \Delta) = \frac{dN}{d\varepsilon}(-\Delta) \text{ where } \Delta = \frac{Um}{2}. \tag{2.191}$$

Since the Fermi energy ε_{F} is much larger than Δ, one can set the lower limit to be zero and combine the integrals,

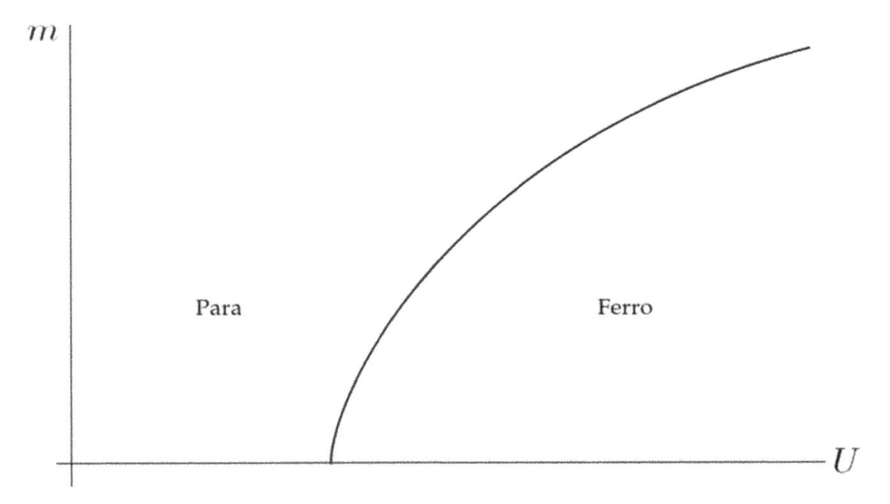

Figure 2.11. m as a function of Hubbard U. The plot emphasizes Stoner criterion and shows the boundary separating the paramagnetic and ferromagnetic phases.

$$m = 2\Delta \int_{\Delta(=0)}^{\varepsilon_{\mathrm{F}}} \frac{dN}{d\varepsilon} d\varepsilon \tag{2.192}$$

$$= 2\Delta N(\varepsilon_{\mathrm{F}}) = mUN(\varepsilon_{\mathrm{F}}). \tag{2.193}$$

Cancelling m from both sides, we arrive at a condition,

$$N(\varepsilon_{\mathrm{F}}) = \frac{1}{U}. \tag{2.194}$$

This is known as Stoner criterion which yields a very stringent condition on the occurrence ferromagnetism. From figure 2.11, it is clear that the interaction term multiplied by the density of states at the Fermi level has to be necessarily greater than unity and the boundary between the paramagnetic and ferromagnetic phases is indicated by the curved line.

It is important to note that only Fe, Ni and Co pass the Stoner criterion. As a toy model, if we consider a two-dimensional square lattice whose density of states diverges at the half-filling, that is, one particle per site ($n = 1$), $U_c \to 0$ which says that the system is unstable against a ferromagnetic ordering even at negligible inter-particle interaction.

2.15 Antiferromagnetism in the Hubbard model

Having discussed ferromagnetism in an itinerant electronic model, we shall discuss antiferromagnetism. For concreteness, we consider a two-dimensional square lattice, consisting of two sublattices, namely A and B sublattices. A sublattice contains predominantly $\uparrow$-spin particles (the $\uparrow$-spin density being large, with a small $\downarrow$-spin density). Similarly, the other sublattice, namely the B sublattice contains predominantly $\downarrow$-spin density (along with a small $\uparrow$-spin density). It may be noted a perfect

Néel order has not been assumed and a small density of the other type is essential for our purpose. Further, every A sublattice site has four neighbours which are those of B sublattice and vice versa. Thus there is an antiferromagnetic ordering with a wave vector $q = (\pi, \pi)$ [12].

Once again we take the Hubbard Hamiltonian, and do an unrestricted Hartree–Fock approximation (with of course ignoring terms, such as, $\langle c^\dagger c^\dagger\rangle$ and cc) on the interaction term, which yields,

$$\mathcal{H}_{\text{int}}^{\text{HF}} = U\sum_i\left[\langle c_{i\uparrow}^\dagger c_{i\uparrow}\rangle c_{i\downarrow}^\dagger c_{i\downarrow} + \langle c_{i\downarrow}^\dagger c_{i\downarrow}\rangle c_{i\uparrow}^\dagger c_{i\uparrow} - c_{i\uparrow}^\dagger c_{i\downarrow}\langle c_{i\downarrow}^\dagger c_{i\uparrow}\rangle - c_{i\downarrow}^\dagger c_{i\uparrow}\langle c_{i\uparrow}^\dagger c_{i\downarrow}\rangle\right] \tag{2.195}$$

where we have neglected a constant term of the form, $U\langle c_{i\sigma}^\dagger c_{i\sigma'}\rangle\langle c_{i\sigma'}^\dagger c_{i\sigma}\rangle$. Rewriting the decoupled interaction term,

$$\mathcal{H}_{\text{int}}^{\text{HF}} = U\sum_i[\langle n_{i\uparrow}\rangle n_{i\downarrow} + \langle n_{i\downarrow}\rangle n_{i\uparrow}]. \tag{2.196}$$

This implies that an $\uparrow$-spin particle feels a potential $U\langle n_{i\downarrow}\rangle$ and a $\downarrow$-spin Hamiltonian feels a potential $U\langle n_{i\uparrow}\rangle$, with the constraint that

$$\langle n_\uparrow^{i(A/B)}\rangle + \langle n_\downarrow^{i(A/B)}\rangle = 1. \tag{2.197}$$

Also, the sublattice magnetization m is defined by,

$$\langle n_\uparrow^{i(A/B)}\rangle - \langle n_\downarrow^{i(A/B)}\rangle = \pm m \tag{2.198}$$

with m being replaced by $(-1)^i m$ implying a change in sign as one goes from A sublattice to B sublattice. Thus in terms of U and m, the potential experienced by $\uparrow$-spin and $\downarrow$-spin particles is given by,

$$\begin{aligned} V_\uparrow(i) = U\langle n_{i\downarrow}\rangle &= \frac{U}{2} - \frac{Um}{2}(-1)^i \\ V_\downarrow(i) = U\langle n_{i\uparrow}\rangle &= \frac{U}{2} + \frac{Um}{2}(-1)^i. \end{aligned} \tag{2.199}$$

Defining, $\Delta = \frac{Um}{2}(-1)^i$ (distinguish it from the definition of Δ used in ferromagnets), one may write,

$$\begin{aligned} V_\uparrow(i \in A) &= \frac{U}{2} - \Delta = V_\downarrow(i \in B) \\ V_\uparrow(i \in B) &= \frac{U}{2} + \Delta = V_\downarrow(i \in A). \end{aligned} \tag{2.200}$$

Thus the mean field Hamiltonians for each spin, say $\uparrow$-spin is written as,

$$\mathcal{H}_\uparrow(i) = -t\sum_{\langle ij\rangle} c_{i\uparrow}^\dagger c_{j\uparrow} + \sum_{i\in A,B} V_\uparrow(i) n_{i\uparrow} \tag{2.201}$$

where $n_{i\uparrow} = c_{i\uparrow}^\dagger c_{i\uparrow}$. Assuming translational invariance, one can Fourier transform the above Hamiltonian,

$$\mathcal{H}_\uparrow(k) = \sum_k \epsilon_k c_{k\uparrow}^{\dagger A} c_{k\uparrow}^{B} + \sum_Q \left[(-\Delta) c_{k\uparrow}^{\dagger A} c_{k-Q\uparrow}^{A} + \Delta c_{k\uparrow}^{\dagger B} c_{k-Q\uparrow}^{B}\right]. \tag{2.202}$$

Writing $\mathcal{H}_\uparrow(k)$ in the sublattice basis renders a 2×2 form,

$$\mathcal{H}_\uparrow(k) = \begin{bmatrix} -\Delta & \varepsilon_k \\ \varepsilon_k & \Delta \end{bmatrix} \begin{bmatrix} c_{k\uparrow}^{A} \\ c_{k-q\uparrow}^{B} \end{bmatrix}. \tag{2.203}$$

Diagonalizing, one gets the eigenvalues as,

$$E_{k\uparrow} = \pm\sqrt{\Delta^2 + \varepsilon_k^2}. \tag{2.204}$$

Similarly for the $\downarrow$-spin,

$$\mathcal{H}_\downarrow(k) = \begin{bmatrix} \Delta & \varepsilon_k \\ \varepsilon_k & -\Delta \end{bmatrix}. \tag{2.205}$$

Again one gets, $E_{k\downarrow} = \pm\sqrt{\Delta^2 + \varepsilon_k^2} = E_{k\uparrow}$, that is, the same as the $\uparrow$-spin.

The eigenvectors corresponding to the filled band (negative eigenvalue) for the $\uparrow$-spin are,

$$\begin{bmatrix} \alpha_{k\uparrow} \\ \beta_{k\uparrow} \end{bmatrix} = \begin{bmatrix} \sqrt{1 - \beta_{k\uparrow}^2} \\ \dfrac{\varepsilon_k}{(E_{k\uparrow} - \Delta)^2 + \varepsilon_k^2} \end{bmatrix}. \tag{2.206}$$

The magnetization, m can now be computed using,

$$m = n_\uparrow - n_\downarrow = \frac{1}{N/2} \sum_{k \in \text{LHB}} (\alpha_{k\uparrow}^2 - \alpha_{k\downarrow}^2) \tag{2.207}$$

where LHB denotes the lower (filled) Hubbard band. Also, $\frac{N}{2}$ arises because we are summing over number of unit cells which are $N/2$ in number corresponding to N sites. This yields a self-consistent value of m,

$$m = \frac{2}{N} \sum_k \frac{\Delta}{E_k} = \frac{1}{N} \sum_k \frac{mU}{E_k} \tag{2.208}$$

where, $mU = 2\Delta$ (ignoring the sign that it picks up in going from one lattice to the next). The band gap 2Δcan be seen from the upper panel of figure 2.12. The band gap scales with U and vanishes as $U \to 0$ (lower panel of figure 2.12).

The condition in equation (2.208) can further be simplified to,

$$\begin{aligned} 1 &= \frac{1}{N} \sum_k \frac{U}{E_k} \\ \frac{1}{U} &= \frac{1}{N} \sum_{k \in \text{LHB}} \frac{1}{E_k}. \end{aligned} \tag{2.209}$$

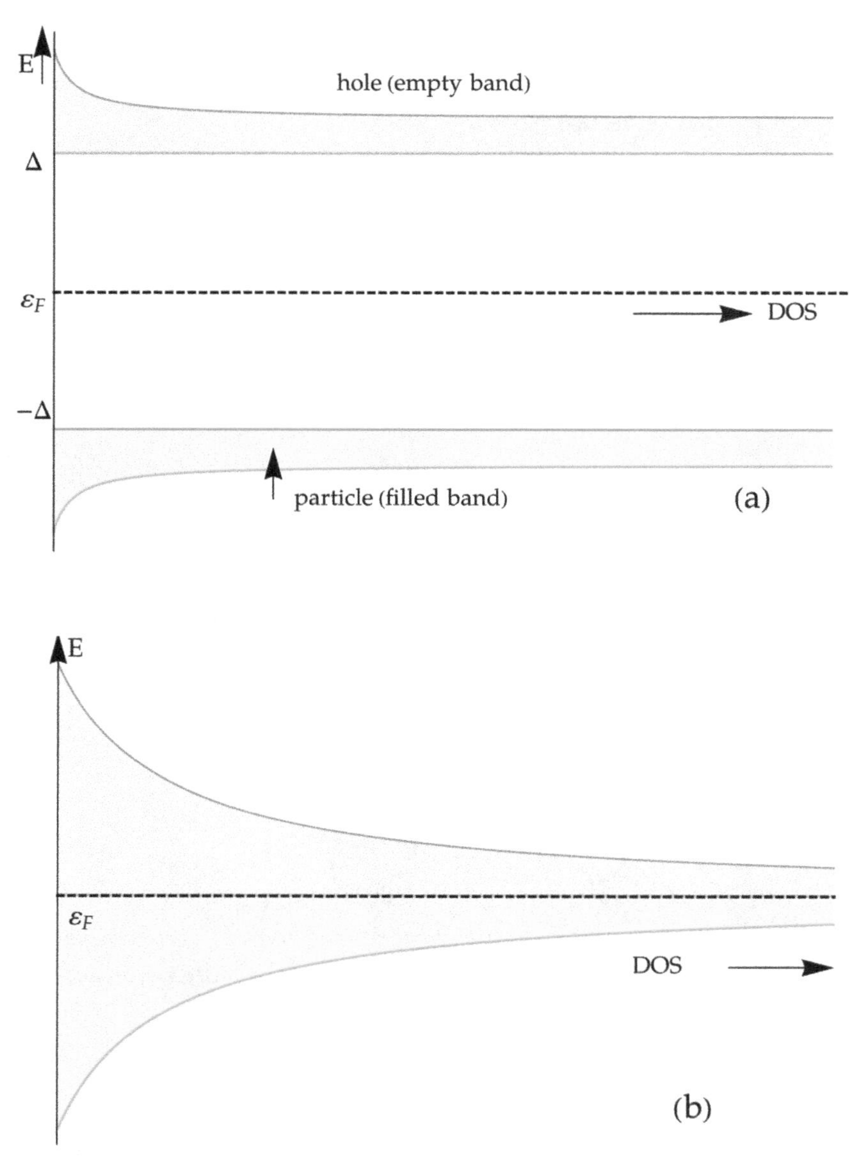

Figure 2.12. Density of states (DOS) along x-axis is shown as a function of energy (E) shown along y-axis. (a) The lower (light yellow) and the upper (light blue) Hubbard bands are shown. The band gap scales with the Hubbard interaction, U. (b) The lower and upper Hubbard bands merge in the limit $U \to 0$.

The above equation can be considered as the self-consistent equation for calculating magnetization, and requires a numerical solution for Δ $(=mU)$. Instead of resorting to the discussion of the numerical solution, let us consider different limits to explore the behaviour of the magnetization.

2.15.1 Strong coupling limit

Here we consider $U \gg t$ or equivalently $U \to \infty$ (that is greater than any other energy scale of the problem) where it is energetically unfavourable for electrons of opposite spins to occupy the same lattice site. Thus, $\Delta = \frac{mU}{2}$ is very large and $\beta_{k\uparrow}$ in equation (2.206) takes the form,

$$\beta_{k\uparrow} = \frac{\varepsilon_k}{(E_{k\uparrow} - \Delta)^2 + \varepsilon_k^2} \simeq \frac{\varepsilon_k}{2\Delta} \simeq 0. \tag{2.210}$$

Thus, $\alpha_{k\uparrow} \approx 1$ (since $\alpha_{k\uparrow} = \sqrt{1 - \beta_{k\uparrow}^2}$). Hence the magnetization becomes,

$$m = \frac{2}{N}\sum_k 1 = \frac{2}{N} \times \frac{N}{2} \simeq 1 \tag{2.211}$$

where we have used $\sum_k 1 = N$.

The same inference can also be obtained from the self-consistent equation,

$$\frac{1}{U} = \frac{1}{N}\sum_k \frac{1}{2\Delta} \tag{2.212}$$

$2\Delta = U$ implies $m = 1$. Thus in the strong coupling limit, the sublattice magnetization attains its saturation value, namely equal to 1. Quantum fluctuations reduce this value. For example the magnetization obtained from quantum Monte Carlo calculations is 0.6 for a two-dimensional square lattice [17].

Now consider the opposite limit, that is, $U/t \to 0$ or equivalently, $\Delta/t \to 0$, then one gets for the self-consistent equation,

$$\frac{1}{U} = \frac{1}{2N}\sum_k \varepsilon_k. \tag{2.213}$$

Converting the sum into an integral,

$$\frac{1}{U} = \int N(\varepsilon)\frac{1}{\varepsilon}d\epsilon. \tag{2.214}$$

As a particular case in two dimensions (2D), the DOS has a divergence, that is,

$$N(\varepsilon) \sim ln\left(\frac{1}{\epsilon}\right) \qquad \text{(van Hove singularity)}$$

Thus one gets,

$$\frac{1}{U} \simeq \lim_{\varepsilon \to 0} ln\frac{\left(\frac{1}{\varepsilon}\right)}{\varepsilon} \simeq \left[ln\left(\frac{1}{\varepsilon}\right)\right]^2. \tag{2.215}$$

One can take a tiny cut off δ to satisfy the divergence of the RHS which renders,

$$\frac{1}{U} \sim \left[ln\left(\frac{1}{\delta}\right)\right]^2 \tag{2.216}$$

yielding,

$$t/U \sim [ln(t/\delta)]^2$$

Thus, one gets, $\Delta/t \sim e^{-t/U}$, implying that the magnetization takes a form,

$$m \sim \frac{t}{U}e^{-t/U} \sim \frac{1}{U}e^{-1/U}. \tag{2.217}$$

Thus as $U \to 0$, $m \to 0$, but in a complex fashion, that is,

$$m \simeq \frac{1}{U}e^{-1/U}.$$

A schematic plot of magnetization is shown in figure 2.13. Thus, at low interaction strengths, the magnetization vanishes as $\frac{1}{U}e^{-1/U}$, while, for obvious reasons, it saturates at a value 1 for large U. Hence, even without a self-consistent numerical solution, we get an idea about the behaviour of magnetization as a function of the Hubbard interaction parameter. While it is hard to guess the exact form at low interaction strengths, it still is possible to reconcile that magnetization vanishes as $t \gg U$.

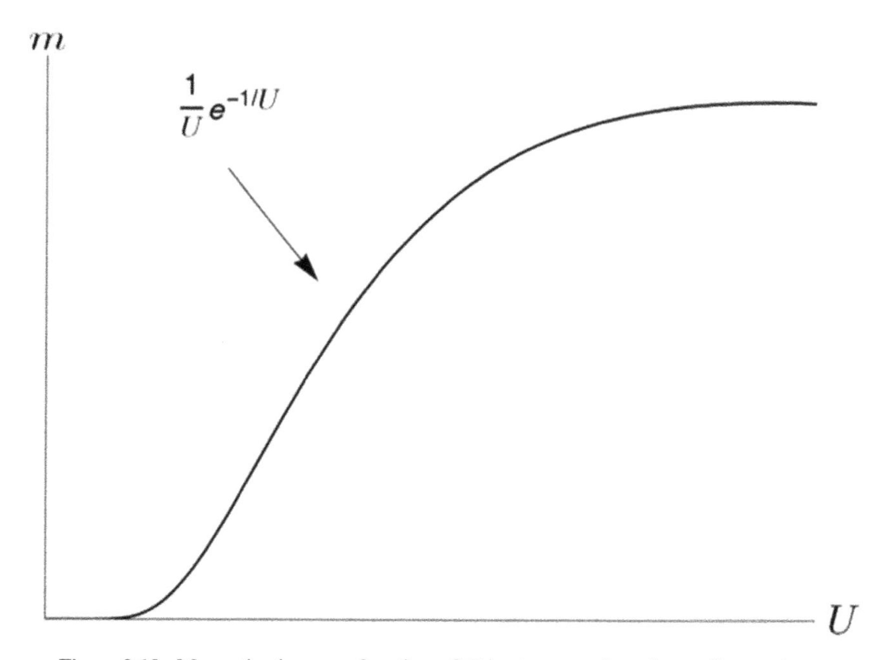

Figure 2.13. Magnetization as a function of U in strong and weak coupling regimes.

2.15.2 Summary and outlook

In this chapter, we have given an overview of different magnetic orders in solids. What follows afterwards is a brief discussion of magnetism in materials with atoms having filled and partially filled shells. Pauli paramagnetism, Curie's law and Landau diamagnetism are hence discussed. We have underscored the role of electron–electron interactions in explaining ferromagnetism and antiferromagnetism. MFT of ferromagnetism is introduced. As an extension, spin wave theory is applied to ferromagnets using Holstein–Primakoff transformation. A simplified version of the Heisenberg model, that is, without the interaction between the z-component of the spins at neighbouring sites, namely, the quantum XY model is solved. Further, antiferromagnetism is studied within a similar formalism. Yet another technique is employed, namely, the transfer matrix approach to study the critical properties of a ferromagnet to a paramagnet phase transition. The magnetic susceptibility is hence studied within a linear response theory. To wind up our discussion, we have introduced an itinerant electronic model to study properties of materials, namely the Hubbard model and discussed its properties. The weak and the strong coupling limits of the model are studied. The model is hence solved using Hartree–Fock approximation. The magnetic properties of both ferromagnets and antiferromagnets are studied, where the magnetization is solved self-consistently as a function of inter-particle (Hubbard) interaction.

Summarizing, apart from studying diamagnetism and paramagnetism which can be comprehensively understood without invoking electron–electron interaction, we have presented interacting models involving spins and electronic degrees of freedom and computed the key magnetic property, that is, magnetization (or equivalently, the magnetic susceptibility) as a function of the inter-particle interaction parameter.

2.16 Appendix

The electronic configuration of elements can be understood by adding one electron at a time to the available energy levels of the atoms. Each electron can be added adhering to the Pauli exclusion principle, and Hund's rule (discussed later). There are a number of factors that decide on the energy of an electron in an orbital. They are, respectively, the mass number, (that is number of protons), average distance of the orbital from the nucleus, and last, but not the least, screening effects. The last one needs a special mention, particularly for many-electron atoms, where the positive charge of the nucleus is partially screened to an electron in the outer shells owing to the presence of intervening electrons. The screening is not uniform for all the orbitals of a given energy level. The orbitals of an atom overlap the region that surrounds the nucleus. The more an orbital penetrates onto the negative cloud of the screening electrons, the more strongly the electron is pulled by the nucleus, and consequently the corresponding energy will be lower. For example, corresponding to the same energy level (same value of the principle quantum number, n), the smaller the value of the orbital quantum number, l, the larger would be the probability of finding the electron near the nucleus. Thus an s-electron ($l = 0$) for any n is more penetrating than the p-electron ($l = 1$). Thus for a given n, the energies are arranged as, $E_s < E_p < E_d < E_f \cdots$ (d corresponds to $l = 2$ and f corresponds to $l = 3$).

Even the above classification is not free of loopholes. For instance, the 4s orbital ($n = 4$, $l = 0$) may have lower energy than a $3d$ orbital ($n = 3$, $l = 2$). Here is where the Aufbau ordering, and Hund's rule come into the picture for deriving the correct electronic configuration of atoms. We shall discuss them below.

2.17 RS coupling

In a many-electron system, the combination of the orbital and the spin degrees of freedom is quite complicated, owing to the presence of spin–spin, orbit–orbit, and spin–orbit couplings. In the Russell–Saunders's (RS) scheme, it is generally assumed that the strength of the spin–orbit coupling is the weakest, followed by the orbit–orbit coupling and the spin–spin coupling, respectively. This gives rise to two principle coupling schemes, namely, RS coupling (same as LS coupling) and jj coupling. The above hierarchy of the coupling strengths is found to be valid for the first row transition series where the coupling of the spin and orbit can be ignored. However, for larger elements (mostly with the ones with atomic number greater than thirty), the spin–orbit coupling becomes increasingly important, and the jj coupling scheme should be used.

The coupling schemes are best described by the selection rules that govern the transition probabilities between different atomic energy levels. Not all possible transitions actually occur. There are constraints imposed by the selection rules. As a particular case, consider the electric dipole transitions. Assuming LS coupling, the selection rules can be stated as in the following.

(i) Only one electron makes a transition at a time.

(ii) The l-value will only alter by one unit, that is, $\Delta = \pm 1$.

This ensures that the parity of the wavefunction must change in an electric dipole transition. To remind readers, the parity of the wavefunction is denoted by the factor $(-1)^l$, which implies that if l is even, the wavefunction has even parity[13] while if l is odd it has odd parity. Since the electric dipole moment ($\mathbf{p} = e\mathbf{r}$) involves a Hamiltonian of the form $\mathcal{H}' \sim r$, which itself has odd parity, the transition matrix elements $\langle f|\mathcal{H}'|i\rangle$ must have even parity. That dictates that the above matrix elements will be non-zero only when the eigenstates ($|i\rangle$ and $|f\rangle$) have opposite parity.

(iii) The quantum numbers for the whole atom must charge as follows. $\Delta s = 0$, $\Delta l = 0, \pm 1$, $\Delta J = 0, \pm 1$ (with the restriction that $J = 0$ to $J = 0$ is prohibited), and finally $\Delta M_J = 0, \pm 1$ (again $M_J = 0$ to $M_J = 0$ is not allowed if $\Delta J = 0$).

2.18 jj Coupling

In a situation (perhaps for large atoms) where there is a strong coupling between l_i and s_i are large, one has to apply the jj coupling scheme. Both l_i and s_i discontinue to

[13] $r \to r, \quad \theta \to \pi - \theta, \quad \phi \to \pi + \phi.$

become good quantum numbers, while they combine to yield a total angular momentum quantum number, j_i to be a good quantum number. The individual j_i's are loosely coupled. Thus there is no definite L or S, making room for J $(= L + S)$ to provide a valid description of the system.

In this case, the selection rules are stated as;

(i) Only one electron makes a transition at a given time.
(ii) The value of l must change by '1', that is $\Delta = \pm 1$. More fundamentally, the parity of the initial and final states should be opposite.
(iii) $\Delta j = 0, \pm 1$.
(iv) For the entire atom, $\Delta J = 0, \pm 1$ (with $J = 0$ to $J = 0$ is forbidden), $\Delta M_J = 0, \pm 1$, (however, $M_J = 0$ to $M_J = 0$ is forbidden for $\Delta J = 0$).

2.19 Hund's rule

The Aufbau scheme discusses how the electrons fill the energy levels stating with the lowest orbitals. They move to the higher orbitals only when the lower orbitals are filled up. For example, 1s orbitals are filled up ahead of the 2s orbitals. However, the question remains, in what order the different 2p orbitals get filled up. The answer to this involves Hund's rule.

Hund's rule can be stated as:

1. Every orbital is first filled up with a single electron, before any double occupancy occurs.
2. All the electrons in the singly occupied orbitals have only one kind of spin. This maximizes the total spin.

The essence of the first rule is, while assigning electrons to orbitals, an electron first attempts to fill the degenerate orbitals, that is, the ones with the same energy, before pairing with another electron in the half- filled orbital. Atoms, in their ground states tend to have as many unpaired electrons as possible. In fact, all the electrons repel each other and try to get as far as possible, before they pair up. The electrons tend to minimize repulsion by occupying their own orbital, rather than sharing an orbital with another electron. The singly occupied orbitals are less efficiently screened from the nuclear charge.

According to the second rule, the unpaired electrons in the singly occupied orbitals have the same spin. Assume the electron which is placed first in an orbital has a spin-↑, the spins of all of the other electrons get fixed, which means the unpaired electrons are all spin-↑.

As an example, consider carbon (C) atom which has six electrons and has an electronic configuration $1s^2 2s^2 2p^2$. The two 2s electrons will occupy the same orbital, while the two 2p electrons will be placed in different orbital and will be aligned in the same direction (say, all spin-↑) according to Hund's rule.

Similarly, the next atom, namely, nitrogen ($z = 7$) has a configuration $1s^2 2s^2 2p^3$. The 1s and the 2s orbitals get completely filled up, which leaves three (unpaired) electrons. Again, according to Hund's rule, the remaining electrons will fill all the empty orbitals.

Finally, consider oxygen which has $z = 8$ and the corresponding configuration is written as $1s^2 2s^2 2p^4$. The $1s$ and $2s$ orbitals get paired up electrons which leaves four electrons to be accommodated in the $2p$ orbitals. According to Hund's rule, again all the orbitals will be singly occupied before at least one (which is the case here) of the $2p$ orbitals get doubly occupied. Thus $2p$ will have one paired, and the other two unpaired levels comprising the same spin.

References

[1] Ashcroft N and Mermin N D 1976 *Solid State Physics* (Philadelphia, PA: Saunders)
[2] Kittel C 2005 *Introduction to Solid State Physics* 8th edn (Hoboken, NJ: Wiley)
[3] Pathria R K and Beale P D 2011 *Statistical Mechanics* 3rd edn (Amsterdam: Elsevier)
[4] Bethe H 1931 Zur Theorie der Metalle. Eigenwerte und Eigenfuntionen der linearen Atomkette *Z. Phys.* **71** 205
[5] Holstein T and Primakoff H 1940 *Phys. Rev.* **58** 1098
[6] Goldstone J 1961 *Nuovo Cimento* **19** 154
[7] Jordan P and Wigner E 1928 *Z. Phys.* **47** 631
[8] Chandler D 1987 *Introduction to Modern Statistical Mechanics* (Oxford: Oxford University Press)
[9] Stauffer D 1992 *Introduction to Percolation Theory* 2nd edn (London: Taylor and Francis)
[10] Hubbard J 1963 *Proc. R. Soc.* A **276** 238
[11] Doniach S and Sondheimer E 1974 *Green's Functions for Solid State Physicists* (Reading, MA: Addison-Wesley)
[12] Singh A and Tesanović Z 1990 *Phys. Rev.* B **41** 11457
[13] Ruderman M A and Kittel C 1954 *Phys. Rev.* **96** 99
Kasuya T 1956 *Prog. Theor. Phys.* **16** 45
Yoshda K 1957 *Phys. Rev.* **106** 893
[14] Kübler J 2000 *Theory of Itinerant Electron Magnetism* (Oxford: Oxford University Press)
[15] Mahan G D 2000 *Many Particle Physics 3rd end* (Berlin: Springer)
[16] Cohen-Tonnoudji C, Diu B and Laloë F 1977 *Quantum Mechanics* vol II (New York: Wiley)
[17] Sorella S, Parola A, Parinello M and Tosatti E 1989 *Int. J. Mod. Phys.* B **3** 1875

IOP Publishing

Condensed Matter Physics: A Modern Perspective

Saurabh Basu

Chapter 3

Transport in electronic systems

3.1 Introduction

The study of electronic transport phenomena in solids started several decades ago and is a well-studied problem by now. Transport evidently involves electronic bias in the form of including a battery in the circuit, thereby making a current flow. However, a battery can be replaced by a thermal gradient, which also results in the manifestation of current–voltage characteristics of the circuit.

Developing a formalism for the electronic transport in the presence of a bias necessitates inclusion of two important aspects, namely a non-equilibrium description, and a full quantum mechanical treatment of the system. A real system in general includes a very large number of particles, which is intractable in terms of computational logistics in any foreseeable future. Thus treating only a few particles (at the most of the order of 100) is manageable using exact techniques, and for numbers larger than that, one has to resort to controlled approximation methods. To treat systems out of equilibrium (but not too far from it), the formalism of non-equilibrium Green's functions is popular (interested readers may see references [1, 2]), since the method retains important correlations in space and time that are very important in the study of transport properties.

Another important technique to address transport is via solving the Boltzmann transport equation (BTE) [3], where the time evolution of the distribution function that involves the positions and the momenta of the constituent particles that enter through the local density and the temperature of the system. The prime importance of this equation is to account for non-equilibrium phenomena, and hence discuss transport beyond the linear regime.

Yet another, and most widely used technique for exploring quantum transport is the Kubo formula, which describes conduction using linear transport theory. The method involves computing the matrix elements of the current operator between the eigenfunctions of the Hamiltonian. A standout advantage of this method lies in the ability to solve the system both in real and momentum spaces, and hence

doi:10.1088/978-0-7503-3031-2ch3

disorder and imperfections, which are indispensable in real materials, can be taken into account. We defer detailed discussion of this formalism for now, and come back to it later.

Transport in lower dimensions, or relevant to nanoscale and mesoscale devices is a subject of prolonged research interest. Depending on certain critical length scales, the transport may either be semi-classical, quantum, or an intermediate between the two. The last one is the most interesting scenario, since the effects of decoherence and dissipation have crucial roles on the transport properties, and quantum effects dominate the energy spectra. We address some of these in the following.

Mesoscopic physics is a relatively new branch of condensed matter physics. It deals with systems having dimensions that are intermediate between the microscopic (a few angstroms) and the macroscopic length (of the order of a micron) scale. A conductor of length L appears to be ohmic if $L \ll \lambda_{\text{db}}$ (λ_{db}: de Broglie wavelength), however, it is termed as non-ohmic or mesoscopic if $L \ll \lambda_{\phi}$ (λ_{ϕ}: phase coherence length, over which the electrons retain phase coherence). This regime is governed by the Landauer formula for conductance [4], namely,

$$G = \left(\frac{e^2}{h}\right) T \tag{3.1}$$

where T is the probability of an electron to be transmitted through the conductor and has been summed over all the channels or modes. The constant, $\frac{e^2}{h} = [(25.8)k\Omega]^{-1}$ represents the quantum of conductance corresponding to a single transverse mode in a conductor.

Several typical characteristic length scales are briefly introduced to define and characterize different transport regimes [1, 5, 6]. They are:

(a) Fermi wavelength (λ_{F}): It is the de Broglie wavelength (λ_{db}) of the electrons near the Fermi energy.

(b) Mean free path (l_e): It is the distance that an electron travels before its initial momentum is destroyed, and describes the elastic scattering caused by the random impurity potential.

(c) Phase coherent length $\left(l_{\phi}\right)$: It is the distance that an electron travels before it looses its initial phase due to scattering from other electrons or lattice vibrations.

(d) Localization length ξ: It is an asymptotic property of the conductance as a function of the linear sample size and can be expressed by the relation, $G = G_0 e^{-2L/\xi}$, where G_0 is the conductance without the disorder.

In terms of the above length scales, one can define three distinct regimes for the mesoscopic transport. These are:

(i) Ballistic regime $\left(\lambda_{\text{db}} < L \ll l_e, l_{\phi}\right)$: In this regime, the electrons propagate through the mesoscopic sample without any elastic or phase breaking scattering in the bulk. The scattering occurs at the contact region separating the leads and the conductor and the impurity scattering can be neglected.

(ii) Diffusive regime ($\lambda_{\rm db} \ll l_e \ll L < \xi$): In this regime, a few impurity atoms or a small concentration of disorder exist in the conducting sample and the electrons traverse coherently. Scattering is sample specific and one observes the full statistical fluctuations which are not suppressed by self-averaging.

(iii) Insulating regime ($\xi < L$): It's also called localized regime in the presence of strong scattering limit where the interference can completely cease the transport inside the disordered mesoscopic conductor.

Many novel phenomena exist that are intrinsic to mesoscopic systems, such as the integer quantum Hall effect, where the Hall resistance is quantized in units of h/e^2, Aharonov–Bohm (AB) oscillations [7] in the conductance spectrum of mesoscopic rings with flux period h/e and 'the dissipationless persistent currents' flowing in such systems in the normal (non-superconducting) resistive states, etc. Since then, attention on the physics governing mesoscopic transport has been growing rapidly, and a wide range of new physical concepts have been realized. Some of them are: mesoscopic resistors in series that do not obey standard resistance addition rules [1, 2], the conductance of very narrow constrictions being quantized [8, 9], the conductance of disordered systems shows sample specific reproducible fluctuations with universal amplitudes, the weak localization phenomena [10, 11], etc.

Over the last two decades, it is possible to experimentally realize dimensions, $L < l_\phi$, using modern electronic material fabrication techniques, such as molecular beam epitaxy (MBE), lithography technique, scanning tunnelling microscope (STM), etc [12]. They are denoted as nanostructures. Familiar examples are GaAs/GaAlAs semiconductor heterostructures, which are formed by two semiconducting samples with different doping levels. At the interface, a thin layer of charges, called a two-dimensional electron gas is realized (2DEG). These heterostructures yield mobilities of the order of $10^6 \text{cm}^2/(\text{V-s})$ that correspond to flow of electrons having a mean free path of about 10 μm.

We shall not prolong this introduction on the transport phenomena at the mesoscale or nanoscale beyond this point, as that forms a subject of intense research interest on its own. Rather, we shall concentrate on transport in two-dimensional electronic systems in the presence of a perpendicular magnetic field. This is known as the Hall effect, which the reader should be familiar with at the beginning phase of his/her undergraduate studies. We shall mainly discuss the quantum version of it, which, for obvious reasons is known as the quantum version of the Hall effect (QHE). QHE laid the foundation of studying condensed matter physics from a whole new perspective, and necessitated deeper introspection to connect the emergent quantum phenomena to the fundamental description of materials.

3.2 Quantum Hall effect

The date of discovery of the QHE is known pretty accurately. It occurred in the night between 4th and 5th February, 1980 at 2:00 AM in the morning at the high magnetic lab in Grenoble, France (see figure 3.1). There was ongoing research on the

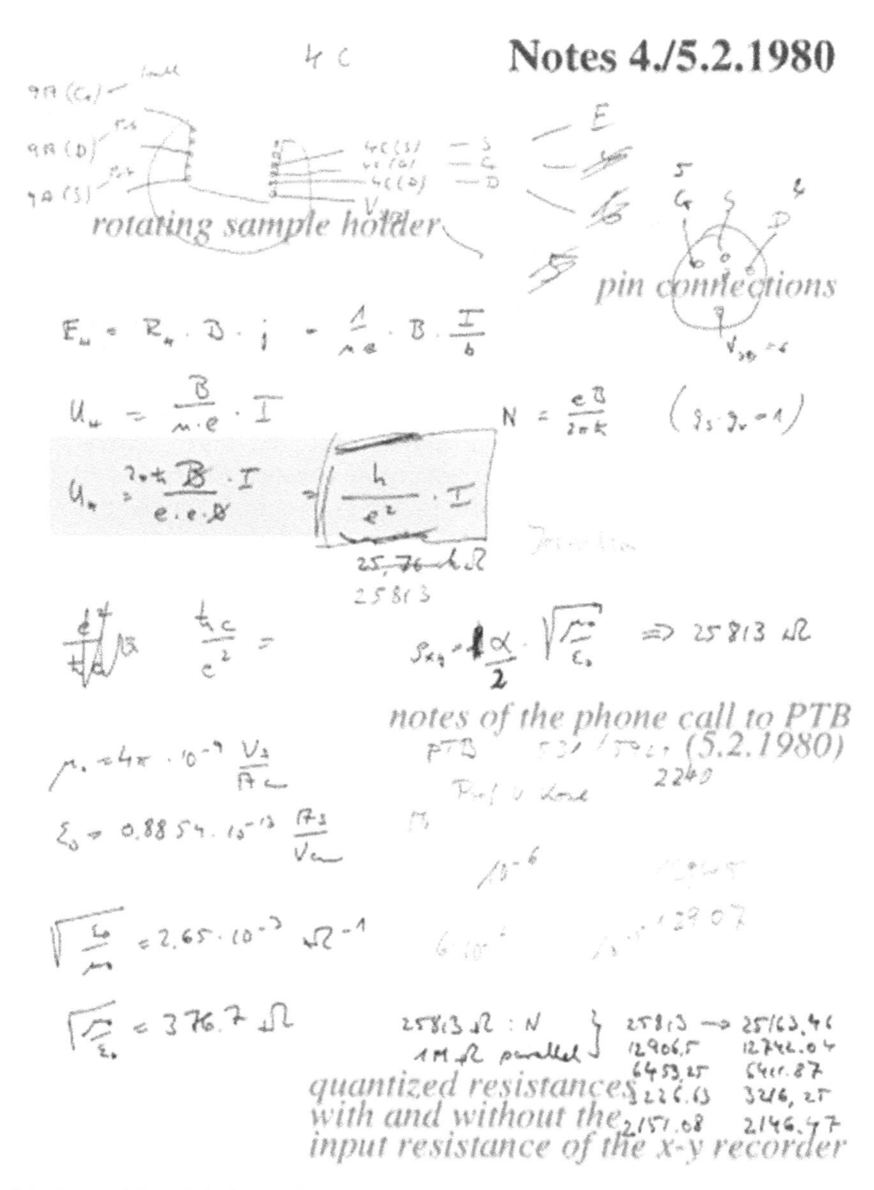

Figure 3.1. Copy of the original notes by Professor K v Klitzing on the discovery of the QHE. It documents that the Hall resistance ($\frac{U_H}{I}$) involves a fundamental constant h/e^2, copyright Klaus von Klitzing.

transport properties of silicon field-effect transistors (FETs). The main motive was to improve the mobility of these FET devices. They were provided by Dorda and Pepper which allowed direct measurement of the resistivity tensor. The system is a highly degenerate two-dimensional electron gas contained in the inversion layer of a metal-oxide semiconductor field-effect transistor (MOSFET) operated at low temperatures and strong magnetic fields. The original notes appear in figure 3.1, where it is clearly stated that the Hall resistivity involves universal constants and hence signals towards the involvement of a very fundamental phenomenon.

In the classical version of the phenomenon discovered by E Hall in 1879, just over a hundred years before the discovery of its quantum analogue, one may consider a sample with a planar geometry so as to restrict the carriers to move in a two-dimensional (2D) plane. Next, turn on a bias voltage so that a current flows in one of the longitudinal directions and a strong magnetic field perpendicular to the plane of the gas (see figure 3.2). Because of the Lorentz force, the carriers drift towards a direction transverse to the direction of the current flowing in the sample. At equilibrium, a voltage develops in the transverse direction, which is known as the Hall voltage. The Hall resistivity, R defined as the Hall voltage divided by the longitudinal current is found to linearly depend on the magnetic field, B and inversely on the carrier density, n through $R = \frac{B}{nq}$ (q is the charge). A related and possibly more familiar quantity is the Hall coefficient, denoted by, $R_H = R/B$ which, via its sign yields information on the type of the majority carriers, that is, whether they are electrons or holes.

At very low temperature, or at very high values of the magnetic field (or at both), the resistivity of the sample assumes quantized values of the form, $\rho_{xy} = \frac{h}{ne^2}$. Initially n was found to be an integer with extraordinary precession (one part in $\sim 10^8$). This is shown in figure 3.3. The quantization of the Hall resistivity yields the name

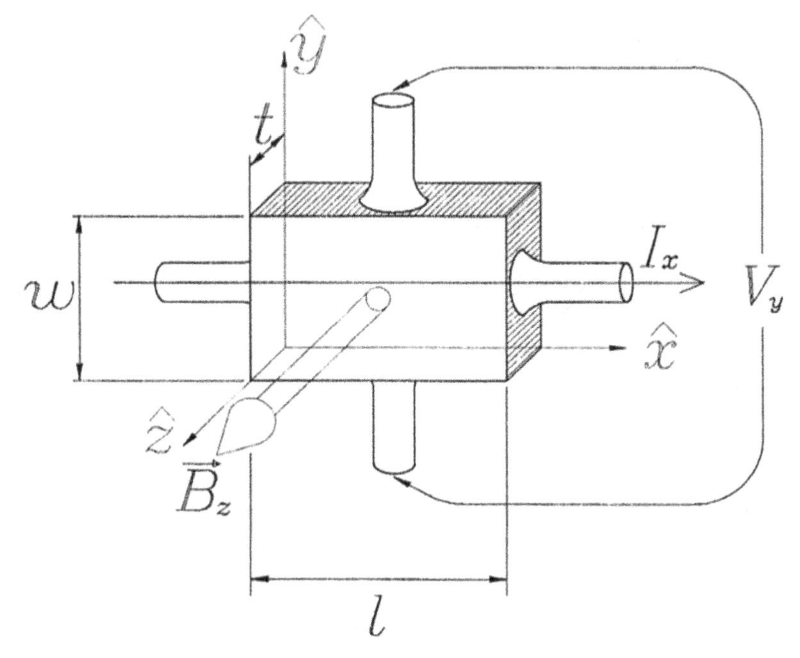

Figure 3.2. Typical Hall experiment setup showing direction of the current, I_x and the magnetic field B_z. V_y denotes the Hall voltage.

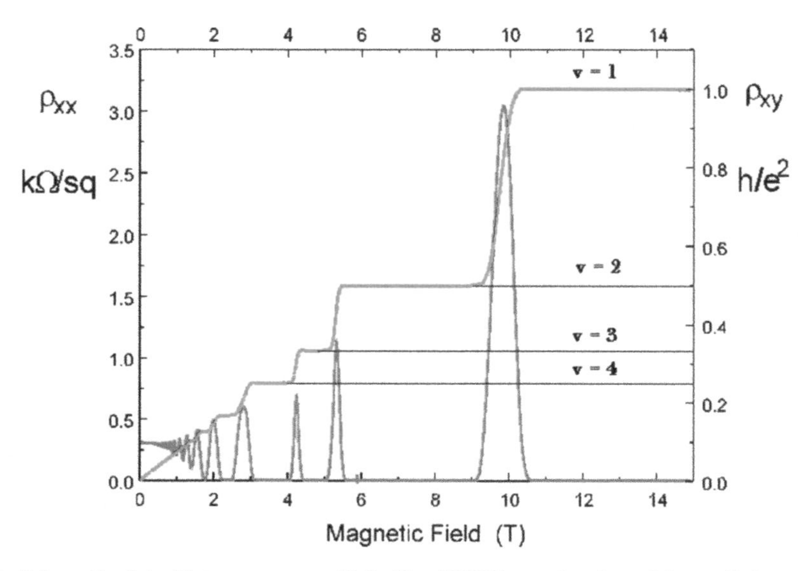

Figure 3.3. Schematic plot of integer quantum Hall effect (IQHE) as a function of the applied magnetic field. The plot in red denotes the Hall resistivity and the one in green shows the longitudinal resistivity (or the magnetoresistivity).

'*quantum*' (or quantized) Hall effect, which we refer to as QHE throughout the chapter.

Klaus von Klitzing and his co-workers [13, 14], while measuring the electrical transport properties of planar systems formed at the interface of two different semiconducting samples in the strong magnetic field facility at Grenoble, France, noted that the Hall resistivity is quantized in units of h/e^2 as a function of the external magnetic field. The flatness of the plateaus occurring at integer or fractional values of h/e^2 has an unprecedented precession and is independent of the geometry of the sample (as long as it is 2D), density of the charge carriers and its purity. The accuracy of the quantization aids in fixing the unit of resistance, namely, $h/e^2 = 25.813\ \mathrm{K}\Omega$, also called as the Klitzing constant. Thus, among other significant properties of QHE that we shall be discussing in due course, an experiment performed at a macroscopic scale that can be used for metrology or yields the values for the fundamental constants used in quantum physics is truly amazing and hence calls for intense scrutiny. The effect occurs when the density of the carriers, n are such that they are encoded in the integers (that come as proportionality constants to the Hall resistivity in terms of h/e^2) as if the charges locked their separation at some particular values. The phenomenon remains resilient to changing the carrier density by a small amount, however, changing it by a large amount does destroy the effect.

The Hall resistivity (in red) in figure 3.3 becomes constant for certain ranges of the external magnetic fields, which are called plateaus. Further the longitudinal resistivity (in green) in the same plot vanishes everywhere. Although it shows peaks wherever there is a jump in the Hall resistivity from one plateau to another. Later on

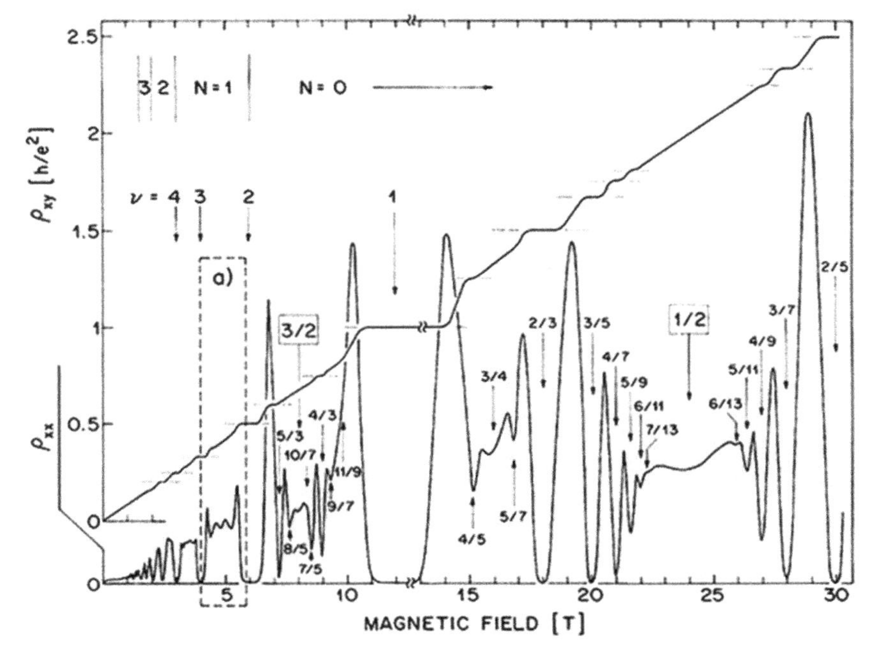

Figure 3.4. The plot shows fractional quantum Hall effect (FQHE). The plateaus are shown at fractional values in units of h/e^2. Reprinted with permission from [15], copyright (1982) by the American Physical Society.

it was found that ν is not only restricted to integer values, but also takes values which are rational fractions, such as $\nu = \frac{1}{5}, \frac{2}{5}, \frac{3}{5}, \frac{3}{7}, \frac{4}{9}, \frac{5}{9}$, etc. There are about 100 fractions (including the improper ones) that have been noted in experiments so far. The corresponding plot appears in figure 3.4.

3.2.1 General perspectives

The charge carriers being confined in 2D wells has a longer history. Since 1966 it has been known that the electrons accumulated at the surface of a silicon single crystal induced by a positive gate voltage form a 2D electron gas (2DEG). The energy of the electrons corresponding to a motion perpendicular to the surface is quantized (box quantization) and on top of it, the free motion of the electrons in 2D becomes quantized when a strong magnetic field is applied perpendicular to the plane (Landau quantization). Thus, QHE has both the quantization phenomena in-built into it.

An important recent development in the study of semiconductors is the achievement of structures in which the electrons are restricted to move essentially in 2D. This immediately says that the carriers are prohibited from moving along the direction transverse to the plane. Hence the motion is quantized. Such 2D behaviour of the carriers can be found in metal-oxide semiconductor (MOS) structures, quantum wells and superlattices. An excellent prototype is the metal-insulator-

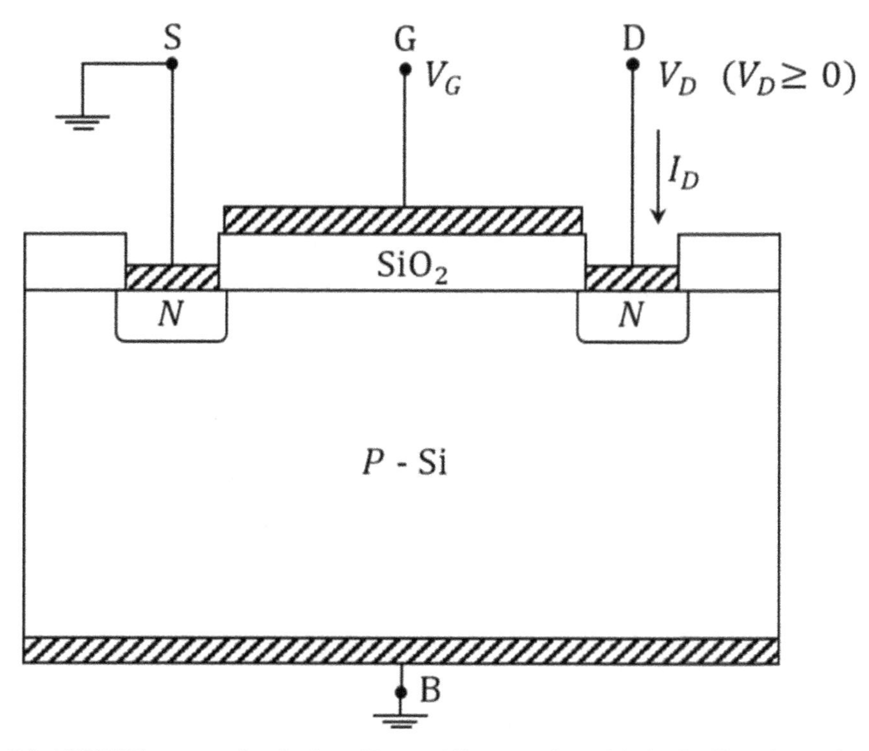

Figure 3.5. MOSFET structure showing base (B), gate (G), source (S) and drain (D). The substrate is a p-type Si. SiO_2 denotes the insulating oxide layer.

semiconductor (MIS) layered structure of which the insulator is usually an oxide, such as Al_2O_3 (thereby making it a MOS structure). In figure 3.5 we show a typical MOS device where the substrate is a doped p-type silicon (Si) which is grounded and is called a base (shown by B in figure 3.5). On the top, there is a metallic layer (shown by the hatched regime) followed by an insulating layer formed by SiO_2. The metallic layer is called the gate, denoted by G which is biased by a voltage V_G. The source (grounded) and the drains are denoted by S and D, respectively, which in figure 3.5 consist of n-type materials. The gate voltage causes the carriers beneath the gate electrode to drift between the source and the drain. The layer of charge carriers below the oxide layer forms the 2DEG which is central to our discussion. The energy dispersion in this case reads,

$$E_n(k_x, k_y) = \frac{\hbar^2 k_x^2}{2m_{xx}^*} + \frac{\hbar^2 k_y^2}{2m_{yy}^*} \tag{3.2}$$

where m_{xx}^* and m_{yy}^* are the components of the effective mass tensor defined by the inverse of the curvature of the band structure, namely,

$$m^*_{\alpha\beta} = \hbar^2 \left(\frac{\partial^2 E(k)}{\partial k_\alpha \partial k_\beta} \right)^{-1}. \tag{3.3}$$

The physical properties of all systems are governed by their density of states (DOS) which plays a crucial role in deciding on the dependence of the temperature, density of carriers, etc. In 2D systems with a parabolic dispersion as elaborated above in equation (3.2), the DOS is a constant and assumes a form[1],

$$g(E) = g_{2D}(E) = \frac{m^*}{\pi \hbar^2}. \tag{3.4}$$

The energy independent DOS is very special to 2D and is in sharp contrast to three-dimensional (3D) where it goes as $E^{1/2}$, and in one-dimensional (1D) where it goes as $E^{-1/2}$. In a general sense, and not restricted to the discussion on Hall effect, the DOS enters while calculating the average quantities, such as the average energy or the average number of particles. For example, the average of a physical observable, O of a fermionic system is computed using,

$$\langle O \rangle = \int_0^\mu Of(E)g(E)dE \tag{3.5}$$

where $f(E)$ is the Fermi distribution function given by,

$$f(E) = \frac{1}{e^{\beta(E-\mu)} + 1}$$

with $\beta = \frac{1}{k_B T}$ and μ denotes the chemical potential. In general this integral is quite challenging to compute analytically because of the Fermi distribution function ($f(E)$) present in the integrand.

Meanwhile, there is a wonderful simplification where $f(E)$ assumes a value unity at all temperatures for which the experiments are performed. Only at temperatures close to the Fermi temperature, T_F defined via, $\epsilon_F = k_B T_F$ (ϵ_F being the Fermi energy), $f(E)$ starts to deviate from unity, and its exact form needs to be incorporated in the integral. However, T_F is usually of the order of tens of thousands of Kelvin for typical metals (such as, Cu, Al, etc) which is too high for them to appear in experimental situations. Moreover, the DOS only depends on energy and is independent of the temperature to a very good approximation. Thus computation of equation (3.5) becomes trivial as the integrand becomes independent of temperature or weakly dependent on temperature [16, 17].

In the following, we mention that there is something interesting about transport properties of 2D systems. In the linear response regime Ohm's law is valid, and says that

$$V_\alpha = R_{\alpha\beta} I_\beta$$

[1] Parabolic dispersion in 3D yields DOS as $E^{1/2}$ (E: energy), in 2D it is constant, while the DOS varies as $E^{-1/2}$ in 1D.

where $R_{\alpha\beta}$ denotes the resistivity tensor, and α, β denote spatial variables, x, y, etc. One can equivalently invert this equation to write, $I_\alpha = G_{\alpha\beta}V_\beta$, where $G_{\alpha\beta}$ represents the conductivity tensor with $G = R^{-1}$. Equivalent relations in terms of the components of the electric field (E) and current density (j) read as,

$$E_\alpha = \rho_{\alpha\beta}j_\beta \qquad \text{and} \qquad j_\alpha = \sigma_{\alpha\beta}E_\beta \tag{3.6}$$

where ρ and σ denote the resistivity and the conductivity tensors, respectively.

An interesting (and useful too) artefact of 2D physics is an accidental similarity that exists in decoding some of the key features of the transport properties. For example, the resistivity, ρ (or the conductivity, σ) is a quantity that is independent of the system geometry, and hence is useful for a theoretical analysis, whereas in experiments, one measures the resistance of a sample, R (or the conductance, G). For a sample in the shape of a hypercube of sides, L, the resistance and the resistivity are related by,

$$R = \rho L^{2-d} \tag{3.7}$$

where d denotes the dimensionality. Only for $d = 2$ the resistance is a scale invariant quantity. This puts the experimentalists and the theorists on the same page, as the geometry of the sample does not enter explicitly in the analysis of its transport properties.

As we dig more into the details of transport properties of 2DEG in the presence of a magnetic field, further useful information emerges. The off-diagonal elements of both the conductivity and the resistivity tensors are antisymmetric with regard to the direction of the applied field, $\mathbf{B}$. Consider a planar sample with dimensions $L_x \times L_y$. The conductivity tensor is of the form,

$$\sigma = \begin{pmatrix} \sigma_{xx} & \sigma_{xy} \\ \sigma_{yx} & \sigma_{yy} \end{pmatrix}. \tag{3.8}$$

Let us try to understand the nature of the tensor, σ in the presence of a magnetic field. A conductor in an external magnetic field obeys,

$$j_\alpha = \sigma_{\alpha\beta}E_\beta. \tag{3.9}$$

Onsager's reciprocity principle does not hold in the presence of a magnetic field, $\mathbf{B}$, which implies [18],

$$\sigma_{\alpha\beta}(\mathbf{B}) \neq \sigma_{\beta\alpha}(\mathbf{B}). \tag{3.10}$$

Instead one has, $\sigma_{\alpha\beta}(\mathbf{B}) = \sigma_{\beta\alpha}(-\mathbf{B})$ to make sure that the time reversal holds only if $\mathbf{B}$ changes sign. Let us write the conductivity tensor as a sum of a symmetric and an antisymmetric tensor (note that this is always possible for a rank two tensor). Thus,

$$\sigma_{\alpha\beta} = S_{\alpha\beta} + A_{\alpha\beta} \tag{3.11}$$

where $\mathbf{S}$ and $\mathbf{A}$ are the symmetric and the antisymmetric tensors which obey the following relations,

$$S_{\alpha\beta}(\mathbf{B}) = S_{\beta\alpha}(-\mathbf{B}) = S_{\alpha\beta}(-\mathbf{B})$$
$$A_{\alpha\beta}(\mathbf{B}) = A_{\beta\alpha}(-\mathbf{B}) = -A_{\alpha\beta}(-\mathbf{B}) \tag{3.12}$$

such that the components of $S_{\alpha\beta}$ are even functions of $\mathbf{B}$, while those of $A_{\alpha\beta}$ are odd functions of $\mathbf{B}$. Putting in equation (3.9),

$$j_{\alpha} = S_{\alpha\beta}E_{\beta} + A_{\alpha\beta}E_{\beta}. \tag{3.13}$$

But owing to the antisymmetry,

$$A_{\alpha\beta} = \epsilon_{\gamma\alpha\beta}A_{\gamma} = -\epsilon_{\alpha\gamma\beta}A_{\gamma}. \tag{3.14}$$

Putting it in equation (3.13),

$$\begin{aligned} j_{\alpha} &= S_{\alpha\beta}E_{\beta} - \epsilon_{\alpha\beta\gamma}A_{\beta}E_{\gamma} \\ &= S_{\alpha\beta}E_{\beta} - (\mathbf{A} \times \mathbf{E})_{\alpha} \\ &= S_{\alpha\beta}E_{\beta} + (\mathbf{E} \times \mathbf{A})_{\alpha}. \end{aligned} \tag{3.15}$$

Assuming that we can expand $\sigma(\mathbf{B})$ in powers of $\mathbf{B}$[2], such that the antisymmetric part contains odd powers of $\mathbf{B}$, then we can write,

$$A_{\alpha} = \eta_{\alpha\beta}B_{\beta} \tag{3.16}$$

and $S_{\alpha\beta}(\mathbf{B})$ consists of even powers of $\mathbf{B}$,

$$S_{\alpha} = (\sigma_0)_{\alpha\beta} + \zeta_{\alpha\beta\gamma\delta}B_{\gamma}B_{\delta}. \tag{3.17}$$

The first term is the zero field conductivity tensor. Thus, putting things together up to terms linear in $\mathbf{B}$,

$$j_{\alpha} = S_{\alpha\beta}E_{\beta} + (\mathbf{E} \times \mathbf{A})_{\alpha}. \tag{3.18}$$

The second term denotes the Hall effect which is linear in $\mathbf{B}$. This implies that the Hall current is perpendicular to the electric field, $\mathbf{E}$ and is proportional to $\mathbf{E}$ and $\mathbf{B}$. Thus an antisymmetric tensor is relevant to the study of the Hall effect, which is why the conductivity and the resistivity tensors are antisymmetric. This is an important result which deviates from the corresponding scenario that arises in the absence of an external magnetic field.

3.2.2 Translationally invariant system: classical limit of QHE

It is quite an irony that the extreme universal signature of the transport properties of a 2DEG are characterized by the flatness of the plateaus, that not only occurs, but survives even in the presence of disorder, impurity and imperfection. In the absence of the magnetic field, Anderson localization would have governed the transport signatures of non-interacting electrons which says that in any dimension less than three, all eigenstates of a system are exponentially localized even for an infinitesimal

[2] Which should be valid for weak magnetic fields, and is not exactly true for QHE, but nevertheless it serves our purpose.

disorder strength. Only in three dimensions is there a critical disorder at which a metal-insulation transition occurs. However, the scenario is strongly altered by the presence of the magnetic field, which yields, as we shall shortly see, a series of unique phase transitions from a perfect conductor to a perfect insulator. No other system demonstrates re-occurrence of the same phases over and over again as the magnetic field is gradually ramped up.

To begin with, we shall consider the case which is free from disorder, or equivalently a translationally (Lorentz) invariant system that possesses no preferred frame of reference. Thus we can think of a reference frame that is moving with a velocity $-\mathbf{v}$ with respect to the lab frame, where the current density is given by, $\mathbf{j} = -ne\mathbf{v}$ (n: areal electron density, $-e$: electronic charge). In this frame, the electric and the magnetic fields are given by[3],

$$\mathbf{E} = -\mathbf{v} \times \mathbf{B} \; ; \qquad \text{and} \qquad \mathbf{B} = B\hat{z}. \tag{3.19}$$

The above transformation ensures that an electric field must exist to balance the Lorentz force $-e\mathbf{v} \times \mathbf{B}$ in order to conduct without deflection. For the electric field, this yields,

$$\mathbf{E} = \frac{1}{ne}\mathbf{J} \times \mathbf{B}. \tag{3.20}$$

This is equivalent to the tensor equation,

$$E^{\mu} = \rho_{\mu\nu} j^{\nu} \tag{3.21}$$

with the resistivity tensor given by,

$$\rho_{\mu\nu} = \frac{B}{ne}\begin{pmatrix} 0 & 1 \\ -1 & 0 \end{pmatrix}. \tag{3.22}$$

Inverting the tensor equation, one can obtain,

$$j^{\mu} = \sigma_{\mu\nu} E^{\nu} \tag{3.23}$$

where the conductivity tensor, $\sigma_{\mu\nu}$ is defined via,

$$\sigma_{\mu\nu} = \frac{ne}{B}\begin{pmatrix} 0 & -1 \\ 1 & 0 \end{pmatrix}. \tag{3.24}$$

There is an interesting paradox which states that, $\sigma_{xx} = \rho_{xx} = 0$ (see above) which are of course contradictory. However, we reserve this rather interesting topic for a discussion immediately afterwards. Here we wish to point out that we get the results for a classical Hall effect, that is, $\sigma_{xy} = \frac{ne}{B}$ (or $\rho_{xy} = \frac{B}{ne}$). It is important to realize that the result is an artefact of Lorentz invariance, where the characteristics of the sample or the 2DEG enter only through the carrier density, n for a translationally invariant system. Thus, in the absence of defect, disorder and impurity, the Hall effect

[3] Remember that $\mathbf{E} = 0$ in the lab frame, though $\mathbf{B}$ remains unchanged.

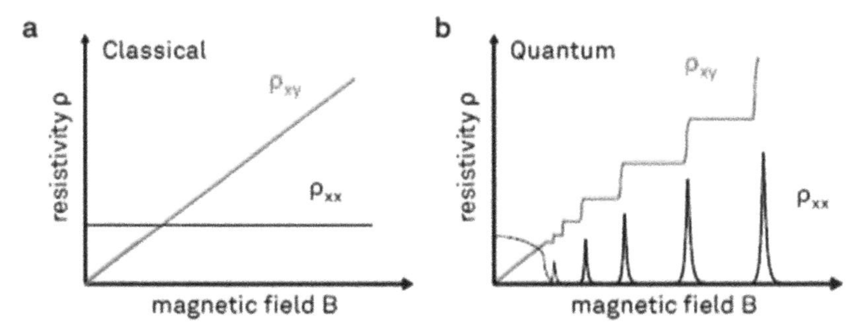

Figure 3.6. Schematic plot showing Hall and magnetoresistivities for both (a) classical and (b) QHE. Copyright Zurich Instruments.

concerns with the carrier density of the sample and nothing else. The Hall resistivity depends linearly on the magnetic field. The QHE which is much more versatile (than merely depending on the density) where disorder, that jeopardizes the translational invariance plays an indispensable part in causing plateaus in the conductivity (or the resistivity) to occur and survive. We show a schematic plot in figure 3.6 to emphasize the difference between the classical and the QHE. Both the Hall and the longitudinal resistivities are significantly different in these two cases.

Writing the equation of motion (EOM) for a charge particle of mass m, moving with a velocity, $\mathbf{v}$ $(= \frac{\mathbf{p}}{m})$ in presence of a longitudinal electric field, $\mathbf{E}$ and a perpendicular magnetic field, $\mathbf{B}$,

$$\frac{d\mathbf{p}}{dt} = -e\mathbf{E} - e\frac{\mathbf{p}}{m} \times \mathbf{B} - \frac{\mathbf{p}}{\tau}. \tag{3.25}$$

The last term is the resistive force arising from electron-impurity scattering with τ being the relaxation time. Using the current density, $\mathbf{j} = -ne\mathbf{p}/m$ and the cyclotron frequency, $\omega_{\mathrm{B}} = eB/m$, at the steady state $(\frac{d\mathbf{p}}{dt} = 0)$,

$$\frac{ne^2\tau}{m}\mathbf{E} + \omega_{\mathrm{B}}\tau\mathbf{j} \times \hat{\mathbf{z}} + \mathbf{j} = 0. \tag{3.26}$$

Assuming the motion of carriers along the x-direction, that is, $\mathbf{j} = j\hat{\mathbf{x}}$, casting it in the form, $\mathbf{j} = \sigma\mathbf{E}$, σ assumes the form,

$$\sigma = \frac{ne^2\tau/m}{1 + \omega_{\mathrm{B}}^2\tau^2}\begin{pmatrix} 1 & -\omega_{\mathrm{B}}\tau \\ \omega_{\mathrm{B}}\tau & 1 \end{pmatrix}. \tag{3.27}$$

This yields the Drude conductivity which can be written as,

$$\sigma_{xx} = \frac{\sigma_0}{1 + \omega_{\mathrm{B}}^2\tau^2} \tag{3.28}$$

where $\sigma_0 = ne^2\tau/m$. In the absence of any scattering by the impurities, the relaxation time, τ is infinitely large, which yields $\sigma_{xx} \to 0$. This induces Ohm's law to assume the form (now writing in terms of the resistivity),

$$\begin{aligned}\mathbf{E} = \begin{pmatrix} E_x \\ E_y \end{pmatrix} &= \begin{pmatrix} 0 & \rho_{xy} \\ -\rho_{xy} & 0 \end{pmatrix}\begin{pmatrix} j_x \\ j_y \end{pmatrix} \\ &= \begin{pmatrix} \rho_{xy} j_y \\ -\rho_{xy} j_x \end{pmatrix}.\end{aligned} \tag{3.29}$$

Hence the electric field $\mathbf{E}$ is perpendicular to the current density, $\mathbf{j}$, which says that, $\mathbf{j} \cdot \mathbf{E} = 0$. The physical significance of $\mathbf{j} \cdot \mathbf{E}$ is the work done that accelerates the charges, which being zero in this case implies that a steady current flows in the sample without requiring any work, and hence causes no dissipation. Thus $\sigma_{xx} = 0$ implies that no current flows in the longitudinal direction, which is actually a signature of a perfectly insulating state. Since the components of the resistivity and the conductivity tensors are related by,

$$\sigma_{xx} = \frac{\rho_{xx}}{\rho_{xx}^2 + \rho_{xy}^2} \quad ; \quad \sigma_{xy} = \frac{-\rho_{xy}}{\rho_{xx}^2 + \rho_{xy}^2}. \tag{3.30}$$

Let us examine the possible scenarios:

(i) If $\rho_{xy} = 0$, one gets $\sigma_{xx} = \frac{1}{\rho_{xx}}$ and $\sigma_{xy} = 0$ which is a familiar scenario.

(ii) If $\rho_{xy} \neq 0$, σ_{xx} and σ_{xy} both exist.

(iii) Now consider $\rho_{xx} = 0$, $\sigma_{xx} = 0$, if $\rho_{xy} \neq 0$. This is truly interesting, since if $\rho_{xx} = 0$, it implies a perfect conductor, and at the same time $\sigma_{xx} = 0$ implies a perfect insulator. This is surprising, but truly occurs in the presence of an external magnetic field. This is reflected in the plots presented in figures 3.3 and 3.4. We shall return for a more thorough discussion later.

3.2.3 Charge particles in a magnetic field: Landau levels

Let us examine the fate of the electrons confined in a 2D plane in the presence of a magnetic field. Consider non-interacting spinless electrons in an external field, $\mathbf{B}$ and write down the Schrödinger equation to solve for the eigenvalues and eigenfunctions. The canonical momentum can now be written as, $\mathbf{p} \to \mathbf{p} - q\mathbf{A} = \mathbf{p} + e\mathbf{A}$ [19] where $q = -e$ is the electronic charge and $\mathbf{A}$ is the vector potential corresponding to the field, $\mathbf{B}$. $\mathbf{B}$ and $\mathbf{A}$ are related via, $\nabla \times \mathbf{A} = \mathbf{B}$. The time-independent Schrödinger equation becomes,

$$\frac{1}{2m}(\mathbf{p} + e\mathbf{A})^2\psi(\mathbf{r}) = E\psi(\mathbf{r}). \tag{3.31}$$

In order to solve this, we need to fix a gauge or a choice of the vector potential. Note that the choice of the gauge will not alter the solution of the equation, which in other words can be stated as the Schrödinger equation being gauge invariant. However, a particular choice is essential for us to go ahead.

Corresponding to a magnetic field, **B** in the z-direction, such that $\mathbf{B} = B\hat{z}$ (just as the case of a 2DEG subjected to a perpendicular magnetic field), the vector potential can be chosen as,

$$A_x = -By, \qquad A_y = A_z = 0. \tag{3.32}$$

This is known as the Landau gauge. It also allows us to assume $A_y = Bx$ and $A_x = A_z = 0$[4]. In the Landau gauge, the Schrödinger equation becomes,

$$\left[\frac{1}{2m}(p_x - eBy)^2 + \frac{p_y^2}{2m} + \frac{p_z^2}{2m}\right]\psi(\mathbf{r}) = \epsilon\psi(\mathbf{r}). \tag{3.33}$$

Clearly, in the z-direction, the particle behaves like a free particle with energy, $\epsilon_z = \frac{p_z^2}{2m}$ with the eigenfunction the same as that of particle in a box in the z-direction. Thus in the x–y plane, the above equation becomes,

$$\begin{aligned}\left[\frac{1}{2m}(p_x - eBy)^2 + \frac{p_y^2}{2m}\right]g(x, y) = \epsilon g(x, y)\\ \mathcal{H}(x, y)g(x, y) = \epsilon g(x, y)\end{aligned} \tag{3.34}$$

where for the 2D case, $\psi(\mathbf{r})$ becomes $g(x, y)$ and $E = \epsilon + \epsilon_z$. It is easy to see that p_x commutes with $\mathcal{H}(x, y)$, that is,

$$[\mathcal{H}(x, y), p_x] = 0. \tag{3.35}$$

Hence p_x is a constant of motion. Thus for a p_x given by, $p_x = \frac{2\pi\hbar}{L_x}n_x$, one can write equation (3.34) as,

$$\left[\frac{p_y^2}{2m} + \frac{1}{2}m\left(\frac{eB}{m}\right)^2(y - y_0)^2\right]f(y) = \epsilon f(y) \tag{3.36}$$

where $y_0 = \frac{p_x}{eB} = k_x l_B^2 = k l_B^2$ (say), and $f(y)$ is only a function of y. y_0 has the dimension of length, and hence l_B is denoted as the magnetic length which is an important quantity in all of subsequent discussions. Moreover, $f(y)$ is the eigenfunction corresponding to the 1D Hamiltonian written above.

Interestingly, the left-hand side of equation (3.36) denotes the Hamiltonian for a simple harmonic oscillator (SHO) which oscillates in the y-direction about a mean position y_0 with a frequency, $\omega_c = \frac{eB}{m}$. ω_c is known as the cyclotron frequency. Taking results from the SHO problem in quantum mechanics, the energy eigenvalues can be found as,

[4] A combination of the two yields a 'symmetric' gauge which we shall introduce and employ later.

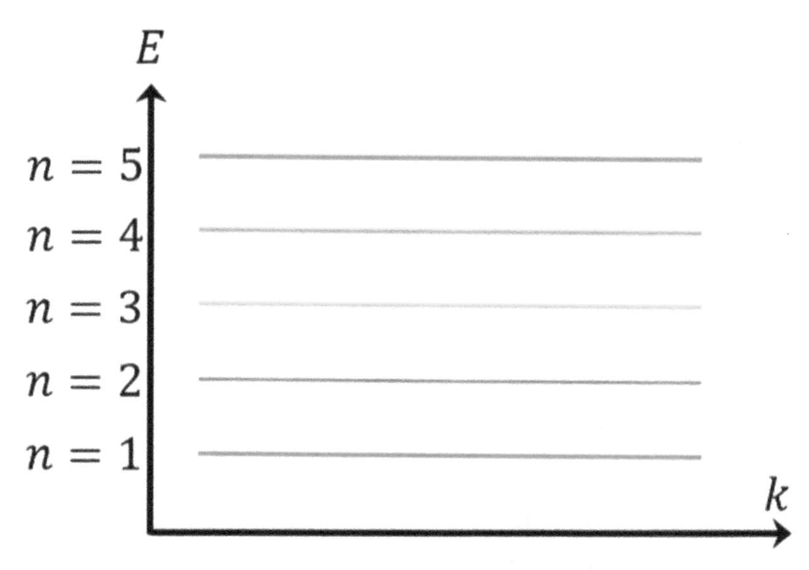

Figure 3.7. Schematic plot showing the Landau levels. Each of these levels has a large degeneracy.

$$\epsilon_n = \left(n + \frac{1}{2}\right)\hbar\omega_c = \left(n + \frac{1}{2}\right)\hbar\frac{eB}{m}. \tag{3.37}$$

The energy levels in equation (3.37) are referred to as the Landau levels. The levels are schematically shown in figure 3.7. They are equidistant for a given value of the magnetic field, however their separation increases with the increase in B. Each of these levels has a large degeneracy (see below).

Further, the eigenfunction, $g(x, y)$ in equation (3.34) corresponding to an oscillatory motion (as an SHO) in the y-direction and a free motion (like a particle in the absence of any potential) in x-direction assumes the form,

$$g(x, y) = \frac{1}{L_x}e^{ik_x x}A_n e^{\frac{-eB(y-y_0)^2}{\hbar}}H_n\left(\frac{eB(y - y_0)}{\hbar}\right) \tag{3.38}$$

where $H_n(\xi)$ with $\xi = \frac{eB(y - y_0)}{\hbar}$ denote the Hermite polynomials that are familiar in the context of SHO and A_n denote the normalization constants. Thus the trajectory of the particle is similar to that of a simple harmonic oscillator centred about a certain value of y (instead of the origin), namely, y_0 and freely propagating along the x-direction. y_0 is controlled by the strength of the magnetic field B, and is inversely proportional to it.

3.2.4 Degeneracy of the Landau levels

The Landau levels given by equation (3.37) are hugely degenerate. Since p_x is a constant of motion, the energy is independent of p_x. Thus all possible values of the quantum numbers corresponding to the motion in the x-direction, namely, n_x that are defined by, $k_x = \frac{2\pi}{L_x} n_x$, ($L_x$ denotes the length of the sample in the x-direction and $n_x = 0, 1, 2, \ldots$) will make the levels degenerate. The degree of degeneracy is only limited by the length of the sample in the y-direction, namely, L_y. Since the magnetic length, $y_0 = kl_B^2 = k\left(\frac{\hbar}{eB}\right)$ about which the simple harmonic motion occurs should not exceed the length L_y of the sample, the maximum degeneracy of the Landau levels can be found by substituting $y_0 = L_y = \left(\frac{\hbar}{eBL_x}\right) n_x$, and hence using the maximum possible value of n_x, that is,

$$(n_x)_{\max} = g = \frac{eBL_xL_y}{h} = \frac{eB\mathcal{A}}{h} \tag{3.39}$$

where $\mathcal{A} = L_xL_y$ is the area of the sample in the x–y plane. This yields the degeneracy, g to be identified as the flux, ϕ ($= B\mathcal{A}$) threading the planar sample, via, $g = \Phi/\Phi_0$ where $\Phi_0 = h/e$[5].

A few comments are in order:

(i) The degeneracy, g is independent of the effective mass of the carriers, and hence independent of the material.
(ii) The degeneracy is proportional to the area of the sample and the value of the magnetic field. Thus the degeneracy can be controlled by the applied magnetic field.

To remind ourselves, we have solved for the properties of a single electron confined in a plane in the presence of a perpendicular magnetic field. The energies are called the Landau levels, and these levels are highly degenerate. The scenario is a prototype of what happens in a Hall effect experiment where the external magnetic field is varied and the resistivities (both the Hall and the longitudinal) are measured. The Hall resistivity shows plateaus in multiples of h/ne^2 whenever the filling fraction, ν of the Landau levels (defined below) is close to an integer n. Consider n_0 to be the density of charge carriers of the sample, then ν is defined using, $\nu = \frac{n_0}{g/\mathcal{A}} = \frac{n_0 h}{eB}$. ν denotes the filling fraction which is also defined as,

$$\nu = \frac{\text{number of electrons}}{\text{flux quantum}}$$

As and when ν assumes a value near an integer, n (or a rational fraction as in the case of FQHE) as the magnetic field is tuned, one observes a plateau in the Hall resistivity, ρ_{xy}. Something else happens at the same time that is equally interesting. The longitudinal resistivity, ρ_{xx} drops to zero whenever ρ_{xy} acquires a plateau. The vanishing of ρ_{xx} makes the system dissipationless. However, the diagonal conductivity (σ_{xx}) also vanishes which makes the system insulating. The paradox is explained in detail elsewhere.

[5] The value of $\phi_0 \simeq 4.13 \times 10^{-15}$ Wb-m^2.

3.2.5 Conductivity of the Landau levels: role of the edge modes

Here we discuss some of the key properties of the Landau levels. Let us now calculate the current carried by the Landau levels. The expression for the current can be found using $\langle \mathbf{J} \rangle = -e\langle \mathbf{v} \rangle$ where the expectation value has to be computed within the Landau states.

$$\langle \mathbf{J} \rangle = -e\langle \psi_k | \mathbf{v} | \psi_k \rangle = -\frac{e}{m}\langle \psi_k | \mathbf{p} + e\mathbf{A} | \psi_k \rangle. \tag{3.40}$$

The longitudinal current in the x-direction carried by the Landau levels is obtained via,

$$\langle J_x \rangle = -\frac{e}{ml_B\sqrt{\pi}} \int dy \; e^{-\frac{1}{l_B^2}(y-y_0)^2}(\hbar k - eBy) = 0. \tag{3.41}$$

The integrand has an even function (the first one) and an odd function (the second one). Thus the integral vanishes. We can get the average velocity, $\langle v \rangle = \frac{1}{\hbar}\frac{\partial \epsilon_k}{\partial k} = 0$ as ϵ does not depend upon k. Thus the Landau wavefunctions carry no current by themselves. They only carry current in the presence of an electric field in the x-direction as we show below.

3.2.6 Spin and the electric field

So far we have been talking about spinless fermions. It is in general a worthwhile exercise to include the spin of the electrons and explore if there is any significant development to the quantization phenomena discussed above. The spin degrees of freedom placed in an external magnetic field introduce a Zeeman energy scale owing to a coupling between the spin and the magnetic field. The Zeeman term is written as $\Delta_Z = g\mu_B B$, where $\mu_B = e\hbar/2m$ is the Bohr magneton and $g = 2$. Now the splitting between the Landau levels originating from the orbital effect ($\mathbf{p} \to \mathbf{p} - q\mathbf{A}$) is $\Delta = \hbar\omega_B = \frac{e\hbar B}{m} = \Delta_B$(say). Δ_B is the so called cyclotron energy. But for electrons, this precisely coincides with the Zeeman splitting $\Delta = g\mu_B B$ between the $\uparrow$ and the $\downarrow$-spins. Thus it looks that the spin-$\uparrow$ particles in a Landau level with index, n have exactly the same energy as the spin-$\downarrow$ particles in the next higher Landau level index, that is, $n + 1$. However, in real materials this does not occur. For example in GaAs, the Zeeman energy is typically about 70 times smaller than the cyclotron energy.

Now consider an external electric field, E applied in the x-direction. This creates an electric potential of the form, $\phi = -Ex$. Thus the resulting Hamiltonian including external electric and magnetic fields can be written as,

$$\mathcal{H} = \frac{1}{2m}\left[p_x^2 + (p_y + eBx)^2\right] + eEx. \tag{3.42}$$

It is important to note that we have considered a different choice of the gauge here, namely, $A_y = Bx$. Instead of a rigorous derivation, we can complete square in the expression for energy of the particles,

Figure 3.8. Schematic plot showing the Landau levels (n = 1, 2, 3, 4) in the presence of an electric field. The levels are tilted because of the electric field.

$$E_{n,\,k} = \hbar\omega_B\left(n + \frac{1}{2}\right) - eE\left(kl_B^2 + \frac{eE}{m\omega_B^2}\right) + \frac{1}{2}m\frac{E^2}{B^2}. \tag{3.43}$$

This is interesting because the degeneracy of the Landau level has now been lifted. The energy in each level now depends linearly on k as shown in figure 3.8, which was earlier independent of k. The eigensolution is simply that of a harmonic oscillator shifted from the origin and displaced along the x-axis by an amount mE/eB^2 and is written as,

$$\psi(x,\, y) = \psi_{n,k}\left(x + \frac{mE}{eB^2},\, y\right). \tag{3.44}$$

Among the other properties, the group velocity is given by,

$$v_y = \frac{1}{\hbar}\frac{\partial E_{n,\,k}}{\partial k} = \frac{e}{\hbar}El_B^2. \tag{3.45}$$

Putting $l_B = \sqrt{\frac{\hbar}{eB}}$ (as said earlier, l_B is an important length scale of the problem which we shall see throughout the discussion),

$$v_y = \left(\frac{eE}{\hbar}\right)\cdot\left(\frac{\hbar}{eB}\right) = \frac{E}{B}. \tag{3.46}$$

Thus the energy has three terms in equation (3.43),

(i) the first one is that of a harmonic oscillator,

(ii) the second one is the potential energy, of a wave packet localized at $x = \left(-kl_B^2 - \frac{mE}{e\omega_B^2}\right)$,

(iii) and finally the last one denotes the kinetic energy of the particle, namely, $\frac{1}{2}mv_y^2$.

3.2.7 Laughlin's argument: Corbino ring

Laughlin intuitively considered the phenomenon of QHE as a quantum pump. Consider a ring where the vacant region admits a magnetic field and hence a flux ϕ. For the argument to be valid, the geometry of the ring is important. Here in addition to the background magnetic field **B** that threads the sample, we can thread an additional flux Φ through the center of the ring (see figure 3.9). This Φ can affect the quantum state of the electrons. In addition, the temperature is low such that the thermal effects can be neglected.

Let us first see what this flux Φ has got to do with the Hall conductivity. Suppose we slowly increase Φ from 0 to $\Phi_0\left(=\frac{h}{e}\right)$, that is, within a time $t_0 \gg \frac{1}{\omega B}$. This induces an emf around the annular region $\varepsilon = \frac{\partial\Phi}{\partial t} = \frac{-\Phi_0}{t_0}$. The purpose of this emf is to transport 'n' electrons from the inner circumference to the outer circumference. This would result in a current in the radial direction, $I_r = -ne/t_0$. Thus the Hall resistivity is,

$$\rho_{xy} = \frac{\varepsilon}{I_r} = -\frac{\Phi_0}{t_0} \cdot \frac{t_0}{(-ne)} = \frac{h}{e^2} \cdot \frac{1}{n}. \tag{3.47}$$

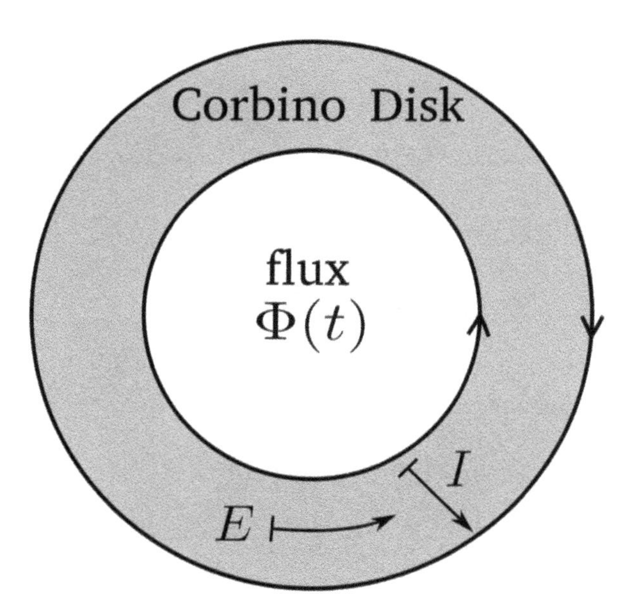

Figure 3.9. Schematic plot showing Corbino ring.

The same arguments hold equally for the IQHE and FQHE, in the former n is an integer, while n is a fraction for the latter. In FQHE, the interpretation is as follows: as we increase the flux from Φ to Φ_0, a charge of magnitude e/m is transported from the inner circumference to the outer one when the flux is increased by Φ_0 units. The resultant Hall conductivity (or equivalently the resistivity) becomes,

$$\sigma_{xy} = \frac{e^2}{h} \cdot \frac{1}{m}. \tag{3.48}$$

Thus a whole electron is transferred only when the flux is increased by $m\Phi_0$ units.

3.2.8 Edge modes and conductivity of the single Landau level

When a particle is restricted to move only in one direction, the motion is said to be chiral where backscattering is prohibited. Thus the particles propagate in one direction at one edge of the sample and move in the other direction at the other end of the sample. Let us understand how the edge modes appear.

An edge can be modelled by a potential $V(x)$ in the y-direction which rises steeply, as shown in figure 3.10. Let us continue working in the Landau gauge, such that the Hamiltonian is given by,

$$\mathcal{H} = \frac{1}{2m}[p_x^2 + (p_y + eBx)^2] + V(x). \tag{3.49}$$

In the absence of the potential $V(x)$, the (lowest) wavefunction is a Gaussian of width $l_B \left(=\sqrt{\frac{\hbar}{eB}}\right)$. If we assume that $V(x)$ is smooth over a distance l_B, and hence assume the center of each of Gaussian to be localized at $x = x_0$, we can Taylor expand the potential $V(x)$ about x_0 in the following fashion,

$$V(x) = V(x_0) + \frac{\partial V}{\partial x}(x - x_0) + \cdots. \tag{3.50}$$

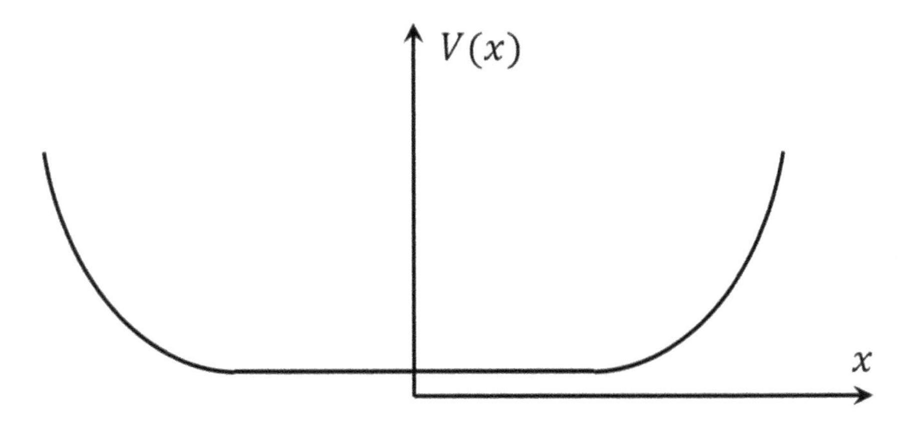

Figure 3.10. Schematic plot showing the potential seen by the charge due to edge of the quantum Hall sample.

Dropping terms after the second one and assuming the constant term (first term) to be zero, the second term looks like the potential due to the electric field. So the particle acquires a drift velocity in the y-direction,

$$v_y = -\frac{1}{eB}\frac{\partial V}{\partial x}. \tag{3.51}$$

Each wavefunction labelled by a momentum, k is located at different x-positions, namely $x = -kl_B^2$, and thus has a different drift velocity. Additionally, the slopes of the two edges drawn above has different slopes, so $\frac{\partial V}{\partial x}$ has different signs at the edges. Thus v_y at the left edge has a different sign with respect to the right edge. Further, because of the drift in y-direction, there will be a current I_y which is known as the Hall current and is calculated as,

$$\begin{aligned} I_y &= -e\int \frac{dk}{2\pi} v_y(k) \\ &= \frac{e}{2\pi l_B^2}\int dx \frac{1}{eB}\frac{dV}{dx}; \quad \text{using } l_B^2 = \frac{\hbar}{eB} \\ &= \frac{e^2}{2\pi\hbar} V_H \end{aligned} \tag{3.52}$$

V_H is the Hall voltage. Now, $\sigma_{xy} = \frac{I_y}{V_H} = \frac{e^2}{2\pi\hbar} = \frac{e^2}{h}$ which is indeed the expected conductivity for a single Landau level.

The above schematic diagram (figure 3.11) shows that the current is entirely carried by the edge states, since the bulk Landau level is absolutely flat (having no k dependence) and hence does not carry any current. The argument is also elegant as it does not depend upon the form of the potential $V(x)$.

Everything that we have discussed so far holds for a single Landau level, however the argument is equally valid for a large number of Landau levels, as long as the Fermi energy lies in between the filled and the unfilled Landau levels. Also, the chiral edge modes are robust to any impurity or disorder as there is no phase space available for scattering. If a left moving electron is to scatter over to a right moving electron, it has to cross the entire sample edge which is not allowed as the probability of scattering would

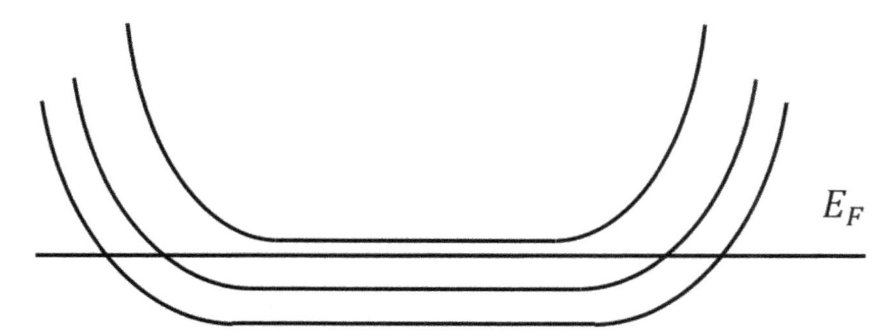

Figure 3.11. Schematic plot showing the appearance of edge modes.

be infinitesimally small owing to the macroscopic physical dimension of the sample. Thus the Hall plateaus are robust to disorder, defects and impurities.

A supremely important issue stands out: how do the plateaus exist in the first place? To see this, let us fix the electron density, n. Then we shall only have filled Landau levels when the magnetic field is exactly $B = \frac{n}{\nu}\Phi_0$ (with $\Phi_0 = \frac{h}{e}$) for some integer ν. But what happens when $B \neq \frac{n}{\nu}\Phi_0$, that is, when the Landau levels are partially filled? Also on top of that there is also a (small) electric field. In the partially filled last Landau level, the longitudinal conductivity will be non-zero, while the Hall conductivity will not be quantized. So how do the plateaus appear and the longitudinal conductivities vanish? The disorder comes to the rescue. It gives a finite width to the Landau levels. So even if the $B \neq \frac{n}{\nu}\Phi_0$, there is a finite regime over which the Hall conductivity, σ_{xy} remains constant and hence the plateaus form so that the Hall resistivity, ρ_{xy} freezes at that value, till the magnetic field is increased further.

3.2.9 Incompressibility and the QH states

The key feature of the QH states is that the states are incompressible. The compressibility, κ is defined by,

$$\kappa = -\frac{1}{A}\frac{\partial A}{\partial P}\bigg|_N \tag{3.53}$$

where A, P and N are the area, pressure and the number of particles, respectively. The system is said to be incompressible when $\kappa = 0$, that is, when the area is insensitive to the pressure applied. The pressure is defined as the change of energy as a result of change of area, that is,

$$P = -\frac{\partial E}{\partial A}.$$

Thus the inverse of the compressibility is defined as,

$$\kappa^{-1} = -A\frac{\partial P}{\partial A}|_N = A\frac{\partial^2 E}{\partial A^2}|_N.$$

Since the energy is an extensive quantity, that is, it depends on the number of particles and hence can be written as,

$$E = N\epsilon(n)$$

where ϵ is energy per particle, and n is particle density, such that the total number of particles is defined by, $N = An$. Hence,

$$\kappa^{-1} = \frac{1}{n} = n^2\left(2\frac{d\epsilon(n)}{dn} + n\frac{d^2\epsilon(n)}{dn^2}\right) = n^2\frac{d^2(n\epsilon)}{dn^2}. \tag{3.54}$$

Also, the chemical potential is given by,

$$\mu = \frac{\partial E}{\partial N}|_V = \frac{d(n\epsilon)}{dn}. \tag{3.55}$$

Thus comparing equations (3.54) and (3.55),

$$\kappa^{-1} = n^2 \frac{d\mu}{dn}. \tag{3.56}$$

The system is incompressible ($\kappa = 0$) when the chemical potential μ increases discontinuously as a function of density, that is, $\frac{\partial n}{\partial \mu} = 0$.

3.2.10 Hall effect in the symmetric gauge

The vector potential **A** in the mixed gauge can be written as (again yields $\mathbf{B} = B\hat{\mathbf{z}}$),

$$\mathbf{A} = -\frac{1}{2}(\mathbf{r} \times \mathbf{B}) = \frac{1}{2}(-By\hat{\mathbf{x}} + Bx\hat{\mathbf{y}}) \tag{3.57}$$

The choice of the symmetric gauge breaks the translation symmetry in both x- and y-directions. However, it preserves rotational symmetry about the origin. This of course means that angular momentum is a good quantum number. This is the most convenient gauge to study FQHE. Further, we can write down the non-canonical momentum as,

$$\boldsymbol{\pi} = \mathbf{p} + e\mathbf{A} = m\dot{\mathbf{r}}. \tag{3.58}$$

This is gauge invariant but non-canonical. One can use this to form the raising and lowering operators,

$$\begin{aligned} a &= \frac{1}{\sqrt{\frac{eBh}{\pi}}}(\pi_x - i\pi_y) = \sqrt{\frac{\pi}{eB}}(\pi_x - i\pi_y) \\ a^\dagger &= \sqrt{\frac{\pi}{eB}}(\pi_x + i\pi_y) \end{aligned} \tag{3.59}$$

where $2e\hbar B = 2e\frac{h}{2\pi}B = \frac{eBh}{\pi}$. Now,

$$\begin{aligned} [a, a^\dagger] = aa^\dagger - a^\dagger a &= \left(\frac{\pi}{eB}\right)(\pi_x - i\pi_y)(\pi_x + i\pi_y) - (\pi_x + i\pi_y)(\pi_x - i\pi_y) \\ &= \left(\frac{\pi}{eBh}\right)[\pi_x^2 + i\pi_x\pi_y - i\pi_y\pi_x + \pi_y^2 - \pi_x^2 + i\pi_x\pi_y - i\pi_y\pi_x - \pi_y^2] \\ &= \left(\frac{2\pi i}{eBh}\right)[\pi_x\pi_y - \pi_y\pi_x] = \left(\frac{i}{eB\hbar}\right)[\pi_x, \pi_y] \end{aligned} \tag{3.60}$$

$\boldsymbol{\pi}$ is not the canonical momenta in the sense, $[x_i, \pi_j] \neq \delta_{ij}$ and $[\pi_i, \pi_j] = \delta_{ij}$. However, they are gauge invariant. The numerical value of $\boldsymbol{\pi}$ does not depend upon the choice of gauge. This can be proved from the commutation relations,

$$\left\{p_i + eA_i, p_j + eA_j\right\} = -e\left(\frac{\partial A_j}{\partial x^i} - \frac{\partial A_i}{\partial x^j}\right) = -e\epsilon_{ijk}B_k. \tag{3.61}$$

Thus, $[a, a^\dagger] = (\frac{i}{eB\hbar})(-ie\hbar B) = 1$

To explore whether the Landau levels yield the expected degeneracy, we can introduce another momentum variable, namely,

$$\tilde{\boldsymbol{\pi}} = \boldsymbol{\pi} - e\mathbf{A}$$

$\tilde{\boldsymbol{\pi}}$ is not gauge invariant, and depends on the gauge potential chosen. The vector potential, $\mathbf{A}$ enjoys the gauge freedom, that is,

$$\mathbf{A}' = \mathbf{A} - \nabla\chi$$

where χ is an arbitrary scalar. This yields,

$$\tilde{\boldsymbol{\pi}}' = \mathbf{p}' - e(\mathbf{A}' - \nabla\chi) = \mathbf{p}' - e\mathbf{A}' + \nabla\chi.$$

The commutation relations for $\tilde{\boldsymbol{\pi}}'$ are,

$$[\tilde{\pi}_x', \tilde{\pi}_y'] = ie\hbar B \tag{3.62}$$

which differ only by a sign with respect to the π momenta. This is also an advantage of this new momenta in that they obey,

$$[\tilde{\pi}_i, \tilde{\pi}_j] = 0.$$

Finally, the Hamiltonian is written in terms of the a and $a^\dagger$ operators as,

$$\mathcal{H} = \hbar\omega_B\left(a^\dagger a + \frac{1}{2}\right) = \hbar\omega_B\left(n + \frac{1}{2}\right) \tag{3.63}$$

where $n = 0, 1, \cdots$ and denote the indices of the Landau levels.

Now as a matter of exercise, in order to explore the degeneracy of the Landau levels, we introduce a second pair of raising and lowering operators, namely,

$$b = \frac{1}{\sqrt{2e\hbar B}}(\tilde{\pi}_x + i\tilde{\pi}_y) \tag{3.64}$$

$$b^\dagger = \frac{1}{\sqrt{2e\hbar B}}(\tilde{\pi}_x - i\tilde{\pi}_y). \tag{3.65}$$

They too obey $[b, b^\dagger] = 1$. These b, $b^\dagger$ will yield the degeneracy of the Landau levels as shown in the following. Thus a general state in the Hilbert space $|n, m\rangle$ is defined by,

$$|n, m\rangle = \frac{(a^\dagger)^n (b^\dagger)^m}{\sqrt{n!m!}}|0, 0\rangle \tag{3.66}$$

where $a|0, 0\rangle = b|0, 0\rangle = 0$ and $\mathcal{H} = \frac{1}{2m}\boldsymbol{\pi}\cdot\boldsymbol{\pi} = \frac{1}{2m}(\mathbf{p} + e\mathbf{A})^2$.

Let us now construct the wavefunction in the symmetric gauge. we are going to focus on the lowest Landau level, $n = 0$ as it is of primary interest for discussing FQHE. The trick is to convert the definition of a into a differential equation,

$$a = \frac{1}{\sqrt{2e\hbar B}}(\pi_x - i\pi_y) = \frac{1}{\sqrt{2e\hbar B}}[p_x - ip_y + e(A_x - iA_y)]$$
$$= \frac{1}{\sqrt{2e\hbar B}}\left[-i\hbar\left(\frac{\partial}{\partial x} - i\frac{\partial}{\partial y}\right) + \frac{eB}{2}(-y - ix)\right] \tag{3.67}$$

using $z = x - iy$ and $\tilde{z} = x + iy$. Remember this is not usually how we define z and z^* (or $\tilde{z}$), however we shall stick to this definition. Also define,

$$\partial = \frac{1}{2}\left(\frac{\partial}{\partial x} + i\frac{\partial}{\partial y}\right) \tag{3.68}$$

and,

$$\tilde{\partial} = \frac{1}{2}\left(\frac{\partial}{\partial x} - i\frac{\partial}{\partial y}\right) \tag{3.69}$$

which obey $\partial z = \tilde{\partial}\tilde{z} = 1$ and $\partial\tilde{z} = \tilde{\partial}z = 0$.

$$\partial z = \frac{1}{2}\left(\frac{\partial}{\partial x} + i\frac{\partial}{\partial y}\right)(x - iy) = \frac{1}{2}(1 + 1) = 1.$$

So a and $a^\dagger$ in terms of the coordinates z can be written as,

$$a = -i\sqrt{2}\left(l_B\tilde{\partial} + \frac{z}{4l_B}\right); \qquad a^\dagger = -i\sqrt{2}\left(l_B\tilde{\partial} - \frac{\tilde{z}}{4l_B}\right). \tag{3.70}$$

Now, the lowest Landau level is found by the one which is annihilated by this operator a.

$$a|0, m\rangle = 0$$

$$-i\sqrt{2}\left(l_B\tilde{\partial} + \frac{z}{4l_B}\right)|0, m\rangle = 0$$

$|0, m\rangle$ is called $\psi_{LLL}(z, \tilde{z})$, where LLL stands for lowest Landau level.

$$\psi_{LLL,\, m=0} \sim e^{-|z|^2/4l_B^2}$$

The ground state is known to be a Gaussian for a linear Harmonic oscillator[6].

One can construct the higher Landau level wavefunctions by employing $b^\dagger$ successively to the $m = 0$ state. This yields,

$$\psi_{LLL,m} = \sim\left(\frac{z}{l_B}\right)^m e^{-|z|^2/4l_B^2}. \tag{3.71}$$

It is straightforward to ascertain that $\psi_{LLL,m}$ are eigenfunctions of J_z, defined by,

$$J_z = \hbar(z\partial - \tilde{z}\tilde{\partial}) \tag{3.72}$$

and obey, $J_z\psi_{LLL,m} = m\hbar\psi_{LLL,m}$.

[6] $(y + \frac{\partial}{\partial y})u(y) = 0$, or $\frac{\partial u}{u} = -ydy$. The solution is $u = e^{-y^2/2}$.

Let us explore the degeneracy associated with the Landau levels $\psi_{LLL,m}$, which is obtained by noting that the wavefunction with angular momentum, m is peaked on a circular ring of radius, $r = l_B\sqrt{2m}$. The number of states in an area, $A = \pi R^2$ is $\mathcal{N} = \pi R^2/\pi r^2 \simeq eBA/2\pi\hbar$ which is a result that we have seen earlier.

3.3 Kubo formula and the Hall conductivity

The important question at this stage is what protects the plateaus in the Hall conductivity (or the resistivity)? Why are they so flat and robust at the integer values (or at certain rational fractions)? Remember the system does not have either translational invariance (because of the presence of disorder) or time reversal invariance (because of the presence of a magnetic field). Thus two known symmetries are lost and still the plateaus persist[7]. We shall now show that the Hall conductivity assumes $\sigma_{xy} = \frac{\nu e^2}{h}$ (or equivalently $\rho_{xy} = \frac{h}{\nu e^2}$) under these conditions.

To compute the conductivity, we resort to the Kubo formula which arises from a more generalized concept, known as the '*linear response theory*'. We shall derive the Kubo formula under a few conditions for the sake of simplicity. They are,

(i) Before the fields are applied, at $t = -\infty$, the system is in a non-interacting many-particle state which obeys $\mathcal{H}_0|\psi_m\rangle = E_m|\psi_m\rangle$, where $\{E_m, \psi_m\}$ are the eigensolutions of $\mathcal{H}_0$.

(ii) Even if we actually apply a constant electric field, $\mathbf{E}$, it is helpful to consider an alternating field with frequency, ω of the form, $\mathbf{E}(t) = \mathbf{E}e^{-i\omega t}$ and at the end of the calculation take the zero frequency limit, that is, $\omega \to 0$.

(iii) Consider a gauge in which the transverse components of the vector potential are zero, that is, $A_t = 0$. In other words, $\mathbf{E} = -\frac{\partial \mathbf{A}}{\partial t}$ with no $\nabla\phi$ term, ϕ being the scalar potential. Equivalently one can assume that ϕ = constant.

Now let us write the full Hamiltonian as,

$$\mathcal{H} = \mathcal{H}_0 + \mathcal{H}' \tag{3.73}$$

where $\mathcal{H}_0$ is the non-interacting Hamiltonian whose exact solutions are known (as described earlier) and $\mathcal{H}'$ is the interaction term due to the coupling of the electrons to the external field. Thus $\mathcal{H}'$ involves the current due to the motion of the electrons coupling with the vector potential arising due to the presence of the magnetic field. Thus,

$$\mathcal{H}' = -\mathbf{J} \cdot \mathbf{A} \tag{3.74}$$

[7] In fact, naively it seems ironical that a broken time reversal symmetry is solely responsible for quantization of the Hall plateaus.

where $\mathbf{J}$ and $\mathbf{A}$ are the electric current density and the vector potential, respectively. $\mathbf{J}$ is related to the mechanical momentum $\mathbf{p} + e\mathbf{A}$. Using $\mathbf{E} = -\frac{\partial \mathbf{A}}{\partial t}$, one can write,

$$\mathbf{A} = \frac{\mathbf{E}}{i\omega} e^{-i\omega t}. \tag{3.75}$$

The aim is to compute the expectation value of the current density and find out how it depends on the applied electric field, such that the proportionality constant yields the conductivity. In particular we are interested in computing the Hall conductivity, σ_{xy}.

Here we shall consider the interaction picture where the time evolution of an arbitrary operator $\hat{O}$ is written as,

$$\hat{O}(t) = e^{i\mathcal{H}_0 t/\hbar} \hat{O}(0) e^{-i\mathcal{H}_0 t/\hbar}. \tag{3.76}$$

Here $\hat{O}$ can be any operator, such as $\mathbf{J}$ or $\mathcal{H}'$. Further, the eigenstates in the interaction picture evolve with time according to,

$$|\psi(t)\rangle = U(t,\, t_0)|\psi(t_0)\rangle \tag{3.77}$$

where t_0 refers to an earlier time when the interaction is switched on and t denotes a later time. The time evolution operator, $U(t,\, t_0)$ is a unitary operator having a form,

$$U(t,\, t_0) = T \exp\left(-\frac{i}{\hbar} \int_{t_0}^{t} \mathcal{H}'(t')dt'\right) \tag{3.78}$$

T denotes time ordering in the above equation [20]. If the interval $[t\colon\, t_0]$ is split into several time steps, T keeps the earliest time to the right. Now let us consider that as $t_0 \to -\infty$, that is, before the perturbation is switched on, the system was in the ground state $|\psi_0(t)\rangle$. Hence the time evolution operator can be written as,

$$U(t,\, t_0 = -\infty) = T \exp\left(-\frac{i}{\hbar} \int_{-\infty}^{t} \mathcal{H}'(t')dt'\right) = U(t) \quad \text{say}. \tag{3.79}$$

The ground state expectation value of the current operator is given by,

$$\begin{aligned} \langle J(t)\rangle &= \langle \psi_0(t)|J(t)|\psi_0(t)\rangle = \langle \psi_0|U^{-1}(t)J(t)U(t)|\psi_0\rangle \\ &= \langle \psi_0|\left[Te^{\frac{i}{\hbar}\int_{-\infty}^{t} \mathcal{H}'(t')dt'} J(t) Te^{-\frac{i}{\hbar}\int_{-\infty}^{t} \mathcal{H}'(t')dt'}\right]|\psi_0\rangle. \end{aligned} \tag{3.80}$$

An expansion of the exponentials (assuming the interaction term to be weak) and retaining terms up to first order in $\mathcal{H}'$ yields,

$$\langle \mathbf{J}(t)\rangle \approx \langle \psi_0|\left[\mathbf{J}(t) + \frac{i}{\hbar}\int_{-\infty}^{t} dt'[\mathcal{H}'(t'),\, \mathbf{J}(t)]\right]|\psi_0\rangle. \tag{3.81}$$

The second term inside the bracket of the RHS involves a commutator of $\mathcal{H}'$ and $\mathbf{J}$. It is to be kept in mind that the commutator does not vanish at two arbitrary times t and t'. Using $\mathbf{A}(t) = \frac{\mathbf{E}}{i\omega}e^{-i\omega t}$ and $\mathcal{H}'(t) = -\mathbf{J} \cdot \mathbf{A}$, the $\langle \mathbf{J}(t)\rangle$ takes the form,

$$\langle \mathbf{J}(t)\rangle \approx \langle \psi_0 | \left[\mathbf{J}(t) + \frac{i}{\hbar}\int_{-\infty}^{t} dt' [-\mathbf{J}(t')\cdot \frac{\mathbf{E}}{i\omega} e^{-i\omega t'}, \mathbf{J}(t)] \right] |\psi_0\rangle. \tag{3.82}$$

The first term inside the bracket in the RHS is the current due to the absence of an external electric field which can safely be ignored in our case. So only the second term survives. Thus, writing for components α, β ($\alpha, \beta \in x, y, z$)

$$\langle J_\alpha(t)\rangle = \frac{1}{\hbar\omega}\int_{-\infty}^{t} dt' \langle \psi_0 | [J_\beta(t'), J_\alpha(t)] |\psi_0\rangle E_\beta e^{-i\omega t'} \tag{3.83}$$

where $\boldsymbol{J}\cdot\mathbf{E}$ is written as $J_\beta E_\beta$.

Since the system is invariant under time translation, the above correlation depends on $t - t'$ and not on t and t' individually. Introducing a new variable $t - t' = \tilde{t}$,

$$\langle J_\alpha(t)\rangle = \frac{1}{\hbar\omega}\left(\int_0^{\infty} d\tilde{t}\ \ e^{i\omega\tilde{t}} \langle \psi_0 | [J_\beta(0), J_\alpha(\tilde{t})] |\psi_0\rangle\right) E_\beta e^{-i\omega t} \tag{3.84}$$

where the term inside the bracket in the RHS can be written as, $\sigma_{\alpha\beta}E_\beta e^{-i\omega t}$, $\sigma_{\alpha\beta}$ being the components of the conductivity tensor. Note that the time t dependence is outside the integral and appears as $e^{-i\omega t}$. Thus in the linear response regime, if an electric field of frequency, ω is applied, the current responds to the external field by oscillating with the same frequency as the external field.

The Hall conductivity is the off-diagonal component, and can be computed using,

$$\sigma_{xy}(\omega) = \frac{1}{\hbar\omega}\int_0^{\infty} dt\ \ e^{i\omega t} \langle \psi_0 | [J_y(0), J_x(t)] |\psi_0\rangle \tag{3.85}$$

$J_x(t)$ can be written as,

$$J_x(t) = e^{i\mathcal{H}_0 t/\hbar} J_x(0) e^{-i\mathcal{H}_0 t/\hbar} \tag{3.86}$$

in the above expression, and using completeness relation of the states, namely, $|\psi_n\rangle\langle\psi_n| = \mathbf{1}$, one obtains,

$$\begin{aligned}\sigma_{xy}(\omega) = \frac{1}{\hbar\omega}\int_0^{\infty} dt\ \ e^{i\omega t}\sum_n \langle \psi_0|J_y|\psi_n\rangle\langle\psi_n|J_x|\psi_0\rangle e^{i(E_n-E_0)t/\hbar} \\ - \langle \psi_0|J_x|\psi_n\rangle\langle\psi_n|J_y|\psi_0\rangle e^{i(E_0-E_n)t/\hbar}.\end{aligned} \tag{3.87}$$

Now we shall perform the integral over time, t. Let us write $\frac{E_n}{\hbar} = \omega_n$ and introduce $\omega \to \omega + i\epsilon$ where ϵ is infinitesimal quantity considered for the convergence of the integral. Then the Hall conductivity is obtained as,

$$\sigma_{xy}(\omega) = -\frac{i}{\omega}\sum_{E_n \neq E_0}\left[\frac{\langle \psi_0|J_y|\psi_n\rangle\langle\psi_n|J_x|\psi_0\rangle}{\hbar\omega + E_n - E_0} - \frac{\langle \psi_0|J_x|\psi_n\rangle\langle\psi_n|J_y|\psi_0\rangle}{\hbar\omega + E_0 - E_n}\right]. \tag{3.88}$$

Finally, we shall take the limit $\omega \to 0$ to account for the constant field. For that let us expand the denominator as follows,

$$\begin{aligned}\frac{1}{\hbar\omega + E_n - E_0} &\approx \frac{1}{E_n - E_0} - \frac{\hbar\omega}{(E_n - E_0)^2} + O(\omega^2)\\ \frac{1}{\hbar\omega + E_0 - E_n} &\approx \frac{1}{E_0 - E_n} - \frac{\hbar\omega}{(E_0 - E_n)^2}.\end{aligned} \tag{3.89}$$

The first term looks divergent and such divergence is responsible for the peak in the longitudinal conductivity which, by now we are familiar with. Moreover, $\sigma_{xy}(\omega)$ should not contain a term which is independent of ω. This is the DC conductivity ($\omega = 0$) which is absent in a translationally invariant system. Thus finally one arrives at,

$$\sigma_{xy}(\omega) = i\hbar \sum_{n\neq 0} \frac{\langle\psi_0|J_y|\psi_n\rangle\langle\psi_n|J_x|\psi_0\rangle - \langle\psi_0|J_x|\psi_n\rangle\langle\psi_n|J_y|\psi_0\rangle}{(E_n - E_0)^2}. \tag{3.90}$$

This is the Kubo formula for the Hall conductivity.

To proceed further, let us assume a specific case of perturbing a system below [21]. Consider a quantum Hall sample in the form of a torus (or a donut). Let us thread two fluxes Φ_x and Φ_y (instead of one) as shown in figure 3.12. Owing to this, the gauge potentials can be written as,

$$A_x = \frac{\Phi_x}{L_x},\ A_y = \frac{\Phi_y}{L_y} + Bx. \tag{3.91}$$

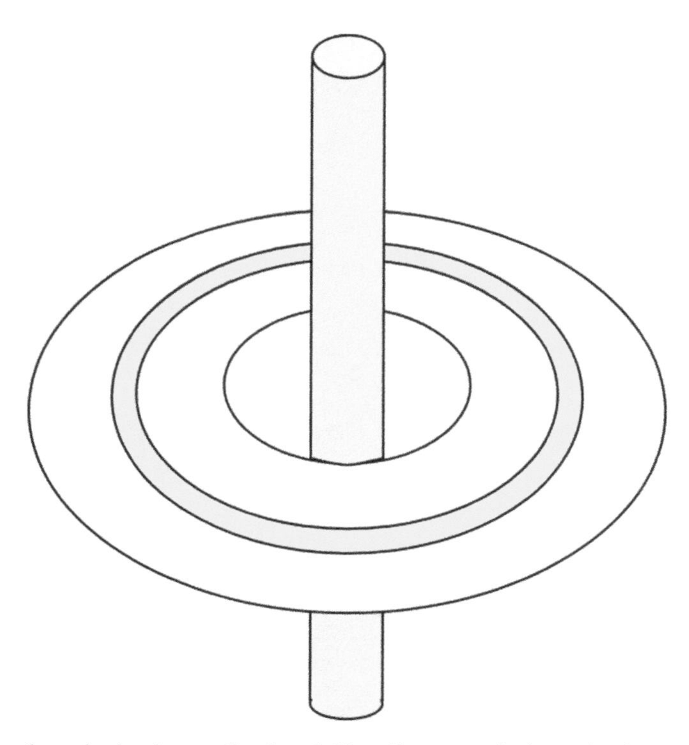

Figure 3.12. The schematic plot shows a disc threaded by a flux perpendicular to the plane and another one in the plane of the disc.

It is clear that the states of the quantum system are sensitive to non-integer values of Φ_i/Φ_0 $(i = x, y)$ where, $\Phi_0 = \frac{h}{e}$. Specifically, if we increase either Φ_x or Φ_y from 0 to Φ_0 then the spectrum of the quantum system must remain invariant. Hence writing the perturbation Hamiltonian in terms of the fluxes,

$$\mathcal{H}' = -\sum_{i\in 1,2} \frac{J_i\Phi_i}{L_i}. \tag{3.92}$$

To the first order corresponding to this perturbation term, the modified ground state becomes,

$$|\psi_0'\rangle = |\psi_0\rangle + \sum_{\substack{\psi_n \neq \psi_0 \\ E_n \neq E_0}} \frac{\langle\psi_n|H'|\psi_0\rangle}{E_n - E_0}|\psi_n\rangle. \tag{3.93}$$

Thus writing,

$$\Delta|\psi\rangle = |\psi_0'\rangle - |\psi_0\rangle = \sum_{\substack{\psi_n \neq \psi_0 \\ E_n \neq E_0}} \frac{\langle\psi_n|H'|\psi_0\rangle}{E_n - E_0}|\psi_n\rangle. \tag{3.94}$$

Considering infinitesimal changes in Φ_i we can write,

$$\left|\frac{\partial|\Delta\psi\rangle}{\partial\Phi_i}\right\rangle = -\frac{1}{L_i}\sum_n \frac{\langle\psi_n|J_i|\psi_0\rangle}{E_n - E_0}|\psi_n\rangle. \tag{3.95}$$

Terms like those in the RHS of the above equation appeared in the Hall conductivity. Let us write the total Hall conductivity including the area factor L_xL_y of the sample, which can be written as,

$$\begin{aligned}\sigma_{xy} &= i\hbar L_xL_y \sum_{\psi_n \neq \psi_0 E_n \neq E_0} \frac{\langle\psi_0|J_y|\psi_n\rangle\langle\psi_n|J_x|\psi_0\rangle - \langle\psi_0|J_x|\psi_n\rangle\langle\psi_n|J_y|\psi_0\rangle}{(E_n - E_0)^2} \\ &= i\hbar\left[\left\langle\frac{\partial\psi_0}{\partial\Phi_y}\middle|\frac{\partial\psi_0}{\partial\Phi_x}\right\rangle - \left\langle\frac{\partial\psi_0}{\partial\Phi_x}\middle|\frac{\partial\psi_0}{\partial\Phi_y}\right\rangle\right] = i\hbar\left[\frac{\partial}{\partial\Phi_y}\left\langle\psi_0\middle|\frac{\partial\psi_0}{\partial\Phi_x}\right\rangle - \frac{\partial}{\partial\Phi_x}\left\langle\psi_0\middle|\frac{\partial\psi_0}{\partial\Phi_y}\right\rangle\right]\end{aligned} \tag{3.96}$$

$\langle\psi_0|\frac{\partial\psi_0}{\partial\Phi_x\partial\Phi_y}\rangle$ will cancel from both the terms This is the expression for Hall conductivity.

3.3.1 Hall conductivity and the Chern number

Remember the spectrum of the Hamiltonian depends upon Φ_i mod Φ_0[8]. If there is a remainder, then the division does not yield an integer, and if there is none, then $\Phi_i/\Phi_0 =$ integer. Φ_i being parameters of $\mathcal{H}'$, Φ_i's are periodic functions. To emphasize the periodicity, we shall introduce angular variables, θ_i such that,

[8] Φ_i mod Φ_0 refers to the remainder when Φ_i is divided by Φ_0.

$$\theta_i = \frac{2\pi\Phi_i}{\Phi_0} \qquad \text{where } \theta_i \in [0, 2\pi]. \tag{3.97}$$

As θ_i increases from 0 to 2π, Φ_i increases from $0 \to \Phi_0$. Now rewrite $\frac{\partial}{\partial\Phi_i}$ as $\frac{\partial}{\partial\theta_i}$ and introduce a quantity called Berry connection which is defined on the surface of the torus as,

$$\mathcal{A}_i(\Phi) = -i\langle\psi_0|\frac{\partial}{\partial\theta_i}|\psi_0\rangle. \tag{3.98}$$

Further, we define a quantity called the Berry curvature (analogous to the magnetic field),

$$\mathcal{F}_{xy} = \frac{\partial\mathcal{A}_x}{\partial\theta_y} - \frac{\partial\mathcal{A}_y}{\partial\theta_x} = (\nabla_\theta \times \mathcal{A}) = -i\left[\frac{\partial}{\partial\theta_y}\left\langle\psi_0\middle|\frac{\partial\psi_0}{\partial\theta_x}\right\rangle - \frac{\partial}{\partial\theta_x}\left\langle\psi_0\middle|\frac{\partial\psi_0}{\partial\theta_y}\right\rangle\right]. \tag{3.99}$$

Note that the last term above in equation (3.99) is the Hall conductivity which is formally written in terms of the Berry curvature as,

$$\sigma_{xy} = -\frac{e^2}{h}\mathcal{F}_{xy}. \tag{3.100}$$

We are still left with the task of understanding the quantization of σ_{xy} which is central to the discussion on QHE. We can now integrate over the surface of the torus to get the total conductivity,

$$\sigma_{xy} = -\frac{e^2}{h}\int_{\text{torus}}\frac{d^2\theta}{(2\pi)^2}\mathcal{F}_{xy}. \tag{3.101}$$

The quantity $C = \frac{1}{2\pi}\int d^2\theta\ \mathcal{F}_{xy}$ is called the first Chern number. Thus if we average over the fluxes the conductivity assumes a form,

$$\sigma_{xy} = -\frac{e^2}{h}C \tag{3.102}$$

C is necessarily an integer. It is also referred to as the TKNN invariant after Thouless, Kohmoto, Nightingale and den Nijs [22].

Here we provide an argument that the Chern number which is the integral over the Berry curvature is indeed an integer. For simplicity, let us assume a translationally invariant system in which the eigenstates can be represented by the Bloch functions. That is, $|\psi\rangle_0$ appearing above can be written as $u_k e^{i\phi_k}$ where u_k captures the periodicity of the lattice. Since the Berry connection requires a derivative to be taken, namely, $\langle\psi_0|\frac{\partial}{\partial\theta}|\psi_0\rangle$, which would be equivalent to,

$$\frac{\partial}{\partial\phi_k}(u_k e^{i\phi_k}) \approx \nabla_k\phi_k. \tag{3.103}$$

Now when one takes an integral over the Brillouin zone (which is equivalent to the surface of the torus in real space), then $(\nabla_k \phi_k) \cdot dk$ over a closed surface is zero.

$$\oint \nabla_k \cdot \phi_k dk = 0. \tag{3.104}$$

This says that the measurable quantity, $e^{i\phi}$ obeys[9],

$$e^{i\phi(0)} = e^{i\phi(2\pi)}. \tag{3.105}$$

Thus,

$$\begin{aligned} |\phi(0) - \phi(2\pi)| &= 2\pi \times \text{(some integer)} \\ &= 2\pi C. \end{aligned} \tag{3.106}$$

Thus the integral over the curvature,

$$\begin{aligned} \oint \mathcal{F}\frac{d^2k}{2\pi} &= \oint \nabla_k \phi_k dk \\ &= \frac{2\pi C}{2\pi} = C. \end{aligned} \tag{3.107}$$

Thus Chern number is an integer and hence we get the Hall conductivity to be quantized in unit of e^2/h. The above calculations are of course applicable to IQHE.

3.4 Quantum Hall effect in graphene

Having studied the quantized Hall effect in a 2DEG in detail, we focus on another system which is of topical interest, namely, graphene. Apart from reviewing the basic electronic properties of graphene, we discuss the properties of the Landau levels. The unequal spacing between the successive Landau levels is a feature that shows up in many experiments. We discuss some of them.

Graphene is made of a single layer of carbon (C) atoms arranged in a honeycomb lattice structure. Each C atom has six electrons with an electronic configuration $1s^2 2s^2 2p^2$. Four electrons in the 2s and the 2p orbitals create a hybridized sp^2 bonding orbital. The orbital connects to the three nearest neighbours in the plane of the crystal lattice, while the fourth electron occupies a π orbital that projects out above and below the plane. This π orbital has a significant overlap with those from the neighbouring C atoms. This is responsible for rendering excellent mobility of graphene, while the σ bonds that connect a C atom to its three neighbours yield the stability of the crystal structure. The discovery of such one-atom thick planar material (graphene is a perfect example of a 2D material realized so far) earned a Nobel prize for A Geim and K Novoselov, both from the University of Manchester in UK in 2010. Before we embark on the Hall effect in graphene, let us review the electronic properties.

[9] It should be remembered that ϕ is not a measurable quantity, while $e^{i\phi}$ is a measurable quantity.

3.4.1 Basic electronic properties of graphene

Owing to the large mobility of π electrons in graphene, a nearest neighbour tight binding Hamiltonian of the following form is most suitable, namely,

$$\mathcal{H} = -t \sum_{\langle ij \rangle, \sigma} (a_{i\sigma}^{\dagger} b_{j\sigma} + \text{h. c.}) \tag{3.108}$$

where $a_i^{\dagger}(b_i)$ denote creation (annihilation) operators for electrons at the A(B) sublattice sites. σ denotes the spin of the electrons, however, that will be suppressed in the next step onwards owing to no active role played by the spin either in the band structure or in QHE. We shall make the spin degrees of freedom apparent only when it is needed. Here $t \simeq 2.7$ eV which is considerably large and allows us to ignore electron–electron interaction. The vectors connecting the nearest neighbours, $\boldsymbol{\delta}_i$, direct, $\mathbf{a}_i$ and the reciprocal $\mathbf{b}_i$ lattice vectors (see figure 3.13) are written as,

$$\begin{aligned}
\boldsymbol{\delta}_1 &= \frac{a}{2}\left(\sqrt{3}\hat{x} + \hat{y}\right); \quad \boldsymbol{\delta}_2 = \frac{a}{2}\left(-\sqrt{3}\hat{x} + \hat{y}\right); \quad \boldsymbol{\delta}_3 = -a\hat{x} \\
\mathbf{a}_1 &= \frac{a}{2}\left(\sqrt{3}\hat{x} + 3\hat{y}\right); \quad \mathbf{a}_2 = \frac{a}{2}\left(-\sqrt{3}\hat{x} + 3\hat{y}\right) \\
\mathbf{b}_1 &= \frac{2\pi}{3a}\left(\sqrt{3}\hat{k}_x + \hat{k}_y\right); \quad \mathbf{b}_2 = \frac{2\pi}{3a}\left(-\sqrt{3}\hat{k}_x + 3\hat{k}_y\right)
\end{aligned} \tag{3.109}$$

with the lattice constant, $a = 1.42$Å. Using the nearest neighbour vectors, $\boldsymbol{\delta}_i$, we explicitly write the tight binding Hamiltonian as,

$$\mathcal{H} = -t \sum_{\mathbf{R}, \delta} [b^{\dagger}(\mathbf{R} + \boldsymbol{\delta}_i) a(\mathbf{R}) + a^{\dagger}(\mathbf{R}) b(\mathbf{R} + \boldsymbol{\delta}_i)]. \tag{3.110}$$

The lattice vector $\mathbf{R}$ at an arbitrary site is given by,

$$\mathbf{R} = n\mathbf{a}_1 + m\mathbf{a}_2 \qquad n, m \in N. \tag{3.111}$$

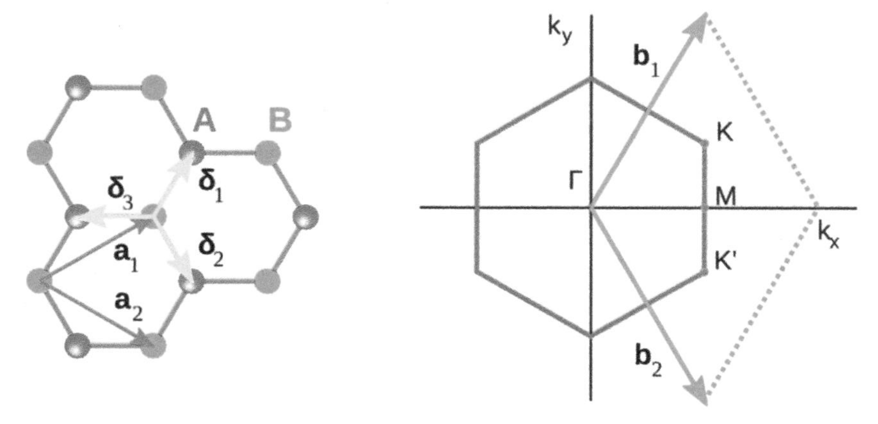

Figure 3.13. Plot showing $\boldsymbol{\delta}_i$, direct, $\mathbf{a}_i$ and the reciprocal $\mathbf{b}_i$ lattice vectors for graphene. Copyright Niels Walet.

The Fourier transform for these operators is done using,

$$a_{\mathbf{k}} = \frac{1}{\sqrt{N}} \sum_{\mathbf{R}} e^{-i\mathbf{k}\cdot\mathbf{R}} a(\mathbf{R}). \tag{3.112}$$

This yields the Hamiltonian in the momentum space as follows,

$$\mathcal{H} = -\frac{t}{N} \sum_{\mathbf{k},\mathbf{q}} \sum_{\mathbf{R}} \sum_{i=1}^{3} e^{i(\mathbf{k}-\mathbf{q})\cdot\mathbf{R}} \left[e^{i(\mathbf{q}-\mathbf{k})\cdot\mathbf{R}} e^{i\mathbf{q}\cdot\boldsymbol{\delta}_i} b_{\mathbf{q}}^\dagger a_{\mathbf{k}} + e^{i(\mathbf{k}-\mathbf{q})\cdot\mathbf{R}} e^{-i\mathbf{q}\cdot\boldsymbol{\delta}_i} a_{\mathbf{q}}^\dagger b_{\mathbf{k}} \right]. \tag{3.113}$$

Using the following definition of the Kronecker delta,

$$\delta_{\mathbf{k},\mathbf{q}} = \frac{1}{N} \sum_{\mathbf{R}} e^{i(\mathbf{k}-\mathbf{q})\cdot\mathbf{R}} \tag{3.114}$$

one gets,

$$\begin{aligned}
\mathcal{H} &= -t \sum_{\mathbf{k}} \sum_{i=1}^{3} [e^{-i\mathbf{k}\cdot\boldsymbol{\delta}_i} b_{\mathbf{k}}^\dagger a_{\mathbf{k}} + e^{i\mathbf{k}\cdot\boldsymbol{\delta}_i} a_{\mathbf{k}}^\dagger b_{\mathbf{k}}] \\
&= -t \sum_{\mathbf{k}} \sum_{i=1}^{3} (a_{\mathbf{k}}^\dagger \quad b_{\mathbf{k}}^\dagger) \begin{pmatrix} 0 & e^{-i\mathbf{k}\cdot\boldsymbol{\delta}_i} \\ e^{i\mathbf{k}\cdot\boldsymbol{\delta}_i} & 0 \end{pmatrix} \begin{pmatrix} a_{\mathbf{k}} \\ b_{\mathbf{k}} \end{pmatrix} \\
&= -t \sum_{\mathbf{k}} \sum_{i=1}^{3} (a_{\mathbf{k}}^\dagger \quad b_{\mathbf{k}}^\dagger) h(\mathbf{k}) \begin{pmatrix} a_{\mathbf{k}} \\ b_{\mathbf{k}} \end{pmatrix}
\end{aligned} \tag{3.115}$$

where $h(\mathbf{k})$ is the Hamiltonian matrix defined by,

$$h(\mathbf{k}) = -t \begin{pmatrix} 0 & (e^{i\mathbf{k}\cdot\boldsymbol{\delta}_1} + e^{i\mathbf{k}\cdot\boldsymbol{\delta}_2} + e^{i\mathbf{k}\cdot\boldsymbol{\delta}_3}) \\ (e^{-i\mathbf{k}\cdot\boldsymbol{\delta}_1} + e^{-i\mathbf{k}\cdot\boldsymbol{\delta}_2} + e^{-i\mathbf{k}\cdot\boldsymbol{\delta}_3}) & 0 \end{pmatrix}. \tag{3.116}$$

Since the difference between two nearest neighbour lattice vectors, $\boldsymbol{\delta}_i$ and $\boldsymbol{\delta}_j$ must yield a lattice vector, $\mathbf{R}$, we can do a transformation,

$$a_{\mathbf{k}} \to e^{i\mathbf{k}\cdot\boldsymbol{\delta}_3} a_{\mathbf{k}}, \qquad \text{and} \qquad a_{\mathbf{k}}^\dagger \to e^{-i\mathbf{k}\cdot\boldsymbol{\delta}_3} a_{\mathbf{k}}^\dagger. \tag{3.117}$$

The above transformation yields a new Hamiltonian matrix,

$$\tilde{h}(\mathbf{k}) = -t \begin{pmatrix} 0 & (e^{i\mathbf{k}\cdot(\boldsymbol{\delta}_1 - \boldsymbol{\delta}_3)} + e^{i\mathbf{k}\cdot(\boldsymbol{\delta}_2 - \boldsymbol{\delta}_3)} + 1) \\ (e^{-i\mathbf{k}\cdot(\boldsymbol{\delta}_1 - \boldsymbol{\delta}_3)} + e^{-i\mathbf{k}\cdot(\boldsymbol{\delta}_2 - \boldsymbol{\delta}_3)} + 1) & 0 \end{pmatrix}. \tag{3.118}$$

Using the definitions of $\boldsymbol{\delta}_i$, one gets,

$$\tilde{h}(\mathbf{k}) = -t \begin{pmatrix} 0 & -(e^{i\mathbf{k}\cdot\mathbf{a}_1} + e^{i\mathbf{k}\cdot\mathbf{a}_2} + 1) \\ -(e^{-i\mathbf{k}\cdot\mathbf{a}_1} + e^{-i\mathbf{k}\cdot\mathbf{a}_2} + 1) & 0 \end{pmatrix}. \tag{3.119}$$

One can check that $\tilde{h}(\mathbf{k})$ obeys $\tilde{h}(\mathbf{k}) = \tilde{h}(\mathbf{k} + \mathbf{G})$, where $\mathbf{G}$ is the reciprocal lattice vector, defined as, $\mathbf{G} = p\mathbf{b}_1 + q\mathbf{b}_2$, with p and q being integers. Thus,

$$\tilde{h}(\mathbf{k}) = -t\begin{pmatrix} 0 & f(\mathbf{k}) \\ f^*(\mathbf{k}) & 0 \end{pmatrix} \tag{3.120}$$

where

$$f(\mathbf{k}) = -t\left(e^{-ik_xa} + 2e^{ik_xa/2}\cos\left(\frac{k_y\sqrt{3}\,a}{2}\right)\right).$$

The tight binding energy is obtained by diagonalizing $\tilde{h}(\mathbf{k})$ which yields,

$$\epsilon_{\mathbf{k}} = \pm t\sqrt{3 + 2\cos(\sqrt{3}\,ak_y) + 4\cos(\sqrt{3}\,ak_y/2)\cos(3ak_x/2)}\,. \tag{3.121}$$

The two bands described by the '+' and the '−' signs in the above dispersion touch at six points in the Brillouin zone. Since graphene has one accessible electron per C atom, one can assume a half-filled system where the lower band is completely filled. Further, we wish to discuss the low lying excitations just above the ground state of the system.

This necessitates exploring the low energy theory of graphene. To achieve that we have to identify the band touching points which can be obtained from the condition, $f(\mathbf{k}) = 0$. Separately putting the real and the imaginary parts equal to zero yield,

$$\begin{aligned} \cos(k_xa) + 2\cos(k_xa/2)\cos(\sqrt{3}\,k_ya/2) &= 0 \\ -\sin(k_xa) + 2\sin(k_xa/2)\cos(\sqrt{3}\,k_ya/2) &= 0. \end{aligned} \tag{3.122}$$

equation (3.122) can be manipulated as follows,

$$\sin(k_xa/2)\left[-\cos(k_xa/2) + \cos(k_ya\sqrt{3}/2)\right] = 0. \tag{3.123}$$

Thus one is left with two options, namely,

either (*i*) $\sin(k_xa/2) = 0$; which means $\cos(k_xa/2) = \pm 1$;

or, (*ii*) $\cos(k_xa/2) = \cos(\sqrt{3}\,k_ya/2)$.

Option (i) gives us,

$$1 + 2\cos(k_y\sqrt{3}\,a/2) = 0$$

which yields the points $\left(0, \pm\frac{4\pi}{3\sqrt{3}\,a}\right)$ (plus or minus the reciprocal lattice vector, **G**), whereas, option (ii) can be written as,

$$\cos(k_ya\sqrt{3}) + 2\cos^2(k_ya\sqrt{3}/2) = 0.$$

Thus we get four more points, which are, $\pm\frac{2\pi}{3a}\left(1, \frac{1}{\sqrt{3}}\right)$, and $\pm\frac{2\pi}{3a}\left(1, -\frac{1}{\sqrt{3}}\right)$ (again plus or minus the reciprocal lattice vector, **G**).

A closer inspection yields all the six points are not independent. For example, the set of vectors, namely, $\left(0, \pm\frac{4\pi}{3\sqrt{3}a}\right)$, $\frac{2\pi}{3a}\left(1, -\frac{1}{\sqrt{3}}\right)$ and $\frac{2\pi}{3a}\left(-1, -\frac{1}{\sqrt{3}}\right)$ can be connected to each other via the combination of the reciprocal lattice vectors, $\mathbf{b}_1$ and $\mathbf{b}_2$. For example,

$$\left(0, \frac{4\pi}{3\sqrt{3}a}\right) + \mathbf{b}_2 = \frac{2\pi}{3a}\left(1, -\frac{1}{\sqrt{3}}\right)$$
$$\left(0, \frac{4\pi}{3\sqrt{3}a}\right) - \mathbf{b}_1 = \frac{2\pi}{3a}\left(-1, -\frac{1}{\sqrt{3}}\right). \tag{3.124}$$

The same is true for the other vectors,

$$\left(0, \frac{4\pi}{3\sqrt{3}a}\right), \quad \frac{2\pi}{3a}\left(-1, \frac{1}{\sqrt{3}}\right), \quad \frac{2\pi}{3a}\left(1, \frac{1}{\sqrt{3}}\right).$$

Thus, only two of them are found to be independent. Traditionally, they are called $\mathbf{K}$ and $\mathbf{K}'$ and can be written as,

$$\mathbf{K} = \frac{2\pi}{3a}\left(1, \frac{1}{\sqrt{3}}\right), \quad \text{and } \mathbf{K}' = \frac{2\pi}{3a}\left(1, -\frac{1}{\sqrt{3}}\right).$$

Any other independent pair is also a valid choice for $\mathbf{K}$ and $\mathbf{K}'$.

It may be kept in mind that the two bands touch at these points and the gap between the conduction and the valence band closes. Thus there are two branches of low energy excitations, namely one of them with momentum close to $\mathbf{K}$ and the other close to $\mathbf{K}'$. Since $f(\mathbf{k})$ becomes zero at $\mathbf{k} = \mathbf{K}$. Defining $\mathbf{q} = \mathbf{k} - \mathbf{K}$, one can expand $f(\mathbf{k})$ near $\mathbf{K}$ in Taylor series about $\mathbf{q} = 0$,

$$f'(\mathbf{q}) = \frac{\partial f(\mathbf{k})}{\partial k_x}|_{(k_x - K_x)}(k_x - K_x) + \frac{\partial f(\mathbf{k})}{\partial k_y}|_{(k_y - K_y)}(k_y - K_y)$$
$$= \frac{3at}{2}(q_x + iq_y).$$

Thus the energy spectrum assumes the form,

$$\epsilon_{\mathbf{K}}(\mathbf{q}) = \hbar v_{\mathrm{F}}(q_x + iq_y) \tag{3.125}$$

where v_{F} is the Fermi velocity defined by, $v_{\mathrm{F}} = \frac{3at}{2\hbar} \simeq 10^6\,\mathrm{ms}^{-1}$. Similarly, if we expand around $\mathbf{K}'$, one gets,

$$\epsilon_{\mathbf{K}'}(\mathbf{q}) = \hbar v_{\mathrm{F}}(q_x - iq_y). \tag{3.126}$$

Thus in a general notation, we can write,

$$\epsilon_{\mathbf{K},\mathbf{K}'} = \hbar v_{\mathrm{F}}\mathbf{q} \cdot \boldsymbol{\sigma} \tag{3.127}$$

where $\mathbf{q}$ is a planar vector (q_x, q_y) and $\boldsymbol{\sigma}$ is the Pauli matrix vector (σ_x, σ_y). The electrons close to the $\mathbf{K}$ and $\mathbf{K}'$ points are called massless Dirac fermions, as they obey the Dirac equation without the 'mass' term[10]. It may be noted that,

$$\epsilon_{\mathbf{K}'}(\mathbf{q}) = \epsilon_{\mathbf{K}}^{*}(\mathbf{q}). \tag{3.128}$$

This implies that (as will be seen later) the 'helicity' of the electrons is opposite at $\mathbf{K}'$ with respect to $\mathbf{K}$.

To sum up our preliminary discussion on graphene, we note that the low energy properties are governed by the dispersion,

$$\epsilon(q) \simeq \pm v_{\mathrm{F}}|q| \tag{3.129}$$

which implies that the eigenvalues are only functions of the magnitude of the wavevector $\mathbf{q}$, and does not depend upon its direction in the 2D plane. The corresponding linear dispersion is shown in figure 3.14. Also the Hamiltonian on a formal note denotes that of a massless $s = 1/2$ particle, such as a neutrino, however the velocity of the particles is reduced by a factor of 300 compared to the speed of light. Further, the handedness (or the helicity) feature of neutrinos is in-built, where the electrons behave similar to a 'left handed' neutrino at the Dirac point $\mathbf{K}$ and as a 'right handed' neutrino at $\mathbf{K}'$ or vice versa.

3.4.2 Experimental confirmation of the Dirac spectrum

When a beam of monochromatic photons with an energy larger than the work function of a particular material interacts with the constituent charges (electrons) by indenting on the surface of the sample, the electrons absorb the photons and thus possess sufficient energy to escape from the sample. By measuring the energy and the momentum of the photoelectrons and using energy–momentum conservations laws, one can derive the properties of the electrons prior to them being incident on the surface and relate them with those getting scattered. Angle resolved photoemission spectroscopy (ARPES) can be a direct probe to resolve the momentum dependent band structure and the topology of the Fermi surface. In ARPES, a photon is employed to eject an electron from the surface of the graphene layer. The intensity of the ARPES is proportional to the transition probability from an initial Bloch state with a crystal momentum, $\mathbf{k}$ and energy, E to a final state, $\mathbf{k}'$. The method conclusively establishes the existence of Dirac fermions seen via linearly dispersing bands in the vicinity of the Dirac points. The experimental setup, ARPES data, and the hexagonal Brillouin zone (which we have discussed before) are shown in figure 3.15.

[10] The Dirac equation is written in conventional notations as, $\mathcal{H} = c\boldsymbol{\alpha} \cdot \mathbf{p} + \beta mc^2$ where α and β are Hermitian operators which do not operate on the space and time variables. In the case of graphene, the second term is absent.

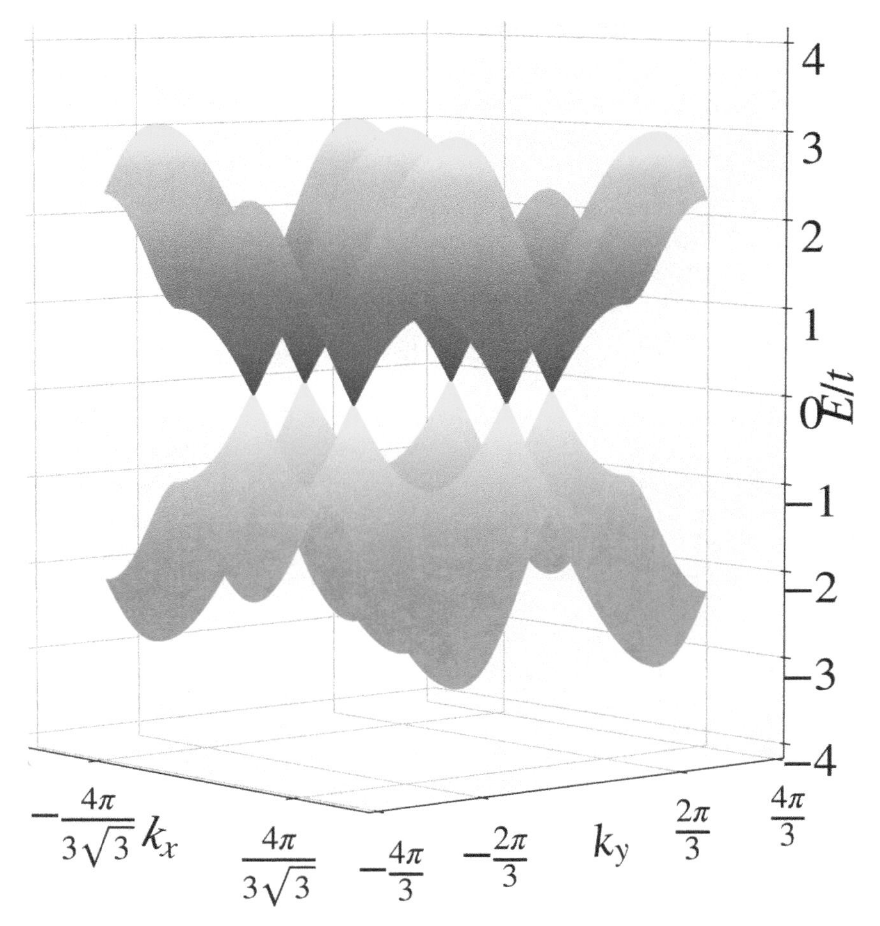

Figure 3.14. Plot showing the two tight binding bands of graphene in the first BZ. The two bands touch at six points. In the vicinity of those points, the bands are linearly dispersing bearing signatures of (pseudo-) relativistic physics.

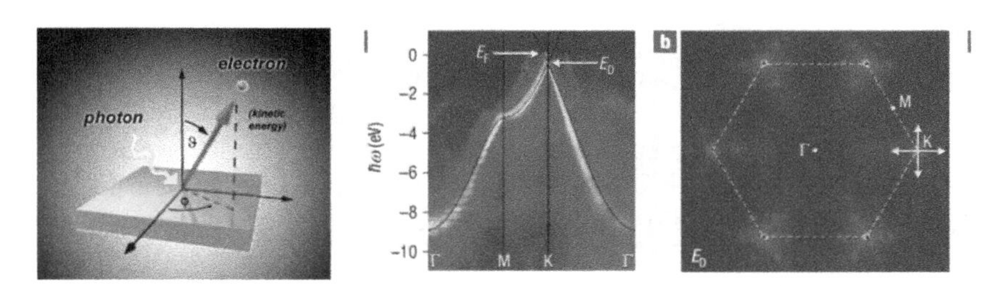

Figure 3.15. Plot showing the experimental setup for ARPES (left), the two linearly dispersing bands (middle) and the Dirac points are shown on the hexagonal BZ of graphene (right). Taken from reference [26], copyright 2006, by permission from Nature Physics: Springer Nature.

3.4.3 Landau levels in graphene

In order to proceed, we can write the Hamiltonian in a unified way that includes the description of both the Dirac points (valleys) **K** and **K**′. So for each valley, we have one 2D spinor Hamiltonian. Thus augmenting the Hilbert space we can write the eigenfunctions as,

$$\psi = (\psi_{\mathbf{K}'}, \psi_{\mathbf{K}})^T \tag{3.130}$$

and the Hamiltonian is given by,

$$\mathcal{H} = \hbar v_{\mathrm{F}}\begin{pmatrix} -\boldsymbol{\sigma}^{*}\cdot\mathbf{k} & 0 \\ 0 & \boldsymbol{\sigma}\cdot\mathbf{k} \end{pmatrix} = v_{\mathrm{F}}\begin{pmatrix} -\boldsymbol{\sigma}^{*}\cdot\mathbf{p} & 0 \\ 0 & \boldsymbol{\sigma}\cdot\mathbf{p} \end{pmatrix}. \tag{3.131}$$

Now we shall be discussing the motion of the massless relativistic electrons in a magnetic field. In an usual 2D electron gas, the Landau quantization produces equidistant levels (see equation (3.37)), which is an artefact of the non-relativistic parabolic dispersion of the free carriers. We need to ascertain how the quantization formula is modified for the case of graphene.

As earlier, we do the Peierls substitution, $\mathbf{p} \rightarrow \mathbf{p} + e\mathbf{A}$. Thus the Hamiltonian becomes,

$$\mathcal{H}_{\mathbf{K},\mathbf{K}'} = v_{\mathrm{F}}\begin{pmatrix} 0 & -(p_x + ip_y) & 0 & 0 \\ -(p_x - ip_y) & 0 & 0 & 0 \\ 0 & 0 & 0 & (p_x - ip_y) \\ 0 & 0 & (p_x + ip_y) & 0 \end{pmatrix}.$$

The wavefunction has now four components, namely,

$$\psi = \begin{pmatrix} \phi_{\mathrm{A}}^{\mathbf{K}'} \\ \phi_{\mathrm{B}}^{\mathbf{K}'} \\ \phi_{\mathrm{A}}^{\mathbf{K}} \\ \phi_{\mathrm{B}}^{\mathbf{K}} \end{pmatrix} \tag{3.132}$$

where $\phi^{\mathbf{K}}_{\mathrm{A,\,B}}$ are the wavefunctions for an electron at momentum values corresponding to the valley **K** at the two sublattice sites A and B. Similar notations carry on for the other valley **K**′.

For a perpendicular magnetic field, $\mathbf{B} = B\hat{z}$, one can choose a Landau gauge, $\mathbf{A} = (-By, 0, 0)$. Since with this choice the Hamiltonian is independent of the spatial variable, x. So $[\mathcal{H}, p_x] = 0$ and hence p_x continues to be a good quantum number.

Further, the Hamiltonian in equation (3.132) is valley decoupled, that is, there are no matrix elements that connect the two valleys, namely, **K** and **K**′. Thus it allows us look at the solutions at each valley separately. For the **K** point, we have a coupled equation for the wavefunctions, ϕ^{A} and ϕ^{B}

$$\epsilon\phi_A^K = v_F(p_x - ip_y)\phi_B^K \tag{3.133}$$

$$\epsilon\phi_B^K = v_F(p_x + ip_y)\phi_A^K. \tag{3.134}$$

One can insert equation (3.133) in equations (3.134) and (3.134) in equation (3.133) to obtain,

$$\epsilon^2\phi_A^K = v_F^2(p_x - ip_y)(p_x + ip_y)\phi_A^K \tag{3.135}$$

$$\epsilon^2\phi_B^K = v_F^2(p_x + ip_y)(p_x - ip_y)\phi_B^K. \tag{3.136}$$

Inserting the Landau gauge such that $p_x \to p_x + eBy$,

$$\begin{aligned}\frac{\epsilon^2}{v_F^2}\phi_B^K &= (p_x + eBy + ip_y)(p_x + eBy - ip_y)\phi_B^K \\ &= \left[(p_x + eBy)^2 - i\left\{(p_x + eBy), p_y\right\} + p_y^2\right]\phi_B^K.\end{aligned} \tag{3.137}$$

Since $[p_x, p_y] = 0$ and $[y, p_y] = i\hbar$, one gets,

$$\frac{\epsilon^2}{v_F^2}\phi_B^K = [(p_x + eBy)^2 + e\hbar B + p_y^2]\phi_B^K. \tag{3.138}$$

Thus, we arrive at,

$$\left(\frac{\epsilon^2}{v_F^2} - e\hbar B\right)\phi_B^K = (\tilde{p}_x^2 + \tilde{p}_y^2)\phi_B^K \tag{3.139}$$

where $\tilde{p}_x^2 = p_x + eBy$ and $\tilde{p}_y^2 = p_y$. Dividing both sides by $2m$,

$$\frac{1}{2m}\left(\frac{\epsilon^2}{v_F^2} - e\hbar B\right)\phi_B^K = \frac{\tilde{p}_x^2 + \tilde{p}_y^2}{2m}\phi_B^K = \frac{1}{2}\tilde{k}(y - y_0)^2 + \frac{\tilde{p}_y^2}{2m}\phi_B^K \tag{3.140}$$

with $\tilde{k} = \frac{e^2B^2}{m}$, $y_0 = \frac{p_x}{eB}$. Thus the RHS is identified as the Hamiltonian for a particle executing SHM in two dimensions about a coordinate point $(0, y_0)$. Thus it is obvious that the energy spectrum is given by, $\epsilon_n = (n + \frac{1}{2})\hbar\omega_B$ with $\omega_B = \frac{eB}{m}$. Hence,

$$\frac{\epsilon^2}{v_F^2} = 2\left(n + \frac{1}{2}\right)\hbar\omega_B - \hbar\omega_B = 2n\hbar\omega_B \qquad \text{where} \quad n = 0, 1, 2, \ldots. \tag{3.141}$$

equation (3.141) allows positive and negative roots for ϵ. So we can obtain the energy spectrum as,

$$\epsilon = sgn(n)\sqrt{n}\; v_F\; (2\hbar eB)^{1/2}. \tag{3.142}$$

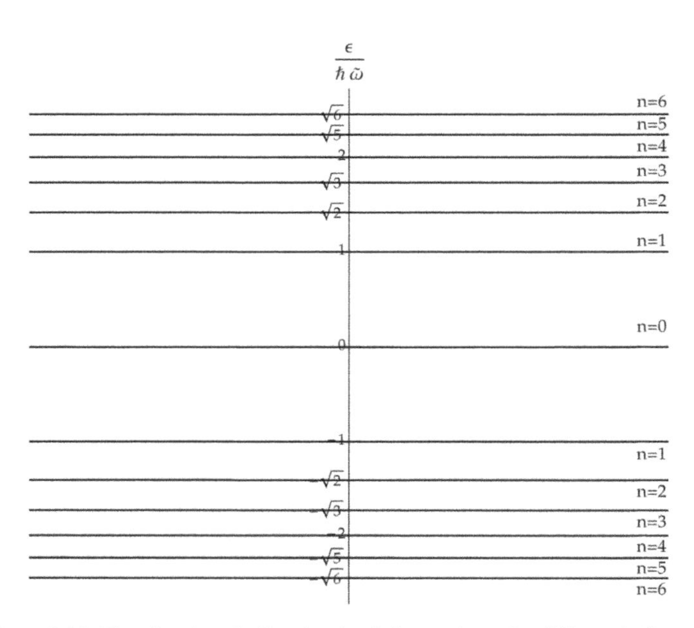

Figure 3.16. Plot showing the Landau levels in graphene for different indices, n.

Let us define another quantity $\tilde{\omega} = v_F \ (2\hbar eB)^{1/2}$, so as to formally write the energy expression as that of a harmonic oscillator. We rewrite the above expression as,

$$\epsilon = \hbar\tilde{\omega} \ sgn(n)\sqrt{|n|}\,. \tag{3.143}$$

Thus, as opposed to the familiar harmonic oscillators when n can take positive integer values (including zero), however, here in graphene all integers, that is, both positive and negative numbers are allowed. The positive integers denote particles (or electrons) in the conduction band and the negative ones denote holes in the valence band. Further, unlike the 2DEG, here the Landau levels are not equidistant. The largest separation occurs between the lowest Landau level ($n = 0$) and the first one ($n = \pm1$). This large gap essentially facilitates observation of QHE in graphene at large temperatures, which is even true for room temperature[11]. The Landau levels for graphene are shown in figure 3.16. Clearly, they are non-equidistant, and hence quite distinct from those for the two-dimensional electron gas.

So far we have been discussing spinless particles. Including the spin, there will be an additional two-fold degeneracy of the Landau levels owing to Zeeman spitting. A hierarchy of energy scales needs to be ascertained here. Let us compare the energy gap between the two lowest Landau levels and the Zeeman splitting corresponding to a typical magnetic field, B, for example, $B = 10$ T.

[11] The title of the paper by Geim and Novoselov is '*Room temperature quantum Hall effect in graphene*'. See [23].

$$\triangle E_{\rm LL} = \frac{\hbar\omega_B/2}{v_{\rm F}} \quad \text{(for the successive Landau levels)} \tag{3.144}$$

$$\triangle E_z = \sqrt{e\hbar B} \quad \text{(for the Zeeman term).} \tag{3.145}$$

For typical $v_{\rm F} \simeq c/300$ (c: speed of light),

$$\frac{\Delta E_z}{\Delta E_{\rm LL}} \simeq 10^2. \tag{3.146}$$

Thus the Zeeman energy scale is much larger than the Landau level splitting, which makes it imperative to include spin degeneracy. Thus including the Zeeman term, the energy can be written as,

$$\frac{\epsilon^2}{v_{\rm F}^2} = 2\hbar eB(n+1) \qquad n = 0, 1, \ldots \tag{3.147}$$

where the additional term in the RHS (denoted by $2\hbar eB$) accounts for the spin. Thus the energy level $\epsilon = 0$ is not present in the spectrum even for $n = 0$. This lowest Landau level is somewhat special in the following sense. The $n = 0$ level receives contribution from only one sublattice at each of the Dirac points. For example, 'A' sublattice contributes to the wavefunction at the Dirac point **K**, and 'B' sublattice contributes at **K**′. However, the $n \neq 0$ Landau levels have non-zero amplitudes at both the A and the B sublattices.

Finally, the wavefunctions corresponding to an arbitrary Landau level index at the two Dirac points are corresponding to the gauge we have chosen are,

$$\psi_{n,k}^{\mathbf{K}} = \frac{c_n}{\sqrt{L}} e^{-ikx} \begin{pmatrix} 0 \\ 0 \\ sgn(n)(-i)\phi_{|n|-1,k} \\ \phi_{|n|,k} \end{pmatrix} \tag{3.148}$$

and

$$\psi_{n,k}^{\mathbf{K}'} = \frac{c_n}{\sqrt{L}} e^{-ikx} \begin{pmatrix} \phi_{|n|,k} \\ sgn(n)(-i)\phi_{|n|-1,k} \\ 0 \\ 0 \end{pmatrix} \tag{3.149}$$

with

$$\begin{aligned} c_n(x) &= 1 && \text{for } n = 0 \\ &= \frac{1}{\sqrt{2}} && \text{for } n \neq 0 \end{aligned}. \tag{3.150}$$

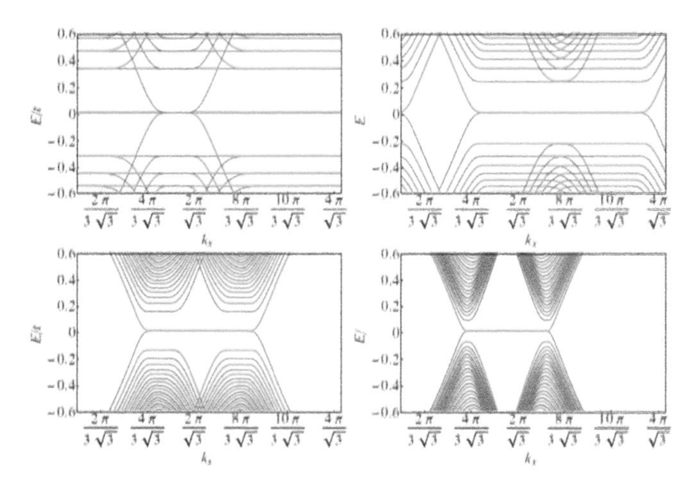

Figure 3.17. Plot showing the Landau levels in graphene for different values of flux, ϕ. The values of the fluxes are $\phi = \frac{\Phi_0}{100}$ (for left upper panel), $\phi = \frac{\Phi_0}{200}$ (for right upper panel), $\phi = \frac{\Phi_0}{500}$ (for left lower panel) and $\phi = \frac{\Phi_0}{1600}$ (for right lower panel). Here $\Phi_0 = \frac{h}{e}$ is the flux quantum.

Further,

$$\begin{aligned} sgn(n) &= 0 && \text{for } n = 0 \\ &= \frac{n}{|n|} && \text{for } n \neq 0 \end{aligned} \tag{3.151}$$

with

$$\phi_{n,k} = \exp\left[-\frac{1}{2}\frac{(y - kl_B^2)^2}{l_B^2}\right] H_n\left[\frac{(y - kl_B^2)}{l_B}\right] \tag{3.152}$$

l_B is the magnetic length $(=\sqrt{\frac{\hbar}{eB}})$ as defined before, and $H_n(x)$ are the Hermite polynomials. $\phi_{n,k}$ denotes the eigenfunctions of an electron in presence of a magnetic field. n refers to the Landau level index. The Landau levels for different values of the flux, ϕ are shown in figure 3.17 as a function of k_x. As the magnitude of the flux is decreased, the width of the flat band appearing at the Fermi energy (at $E/t = 0$) decreases. Further, the flat bands become dissipative in the bulk corresponding to larger values of the Landau level index, n and lower values of the flux ϕ. For a weak magnetic field, such that, $\phi/\Phi_0 = 1/1600$, the energy bands of the bulk regain their Dirac-like structure similar to the single-particle energy levels, while the zero mode flat band continues to exist.

The experimental results for the (room temperature) QHE in graphene is shown in figure 3.18. The Hall conductivity as a function of the gate voltage, V_g shows plateaus at $2e^2/h$, and the longitudinal resistivity (ρ_{xx}) vanishes ($<10\Omega$) at values of the voltage where the Hall conductivity saturates to values that are multiples of e^2/h. The maximum difference between the non-equidistant Landau levels is about

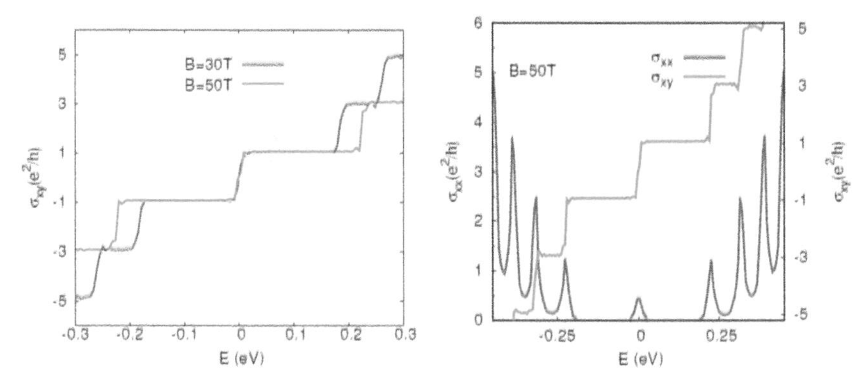

Figure 3.18. Plot showing the variation of Hall conductivity as a function of the bias voltage for a particular value of the magnetic field, such as, $B = 30$ T and 50 T. The plateaus in the Hall conductivity are clearly visible. The longitudinal conductivity (σ_{xx}) is shown for a specific value, namely, $B = 50$ T.

$\Delta E \simeq 600$ K[12] at $B = 29$ T and survival of QHE at very large temperature is owing to a large cyclotron frequency, ω_c. At still larger values of the magnetic field, namely, $B = 45$ T, ΔE is about 2800 K when the Fermi energy lies between the lowest Landau level ($n = 0$) and the first Landau levels ($n = \pm 1$). Thus at this value of the magnetic field, the energy scale $\hbar\omega_c$ exceeds the thermal energy at $T = 300$ K by an order of magnitude. The density of the carriers in graphene is very high ($\sim 10^{13}$ cm^{-2}) such that the lowest Landau level is fully populated even at very large values of the field.

We have included numeric computation of the Hall conductivity as a function of the Fermi energy at comparable values (to those in experiments) of the magnetic field, namely, $B = 30$ T and 50 T using Kubo formula in figure 3.18. The quantization of the Hall conductivity is clearly visible. Further, the longitudinal conductivity, σ_{xx} is shown for one of them, namely, 50 T. σ_{xx} shows vanishingly small values corresponding to plateaus of the Hall conductivities, while it shows spikes when the Hall conductivity jumps from one plateau to another.

3.4.4 Experimental observation of the Landau levels in graphene

There are primarily two experimental techniques for observing the existence of Landau levels. They are (i) the infra-red (IR) spectroscopy [24] and (ii) scanning tunnelling microscopy (STM) [25] experiments. In the following we include a brief discussion on each of them and their utilities in observing the non-equidistant Landau levels in graphene.

In the IR spectroscopy, the optical transitions from one Landau level to another are studied via measuring the cyclotron frequencies. The Landau levels being proportional to $\sqrt{n}$ (n: Landau level index), all the frequencies of the optical transitions are distinct as the energy spacing between each pair of Landau level is

[12] ΔE is called the activation energy in the original paper by [23].

different than the other pairs. These optical transitions are of two types which correspond to transitions between the electron or the hole states in the conduction or the valence bands (intra-band transition), or the transitions between the electron and the hole states pertaining to the valence and conduction bands (inter-band transition), respectively. The photoconductive response and the resistive voltage show the existence of differently spaced Landau levels in graphene at particular values of the magnetic field, longitudinal current, and the IR frequency. The photoconductive intensity as a function of the carrier density, n (not to be confused with the Landau level index) show distinct peaks which are proportional to the energy absorbed from the incident IR radiation.

In the STM experiment, the specific energy levels can be identified by varying the bias voltage between the tip and the surface of the sample, and the tunnelling current generated is proportional to the local density of states. In graphene, the Landau levels are directly observed via the peaks in the tunnelling spectrum. From the positions of the peaks as a function of the sample bias (shown in figure 3.20), the energies of the Landau levels can be extracted. In the STM spectra shown below corresponding to a specific temperature (4.4 K) and magnetic field in the range [0,12 T], the peaks occurring at the zero sample bias (marked as 'A' peak) do not suffer a change in position as the magnetic field is ramped up from 0 to 12 T, while the peaks on either side of the zero bias (marked as 'Z' peak) show a gradual fanning outward with increase in the applied magnetic field. Clearly, the 'A' peaks correspond to the $n = 0$ Landau level, while the 'Z' peaks correspond to $n = \pm 1$ levels. Closer analysis of the other peaks reveals a $\sqrt{n}$ dependence. In the lower left panel of figure 3.20 the STM spectra are shown as a function of $E/\sqrt{B}$. Further, the peak positions conform to a $\sqrt{B}$ dependence where the Landau level indices are shown by different peaks as a function of a scaled energy variable, namely, $E/\sqrt{B}$. In the other figure (right panel of figure 3.20), $E/\sqrt{B}$ is shown as a function of the Landau level index, n and also goes as $sgn(n)\sqrt{|n|}$ where the latter shows a linear behaviour passing through the origin (right bottom lower panel of figure 3.19).

3.4.5 Summary

We begin with a historical overview of the QHE. The experiment and the physical systems are described with an emphasis on the 2D nature of the 'dirty' electronic system in the presence of a strong perpendicular magnetic field at low temperature. The Hall resistivity as a function of the field shows quantized plateaus in unit of h/e^2 with an accuracy of one part in more than a billion. Very surprisingly, the longitudinal resistivity synergetically vanishes at the positions of the plateaus for the Hall resistivity. This indicates emergence of a phase with an inherent ambiguousness of being a perfect conductor and a perfect insulator at the same time. However, such an ambiguity can only be reconciled for an electron gas confined in a plane in the presence of a magnetic field.

Quite intriguingly, the presence of the perpendicular magnetic field introduces '*another*' quantization, which replaces the band structure (energy as a function of the wavevector) of the electronic system. This quantization was shown via solving the

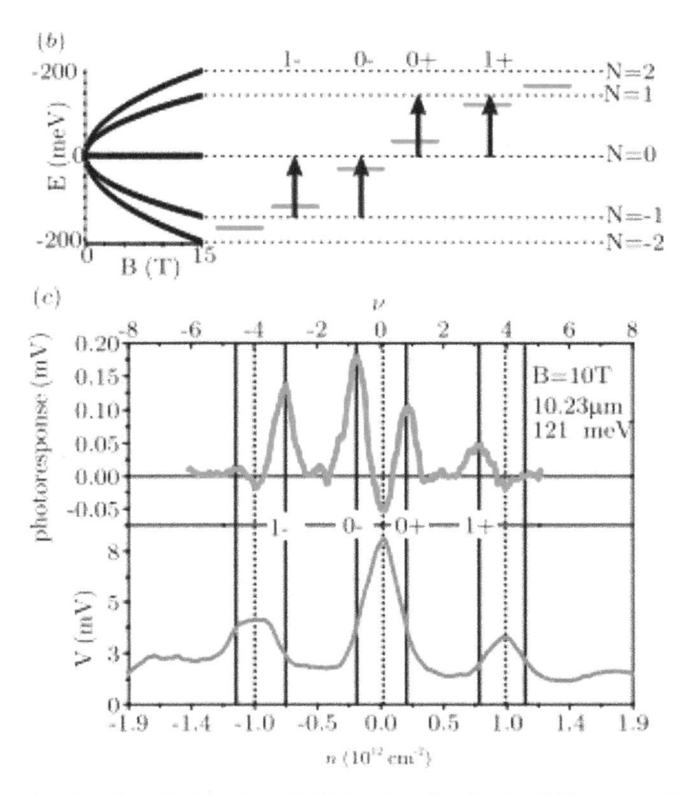

Figure 3.19. Plot showing the identification of distinct Landau levels which go as $\sqrt{n}$ (reprinted with permission from reference [13], copyright (2007) by the American Physical Society).

Schrödinger equation in the presence of a Landau gauge. The resultant energy levels of this problem are the infinitely degenerate Landau levels, which slightly broaden due to the presence of impurity and disorder, but still remain distinct and cause quantization of the Hall conductivity as the magnetic field is ramped up gradually. Further, this quantization is visioned as a quantum pump by Laughlin where an electron gas in a planar disk geometry subjected to a magnetic field shows a transfer of one unit of charge from the inner to the outer edge of the disk as the magnetic flux changes by one quantum ($=\frac{h}{e}$). Further, we have studied the same problem in the circular gauge which is relevant to the study of FQHE.

The quantum Hall state is the first realization of a topological insulator where the transmission of charges occurs through the edges of the electronic system, while the bulk remains insulating. The quantization of the Hall conductivity[13] is shown to be a topological invariant called the Chern number, which can only assume integer values. We have further derived the Kubo formula to compute the Hall conductivity.

[13] It is more appropriate to talk about conductivity, rather than the resistivity.

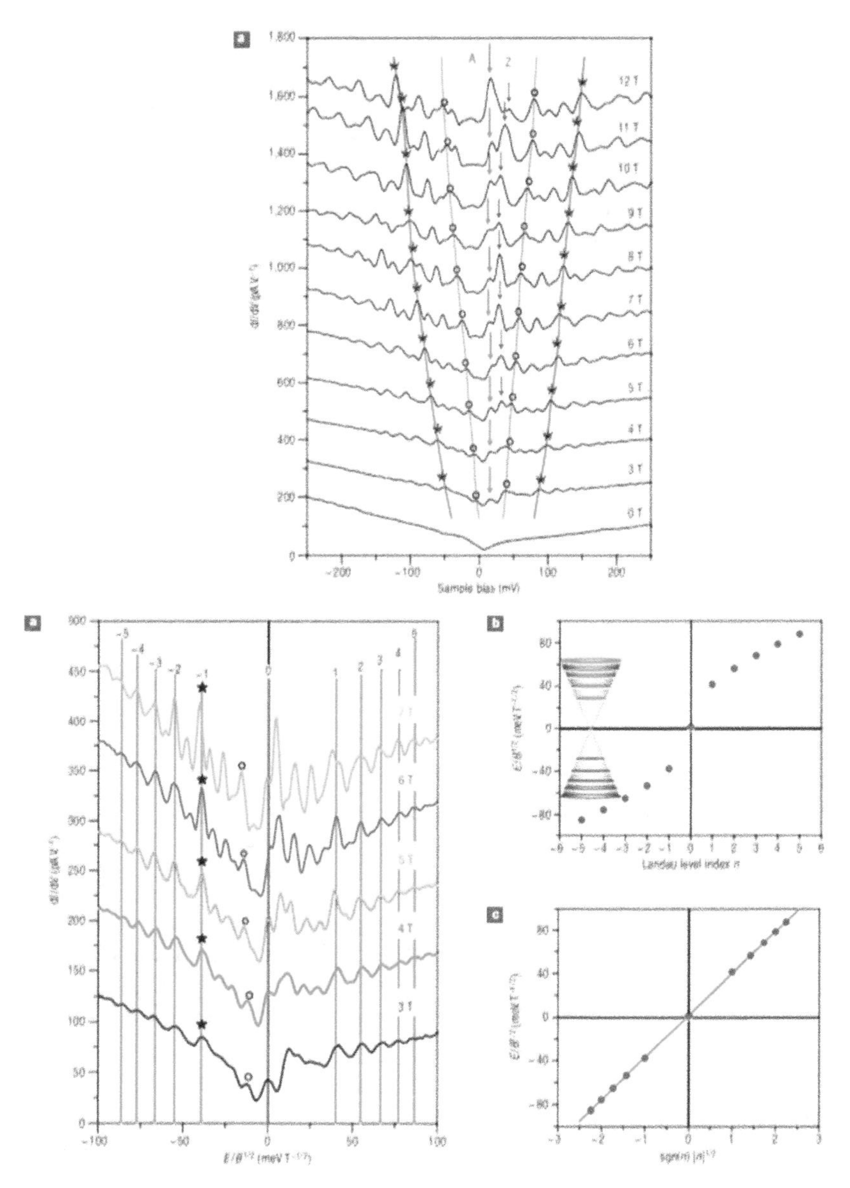

Figure 3.20. Plot of the tunnelling spectra showing Landau levels which go as $\sqrt{n}$ (taken from reference [25], copyright 2007 by permission from Nature Physics: Springer Nature).

Finally, we have chosen graphene as another hobby horse to demonstrate occurrence of QHE. The choice of graphene stems from the fact that the energy scale of the problem allows QHE to be realizable at temperatures as high as room temperature, or even higher than that. The Landau levels of graphene are computed and are shown to be distinct from those for the 2DEG. Lastly, the experimental demonstration and observation of QHE in graphene are discussed.

References

[1] Dutta S 1995 *Electronic Transport Properties in Mesoscopic Systems* (Cambridge: Cambridge University Press)
[2] Imry Y 1997 *Electronic Transport in Mesoscopic Systems* (New York: Oxford University Press)
[3] Hess K 1988 Boltzmann transport equation *The Physics of Submicron Semiconducting Devices* ; H L Grubin
[4] Landauer R 1957 *IBM J. Res. Dev.* **1** 223
[5] Mello N K P R 2004 *Quantum Transport in Mesoscopic Systems: Complexity and Statistical Fluctuations* (New York: Oxford University Press)
[6] Ando T, Arakawa Y, Furuya K, Komiyama S and Nakashima H (ed) 1998 *Mesoscopic Physics and Electronics* (New York: Springer)
[7] Aharonov Y and Bohm D 1959 *Phys. Rev.* **115** 485
[8] van Wees B J, van Houten H, Beenakker C W J, Williamson J G, Kouwen-hoven L P, van der Marel D and Foxon C T 1988 *Phys. Rev. Lett.* **60** 848
[9] Wharam D A *et al* 1988 *J. Phys. C: Solid State Phys.* **21** L209
[10] Bergmann G 1984 *Phys. Rep.* **107** 1
[11] Lee P A and Ramakrishnan T V 1985 *Rev. Mod. Phys.* **57** 287
[12] Garcia R G 1992 *Appl. Phys. Lett.* **60** 1960
[13] Klitzing K v, Dorda G and Pepper M 1980 *Phys. Rev. Lett.* **45** 494
[14] Landwehr G 1986 *Metrologia* **22** 118
[15] Willett R, Eisenstein J P, Störmer H L, Tsui D C, Gossard A C and English J H 1987 *Phys. Rev. Lett.* **59** 1776
[16] Kittel C 2004 *Introduction to Solid State Physics* 8th edn (Hoboken, NJ: Wiley)
[17] Ashcroft N W and Mermin N D 1976 *Solid State Physics* (Fort Worth, TX: Harcourt Brace College Publishers)
[18] Onsager L 1931 *Phys. Rev.* **37** 405
[19] Goldstein H, Poole C P and Safko J 2002 *Classical Mechanics* 3rd edn (Reading, MA: Addison-Wesley)
[20] Mahan G D 2000 *Many Particle Physics* 3rd edn (Berlin: Springer)
[21] Tong D 2016 TIFR, Infosys Lectures on *The Quantum Hall Effect* (arXiv:1606.06687)
[22] Thouless D J, Kohmoto M, Nightingale M P and den Nijs M 1982 *Phys. Rev. Lett.* **49** 405
[23] Novoselov K S *et al* 2007 *Science* **315** 1379
[24] Deacon R S *et al* 2007 *Phys. Rev. B* **76** 081406(R)
[25] Li G and Andrei E Y 2007 *Nat. Phys.* **3** 623
[26] Bostwick A *et al* 2007 *Nat. Phys.* **3** 36

IOP Publishing

Condensed Matter Physics: A Modern Perspective

Saurabh Basu

Chapter 4

Symmetry and topology

4.1 Introduction

'Point set topology is a disease from which the human race will soon recover'
—H Poincaré (1908)

Poincaré conjecture was the first conjecture made on topology which asserts that a 3D manifold is equivalent to a sphere in 3D subject to the fulfilment of a certain algebraic condition of the form $f(x, y, z) = 0$, where x, y and z are complex numbers. G Perelman (arguably) solved the conjecture in 2006 [1]. However, on practical aspects, just the reverse of what Poincaré had predicted, happened. Topology and its relevance to condensed matter physics have emerged in a big way in recent times. The 2016 *Nobel prize* awarded to D J Thouless, J M Kosterlitz, F D M Haldane, C L Kane, and E Mele getting the *Breakthrough Prize* for contribution to fundamental physics in 2019 bear testimony to that.

Topology and geometry are related, but they have a profound difference. Geometry can differentiate between a square and a circle, or between a triangle and a rhombus, however, topology cannot distinguish between them. All it can say is that, individually all these shapes are connected by continuous lines, and hence are identical. However, topology indeed refers to the study of geometric shapes where the focus is on how properties of objects change under continuous deformation, such as stretching and bending, but tearing or puncturing is not allowed. The objective is to determine whether such a continuous deformation can lead to a change from one geometric shape to another. The connection to a problem of deformation of geometrical shapes in condensed matter physics may be established if the Hamiltonian for a particular system can be continuously transformed via tuning of one (or more) of the parameter(s) that the Hamiltonian depends on. Should there be no change in the number of energy modes below the Fermi energy during the process of transformation, then the two systems (that is, before and after the transformation) belong to the same topology class. In the process *'something'*

doi:10.1088/978-0-7503-3031-2ch4

remains invariant. If that *something* does not remain invariant, then there occurs a topological phase transition. This phase transition can occur from one topological phase to another, or from a topological phase to a trivial phase.

In the following we present the geometric aspects of topology, and relate the integral of the geometric properties over closed surfaces to the topological invariants. It turns out that the '*geometric property*' and the '*closed surface*' have smooth connection to physical observables. We shall see soon that in 1982, Thouless, Kohmoto, Nightingale and den Nijs [2] linked the topological invariant to the quantized Hall conductivity.

To test many of the concepts that we are going to discuss in this chapter, we choose two prototype systems, one each in one (1D), and two dimensions (2D). In 1D, we consider a tight binding model, with dimerized hopping, and in 2D, we consider graphene which has been a hobby horse even several years before its experimental discovery. The theme is to discuss the interplay of symmetry and the topological properties. Particularly, in 2D an important highlight in this direction was put forward by Haldane [3], who had proposed a non-trivial topological phase by breaking one of the fundamental symmetries, namely the time reversal symmetry. Finally, after the experimental discovery of graphene, yet another distinct topological state of matter was discovered by Kane and Mele [4] which has culminated in an emerging field of spintronics.

In chapter 3, we saw that the Hall conductivity (or the resistivity) is quantized in the unit of e^2/h (or h/e^2) within a splendid precision, so much so, that the quantity h/e^2 can define the standard of resistance (=25.5kΩ). Clearly the quantization is independent of the details of the Hamiltonian, for example, the nature of the sample, the strength of the magnetic field, and disorder present in the system. It is realized later that the universality of the phenomenon arises due to 'topological' protection of the energy modes that exist at the edges of a quantum Hall sample, which possess completely different character as compared to the ones that exist in the bulk of the sample. Thus an understanding emerges, that says that a physical observable (which is either the resistivity or the conductivity) can be represented mathematically by a topological invariant. This invariant does not change even when the Hamiltonian changes (for example, when the strength of the magnetic field is varied), until and unless a phase transition occurs which will show up via an abrupt change in the value of the topological invariant. There is an elegant explanation of the physics involved with such a universal phenomenon, which brings us to the subject of topology.

Topology in its usual sense deals with geometry of the objects, in the same spirit here we shall study the geometrical properties of the Hilbert space for the system under consideration. The ideas are best demonstrated for a quantum Hall system which undergoes a series of transitions from a conducting to an insulating state as a function of the external magnetic field. In the process, the topological invariant, for example, the Chern number in this case (we shall discuss this later) jumps from one integral value to another. Thus the system repeatedly undergoes a series of topological phase transitions. In the following we describe this topological phase transition in more general terms.

Consider two Hamiltonians, $\mathcal{H}_1$ and $\mathcal{H}_2$ both of which are functions of a tunable parameter, say, β. If the corresponding energy spectra $\epsilon_1(\beta)$ and $\epsilon_2(\beta)$ are such that the number of energy levels below the zero energy (zero energy is usually the Fermi energy) always remain same for all values of β, then the Hamiltonians can be continuously transformed (or deformed as we see in the analogy of a cup and donut later), and there is no phase transition. Now, consider either $\mathcal{H}_1$ or $\mathcal{H}_2$. If, for either of them, the spectrum is such that the number of energy levels vary as a function of β, that is, if any (or more) level crosses the zero energy, then the 'invariant' changes (from one integer value to another), and one encounters the system going from one topological phase to another. A quantum Hall system shows a similar transition where the Hall conductivity changes from ne^2/h to $(n + 1)e^2/h$, where n is strictly an integer.

Thus the study of topology deals with objects (or Hamiltonians) that can be continuously transformed (or deformed) from one to another without puncturing or tearing the object (or without even closing the energy gap for the quantum system). For geometrical objects, being able to transform continuously depends on the number of 'holes' or 'genus' that are preserved during the course of the transformation. For example, a soccer ball can be deformed smoothly into a wine glass since both of them have no hole (zero genus), while a mug (as shown in figure 4.1) can be transformed smoothly into a donut with one hole (genus equal to 1). The first case with zero hole is called topologically trivial, and the second with a finite number of holes (one in this case) is termed as topologically non-trivial.

4.1.1 Gauss–Bonnet theorem

Gauss–Bonnet theorem in differential geometry is about evaluation of the surface integral of a Gaussian surface. Here we state the theorem without proof. In the most general form, for a closed polyhedral surface, the theorem can be stated as,

$$\int_{\partial R} k_g(s)\mathrm{d}s + \iint_R K\mathrm{d}A = 2\pi\chi(R) \tag{4.1}$$

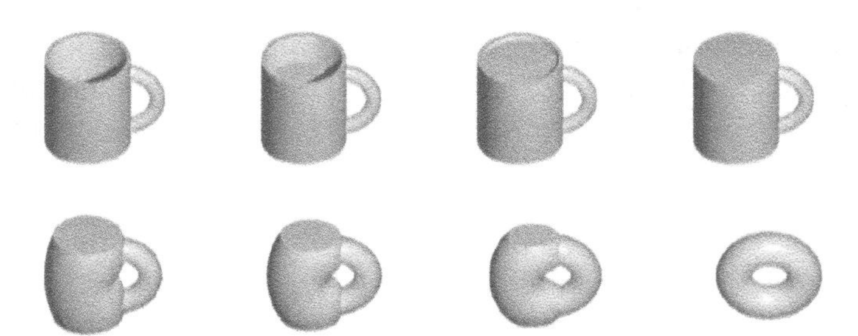

Figure 4.1. A mug can be transformed smoothly into a donut. The handle of the mug remains invariant and emerges as the 'hole' of the donut. Thus the mug and the donut belong to the same universality class. License: The copyright holder of this work (Lucas Vieira), released this work into the public domain. This applies worldwide. Title of the image: Mug and Torus morph.gif. Taken from https://en.wikipedia.org/wiki/File:Mug_and_Torus_morph.gif. Created by Lucas Vieira.

where R denotes a regular region with the boundary ∂R of R, K is the Gaussian curvature, s denotes the arc length of the curves, C_i and the integral is over C_i. Further $\chi(R)$ is called the Euler–Poincaré characteristic. The first term on the left is the integral of the Gaussian curvature over the surface, the second one being the integral of the geodesic curvature of the boundary of the surface. Thus the Gauss–Bonnet theorem simply states that the total curvature of R plus the total geodesic curvature of ∂R is a constant.

As an example we consider the simplest case, that is a sphere of radius R. The Gaussian curvature is $1/r^2$[1] and the corresponding area is,

$$\iint_R K\mathrm{d}A = K \times \text{Area} = \frac{1}{r^2} \times 4\pi r^2 = 4\pi. \tag{4.2}$$

Again,

$$\iint_R K\mathrm{d}A = 2\pi\chi(R). \tag{4.3}$$

Thus,

$$\chi(R) = 2$$

Thus the Euler–Poincaré characteristic of a sphere is 2. Suppose we wish to extend this argument to other closed, however, not necessarily convex surfaces in a three-dimensional space. For that, consider the polar cap of unit radius (see figure 4.2). The area is given by,

$$S = \int_0^\theta 2\pi \sin\theta \mathrm{d}\theta = 2\pi(1 - \cos\theta). \tag{4.4}$$

Thus,

$$\int_R K\mathrm{d}A = 1 \times (\text{Area of } S) = 2\pi(1 - \cos\theta).$$

The geodesic curvature K is $1/\tan\theta$. Thus,

$$\int_s K_g\mathrm{d}s = K_g \times \text{length}(S) = \frac{1}{\tan\theta}. \tag{4.5}$$

Hence

$$\int_s K_g\mathrm{d}s + \int_R K\mathrm{d}A = 2\pi(1 - \cos\theta) + 2\pi\cos\theta = 2\pi = 2\pi\chi(R)$$

thereby yielding,

$$\chi(R) = 1.$$

In fact, an alternate form for the Gauss–Bonnet theorem is more useful for our purpose which states that, for a closed convex surface, the integral over the Gaussian

[1] For a geometry with two different radii of curvature, such as a convex lens, the Gaussian curvature is $\frac{1}{r_1 r_2}$.

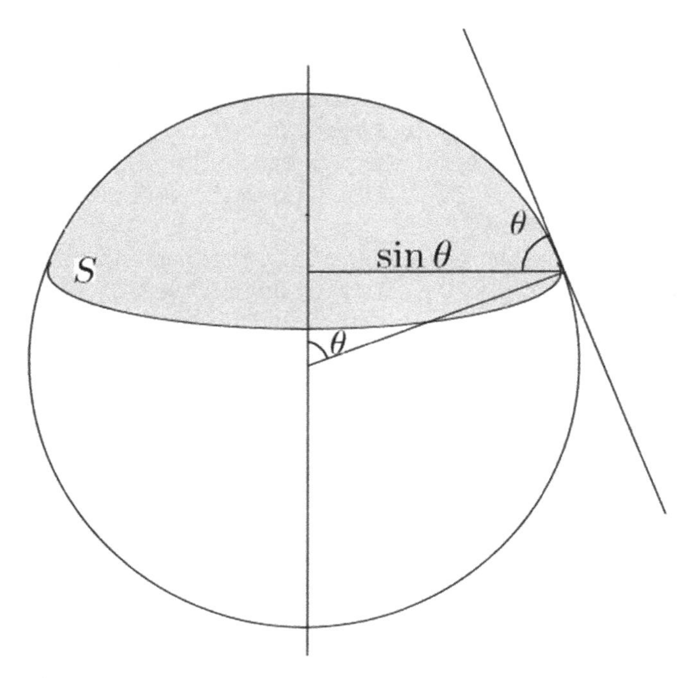

Figure 4.2. A sphere with a polar cap. The figure will aid in calculating the area S of the polar cap.

curvature can be expressed in terms of the number of holes or the genus of the surface. Thus, a simplified (and more relatable for us) version reads,

$$\iint K\mathrm{d}A = 2\pi(2 - 2g) \tag{4.6}$$

Since a sphere has no holes ($g = 0$) (see figure 4.3), the integral of the curvature yields

$$\iint K\mathrm{d}A = 4\pi \tag{4.7}$$

a result that we have seen earlier. Let us look at a case where the genus is non-zero ($g \neq 0$), such as a torus which is topologically equivalent to a mug as we have seen earlier.

For a torus, the Euler–Poincaré characteristic has a value zero. This implies that irrespective of how we bend or deform it, the integrated curvature vanishes. Refer to figure 4.4 (left panel) where there is a positive curvature on the outer surface, and a negative curvature on the inner surface, thereby resulting in a zero total curvature. This is consistent with the Gauss–Bonnet theorem which states that the integral of the Gaussian curvature is $2\pi(2 - 2g)$. Since $g = 1$ here, the integral is zero and so is the Euler–Poincaré characteristic $\chi(R)(= 2 - 2g)$ is zero as well. Similarly, a two-holed donut (see right panel of figure 4.4) will have $\chi(R) = -2$, and hence a negative integrated Gaussian curvature.

Based on the preceding discussion, a sketchy idea emerges on the relationship between topology, and properties of quantum systems. However, it still remains

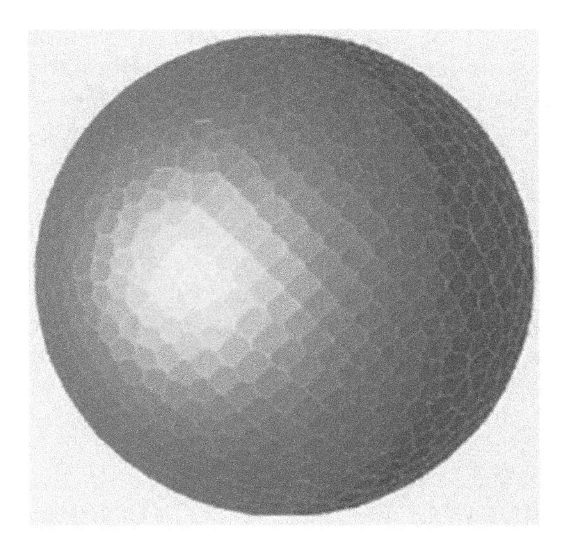

Figure 4.3. A sphere with no hole (or genus). It represents a trivial phase.

Figure 4.4. (Left) A donut with genus (or a hole) equal to 1. (Right) A two-hole object has genus equal to 2. License: Image stated to be in the Public Domain. Title of the image: Double torus illustration.png. Link to image: https://commons.wikimedia.org/wiki/File:Double_torus_illustration.png. Author: Oleg Alexandrov..

unclear how the ideas can relate to properties of materials. At the moment let us talk about crystalline solids for which electron wavefunction is given by Bloch's theorem, namely,

$$\psi(\mathbf{r}) = e^{i\mathbf{k}\cdot\mathbf{r}}u_{\mathbf{k}}(\mathbf{r}) \tag{4.8}$$

where the periodicity of the crystal potential, that is $V(\mathbf{r}) = V(\mathbf{r} + \mathbf{R})$ is captured by the amplitude function $u_{\mathbf{k}}(\mathbf{r})$, such that,

$$u_{\mathbf{k}}(\mathbf{r} + \mathbf{R}) = u_{\mathbf{k}}(\mathbf{r}) \tag{4.9}$$

where $\mathbf{k}$ denotes the crystal momentum, and is distinct from the usual momentum $(= -i\hbar\nabla)$ [5]. The crystal momentum is restricted within the first Brillouin zone (BZ), where the latter is a region in the k-space with periodic boundaries. As the crystal momentum is varied, we map out the energy bands, and one obtains the band structure. The BZ plays the role of the surface over which the integral of the Gaussian curvature is taken which we have discussed earlier.

Now that brings us to the question: *what is the analogue of the Gaussian curvature* for a crystalline solid? To understand this, consider the (non-degenerate) ground state of a Hamiltonian which depends upon a number of parameters that are time dependent. The adiabatic theorem states that if the Hamiltonian is now changed slowly[2] with respect to the parameters, the system remains in its time dependent ground state. However, there is something more to it. As the ground state is evolved in time, in addition to the trivial dynamical phase, there may emerge an irreducible geometric phase that comes into play, namely, the Berry phase put forward by M V Berry in 1984 [6]. In the following we discuss the origin of the Berry phase, and the Berry curvature which is analogous to the Gaussian curvature. The integral of the Berry curvature over the BZ will be shown to yield a constant (or more appropriately an invariant) known as the Chern number, which is analogous to the RHS of equation (4.6) or the Euler–Poincaré characteristic.

4.1.2 Berry phase

Consider a particle in the ground state of a box of length L. Suppose the box slowly expands such that $L(t)$ is a slow function of time. The adiabatic principle says that, if the expansion is slow then the particle always remains in the ground state at any time t. it is true for any state of the system. More generally, consider a Hamiltonian $\mathcal{H}(\lambda(t))$ where λ is a parameter which changes slowly. Now the adiabatic principle says that if the particle starts out in the nth eigenstate of $\mathcal{H}(\lambda(0))$, it will land nth instantaneous eigenstate of $\mathcal{H}(\lambda(t))$ at a time t.

The question is what is the solution of the Schrödinger equation in this approximation? A reasonable guess is,

$$|\psi(t)\rangle = \exp\left(-\frac{i}{\hbar}\int \varepsilon_n(t')\mathrm{d}t'\right)|\phi_n(t)\rangle \tag{4.10}$$

where,

$$\mathcal{H}(t)|\phi_n(t)\rangle = \varepsilon_n(t)|\phi_n(t)\rangle. \tag{4.11}$$

If $\mathcal{H}$ does not vary with time, then clearly the phase is correct. However, it is not so in the case where $\mathcal{H}$ depends on time.

To see what is missing in the above ansatz, let us modify it slightly.

$$|\psi(t)\rangle = C(t)\exp\left[-\frac{i}{\hbar}\int_0^t \varepsilon_n(t')\mathrm{d}t'\right]|\phi_n(t)\rangle. \tag{4.12}$$

In equation (4.10), $C(t) = 1$. Applying Schrödinger equation $(i\hbar\frac{\partial}{\partial t} - \mathcal{H})$ to equation (4.12) and simplifying, for the time dependence of C, one gets,

$$\dot{C}(t) = -C(t)\langle\phi_n(t)|\frac{\mathrm{d}}{\mathrm{d}t}|\phi_n(t)\rangle. \tag{4.13}$$

[2] The time scale for change is larger than the inverse level spacing (level spacing implies a difference between the subsequent energy levels) of the system.

This yields a solution of the form,

$$C(t) = C(0) \exp \underbrace{\left[-\int_0^t \langle \phi_n(t')| \frac{d}{dt'} |\phi_n(t')\rangle dt' \right]}_{e^{i\gamma}} \\ = C(0) e^{i\gamma} \tag{4.14}$$

where,

$$\gamma = i \int_0^t \langle \phi_n(t')| \frac{d}{dt'} |\phi_n(t')\rangle dt'. \tag{4.15}$$

This extra phase is called as the Berry phase or the geometric phase. It is also called the Berry–Panchratnam phase, and is quite a familiar quantity in the field of optics.

In general, phases do not give rise to measurable consequences, since the eigenstates are defined only up to a phase factor. Even here, it may be thought that we can define new Berry states to absorb the phase, namely,

$$|\phi_n'(t)\rangle = e^{i\chi(t)} |\phi_n(t)\rangle \tag{4.16}$$

then,

$$i\langle \phi_n'(t)| \frac{d}{dt} |\phi_n'(t)\rangle = i\langle \phi_n(t)| \frac{d}{dt} |\phi_n(t)\rangle - \frac{d\chi}{dt}. \tag{4.17}$$

Now suppose the parameter $\chi(t)$ changes the Hamiltonian in such a manner that after a complex cycle,

$$\mathcal{H}(0) = \mathcal{H}(t = T)$$

The end result is,

$$i \oint \langle \phi_n'(t)| \frac{d}{dt} |\phi_n'(t)\rangle = i \oint \langle \phi_n(t)| \frac{d}{dt} |\phi_n(t)\rangle - (\chi(T) - \chi(0)). \tag{4.18}$$

The last term on the RHS is an irreducible phase that does not cancel under redefinition of the Berry states. Remember, χ arises from γ which we denote as the Berry phase. Single valuedness of χ demands,

$$\chi(T) - \chi(0) = 2\pi n, \quad \text{where } n \text{ is an integer}$$

The surface integral of the Berry curvature is called the Chern number. This is analogous to Gauss–Bonnet theorem which connects surface integral of radii of curvature. In the Gauss–Bonnet theorem, just like an object with a genus '1' cannot be smoothly transformed into another with genus 'zero' or '2' (unless, of course, something drastic, that is, tearing or puncturing is done to the object), a system with a non-zero Chern number cannot be transformed into that with zero Chern number.

In a quantum Hall system, the Hall conductivity is given by,

$$\sigma_{xy} = Ce^2/h = ne^2/h, \quad \text{where } C = \text{ Chern number}.$$

It can be argued that the Chern number is always an integer. Further, the Berry curvature, $\mathcal{F}$ is defined as curl of the Berry connection, namely,

$$\mathcal{F} = \nabla \times \mathcal{A} \tag{4.19}$$

$\mathcal{F}$ is analogous to the magnetic field. The Chern number is defined as the surface integral of the Berry curvature over a surface enclosed.

4.2 Symmetries and topology

To elucidate more on the topological invariance in materials, we discuss a few discrete symmetries of the Hamiltonian, and how they interplay with the topological properties. In this context we wish to discuss three symmetries, namely, the inversion symmetry (also known as sublattice symmetry), and the time reversal symmetry. A third symmetry that we shall talk about is the contribution of the above two, and is known as the particle–hole symmetry. We shall not worry about the third one here, since, it is not relevant for the present discussion, and it has been discussed in the context of Hubbard model (see chapter 2).

4.2.1 Inversion symmetry

Let us consider an eigenstate in the position basis $|\psi(\mathbf{r})\rangle$, so that,

$$\mathbf{r}|\psi(\mathbf{r})\rangle = |\mathbf{r}||\psi(\mathbf{r})\rangle. \tag{4.20}$$

Now we define the inversion symmetry or the parity operator $\mathcal{P}$ such that,

$$\mathcal{P}|\psi(\mathbf{r})\rangle = |\psi(-\mathbf{r})\rangle. \tag{4.21}$$

Now,

$$\mathbf{r}\mathcal{P}|\psi(\mathbf{r})\rangle = \mathbf{r}|\psi(-\mathbf{r})\rangle = -|\mathbf{r}||\psi(-\mathbf{r})\rangle. \tag{4.22}$$

If we act $\mathcal{P}^{\dagger}$ on both sides (remembering $\mathcal{P}^{\dagger} = \mathcal{P}^{-1} = \mathcal{P}$)

$$\begin{aligned} \mathcal{P}^{\dagger}\,\mathbf{r}\,\mathcal{P}|\psi(\mathbf{r})\rangle &= -\,|\mathbf{r}|\mathcal{P}^{\dagger}|\psi(-\mathbf{r})\rangle \\ &= -\,|\mathbf{r}|\mathcal{P}|\psi(-\mathbf{r})\rangle \\ &= -\,|\mathbf{r}||\psi(\mathbf{r})\rangle = -\mathbf{r}|\psi(\mathbf{r})\rangle. \end{aligned} \tag{4.23}$$

Thus,

$$\mathcal{P}^{\dagger}\,\mathbf{r}\,\mathcal{P} = -\mathbf{r}. \tag{4.24}$$

This yields

$$\begin{aligned} \mathcal{P}^{\dagger}\,\mathbf{r} &= -\,\mathbf{r}\mathcal{P} \\ \text{or, } \{\mathcal{P}, \mathbf{r}\} &= 0. \end{aligned} \tag{4.25}$$

Hence the parity operator anticommutes with the position operator.

Let us now explore the analogous scenario for the momentum operator. For this purpose it is convenient to introduce the transformation operator $T(\mathbf{a})$[3] that translates a state $|\psi(\mathbf{r})\rangle$ to $|\psi(\mathbf{r}+\mathbf{a})\rangle$ where $\mathbf{a}$ denotes a fixed length, for example, $\mathbf{a}$ can be the lattice constant. That is,

$$\begin{aligned} T(\mathbf{a})|\psi(\mathbf{r})\rangle &= |\psi(\mathbf{r}+\mathbf{a})\rangle \\ \mathcal{P}^{\dagger}T(\mathbf{a})\mathcal{P}|\psi(\mathbf{r})\rangle &= T(-\mathbf{a})|\psi(\mathbf{r})\rangle \end{aligned} \tag{4.26}$$

which yields,

$$\mathcal{P}^{\dagger}\ T(\mathbf{a})\ \mathcal{P} = T(-\mathbf{a}). \tag{4.27}$$

This demands that the translation operator is of the form,

$$T(\mathbf{a}) = e^{i\ \mathbf{k}\cdot\mathbf{a}}. \tag{4.28}$$

Expanding for infinitesimal translations,

$$\mathcal{P}^{\dagger}\ \mathbf{p}\ \mathcal{P} = -\ \mathbf{p}. \tag{4.29}$$

Thus, similar to the position operator, the momentum operator too anticommutes with the parity operator.

Since, both $\mathbf{r}$ and $\mathbf{p}$ anticommute, the angular momentum, $\mathbf{L}$ $(= \mathbf{r} \times \mathbf{p})$ commutes with $\mathcal{P}$. In a 3D orthogonal coordinate system, one can invert it about any of the axes. For example in a Cartesian coordinate system,

1. an inversion about the z-axis is denoted as $\sigma_h(xy)$.
2. about the y-axis it is $\sigma_v(xz)$ and
3. $\sigma_v(yz)$ denotes the inversion about x-axis.

Here σ denotes an inversion operation, and has got nothing to do with the Pauli matrices. Under these operations, the position and the angular momentum variable transforms as,

(i) $\sigma_h(xy)$: $x \to x,\ y \to y,\ z \to -z \quad L_x \to -L_x,\ L_y \to -L_y,\ L_z \to L_z$
(ii) $\sigma_v(xz)$: $x \to x,\ y \to -y,\ z \to z \quad L_x \to -L_x,\ L_y \to L_y,\ L_z \to -L_z$
(iii) $\sigma_v(yz)$: $x \to -x,\ y \to y,\ z \to z \quad L_x \to L_x,\ L_y \to -L_y,\ L_z \to -L_z$

4.2.2 Time reversal symmetry

Now we shall discuss time reversal symmetry. It is obvious that under time reversal, the time variable, t changes to $-t$. This makes the position $(\mathbf{r}(t))$ and the momentum variable $(\mathbf{p}(t))$ transform under time reversal as, $\mathbf{r}(-t)$ and $-\mathbf{p}(t)$, respectively. The angular momentum $\mathbf{L}(t)(= \mathbf{r} \times \mathbf{p})$ thus also becomes, $-\mathbf{L}(-t)$ under time reversal. Similar outcomes are expected when $\mathbf{r}(t)$, $\mathbf{p}(t)$ and $\mathbf{L}(t)$ are quantum mechanical operators. Additional inputs to the ongoing discussion can be received from the behaviour of the electric field, $\mathbf{E}(\mathbf{r}, t)$ and the magnetic field $\mathbf{B}(\mathbf{r}, t)$ vectors under time reversal. $\mathbf{E}(\mathbf{r}, t)$ does not change sign under time reversal (refer to the Maxwell's

[3] Distinguish between $T(a)$ for the translation operator, and $\mathcal{T}$ for the time reversal operator.

equations, $\nabla \cdot \mathbf{E} = \frac{\rho}{\epsilon_0}$ where charge density $\rho(\mathbf{r})$ does not change sign), however $\mathbf{B}(\mathbf{r}, t)$ changes sign (owing to, $\nabla \times \mathbf{B} = \mu_0 \mathbf{J}$, where $\mathbf{J}(\mathbf{r}, t)$ is the current density and it changes sign under time reversal).

Now, consider a quantum state $\psi(t)$ that obeys Schrödinger equation,

$$i\hbar \frac{\partial \psi(\mathbf{r}, t)}{\partial t} = \mathcal{H}\psi(\mathbf{r}, t). \tag{4.30}$$

In the following, we suppress the $\mathbf{r}$ dependence of ψ, and simply write $\psi(t)$ which upon the application of the time reversal operator yields $\psi'(-t)$. Mathematically,

$$\mathcal{T}|\psi(t)\rangle = |\psi'(-t)\rangle. \tag{4.31}$$

In order to find $\psi'(-t)$, let us look at the solution of equation (4.30),

$$|\psi(t)\rangle = e^{-i\mathcal{H}t/\hbar}|\psi(0)\rangle. \tag{4.32}$$

For $t = 0$, apply the time reversal operator, that is, $\mathcal{T}|\psi(0)\rangle$. Now, let it evolve forward in time, which means we get a state,

$$e^{-i\mathcal{H}t/\hbar}\mathcal{T}|\psi(0)\rangle.$$

For the Hamiltonian to be invariant under time reversal, this state should be same as $\mathcal{T}\psi(-t)$ which is equivalent to,

$$\mathcal{T}e^{i\mathcal{H}t/\hbar}|\psi(0)\rangle.$$

Thus,

$$\mathcal{T}e^{i\mathcal{H}t/\hbar}|\psi(0)\rangle = e^{-i\mathcal{H}t/\hbar}\mathcal{T}\psi(0).$$

For small time δt we can expand the exponential and write,

$$\mathcal{T}i\mathcal{H} = -i\mathcal{H}\mathcal{T}. \tag{4.33}$$

A natural intuition (albeit wrong as shown later) is to cancel the 'i' from both sides of equation (4.33). This yields

$$\mathcal{T}\mathcal{H} = -\mathcal{H}\mathcal{T} \tag{4.34}$$

which implies,

$$\mathcal{T}\mathcal{H} + \mathcal{H}\mathcal{T} = 0 \quad \text{or,} \quad \{\mathcal{T}, \mathcal{H}\} = 0. \tag{4.35}$$

But that cannot be correct, since we have assumed the time reversal operation to be a valid symmetry operation. This means that cancelling the 'i' from both sides in equation (4.33) was not a legitimate step.

Reconciliation is possible if we understand that time reversal, unlike most other operations in quantum mechanics, is an anti-unitary operation. To remind ourselves a unitary operation U satisfies $UU^{\dagger} = \mathbb{1}$ or a unitary operator acting on a state $\alpha|\psi\rangle$ yields,

$$U(\alpha|\psi\rangle) = \alpha U|\psi\rangle. \tag{4.36}$$

This is also the property of a linear operator. However, for an anti-linear operator, A one gets

$$A(\alpha|\psi\rangle) = \alpha^* A|\psi\rangle \tag{4.37}$$

which means that the anti-linear operator does a complex conjugation. This resolves the dilemma caused by the naive cancellation of 'i' in equation (4.33). Thus the factor 'i' on the LHS of equation (4.33) is complex conjugated when it encounters $\mathcal{T}$ on the way pulling it. This yields an extra minus sign which cancel with the one on the RHS yielding,

$$[\mathcal{T}, \mathcal{H}] = 0. \tag{4.38}$$

This is familiar with the notion that $\mathcal{H}$ is time reversal invariant (we have to deliberately break time reversal invariance of the Hamiltonian, either via an external magnetic field or some other means) and hence the Hamiltonian should commute with the time reversal operator.

Any anti-unitary operator can be written as a product of a unitary operator, U multiplied by a complex conjugation operator, $\mathcal{K}$ such that,

$$\mathcal{T} = U\mathcal{K}.$$

A special case in this regard deserves a mention, that is, the case for a $S = \frac{1}{2}$ particle. Spin being an angular momentum it is odd under time reversal, that is,

$$\mathcal{T}\boldsymbol{\sigma}\mathcal{T}^{-1} = -\boldsymbol{\sigma}. \tag{4.39}$$

This implies, $\mathcal{K}\sigma_x\mathcal{K}^{-1} = \sigma_x$, $\mathcal{K}\sigma_y\mathcal{K}^{-1} = -\sigma_y$ and $\mathcal{K}\sigma_z\mathcal{K}^{-1} = \sigma_z$ which is reasonable as σ_y contains imaginary entries. Thus for the unitary operator, $U\sigma_x U^{-1} = -\sigma_x$, $U\sigma_y U^{-1} = \sigma_y$ and $U\sigma_z U^{-1} = -\sigma_z$. Thus the unitary operator, U commutes with σ_y, but anticommutes with σ_x and σ_z.

Finally, the form of the time reversal operator for a Hamiltonian corresponding to a $S = \frac{1}{2}$ system is given by (without proof, readers are encouraged to try using $\mathcal{T} = Ke^{-\frac{\pi}{2}\sigma_y/\hbar}$),

$$\mathcal{T} = -i\sigma_y\mathcal{K}. \tag{4.40}$$

Further, for spinless particles (or integer spin), $\mathcal{T}^2 = 1$, while for $S = \frac{1}{2}$ particles, $\mathcal{T}^2 = -1$.

4.3 SSH model

4.3.1 Introduction

To make our concepts clear on the topological phase, and whether a model involves a topological phase transition, we apply it to the simplest model available in literature. The Su–Schrieffer–Heeger (SSH) model denotes a paradigmatic 1D model which hosts a topological phase. It also possesses a physical realization in polyacetylene which is a long chain organic polymer (polymerization of acetylene) with a formula $[C_2H_2]_n$ (shown in figure 4.5). The C – C bond lengths are measured

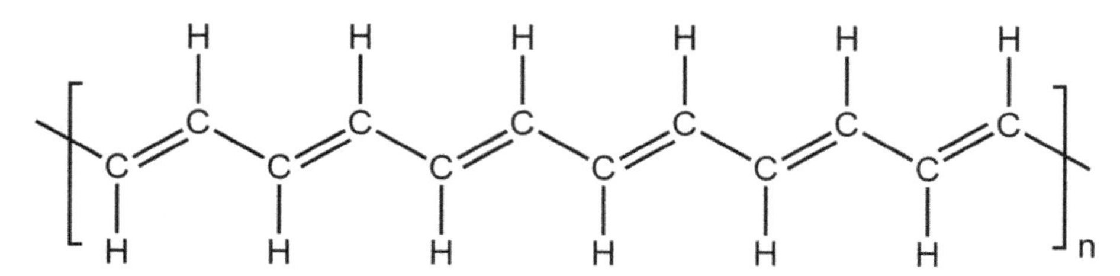

Figure 4.5. A polyacetylene chain with formula $(C_2H_2)_n$ is shown.

by NMR spectroscopy technique and are found to be 1.36 Å and 1.44 Å for the double, and the single bonds, respectively. The chain consists of a number of methyne (=CH–) groups covalently bonded to yield a 1D structure with each C-atom having a π electron. This renders connectivity to the polymer chain.

Possibly intrigued by this bond length asymmetry, one can write down a tight binding Hamiltonian of such a system with two different hopping parameters for spinless fermions hopping along the single, and the double bonds. These *staggered* hopping amplitudes are represented by t_1 and t_2. Let us consider that the chain consists of N unit cells with two sites (that is, two C atoms) per unit cell, and denote these two sites as A and B. The hopping between A and B sites in a cell is denoted by t_1, while those from B to A across the cell are denoted by t_2. Because of the presence of a single π electron at each of the C atoms, the inter-particle interaction effects are completely neglected. We shall show that the staggered hopping or the dimerization has got serious consequences on the topological properties of even such a simple model.

4.4 The SSH Hamiltonian

The above considerations yield the following Hamiltonian,

$$\mathcal{H} = -t_1\sum_{n=1}^{N}(c_{n,\,\mathrm{A}}^{\dagger}c_{n,\mathrm{B}} + \mathrm{h.\ c.}) - t_2\sum_{n=1}^{N-1}(c_{n,\,\mathrm{B}}^{\dagger}c_{n+1,\mathrm{A}} + \mathrm{h.\ c.}). \tag{4.41}$$

For simplicity and concreteness t_1 and t_2 are assumed to be real and non-negative and $c_{n,\,\alpha}^{\dagger}(c_{n,\alpha})$ denotes electron creation(annihilation) operator at site n belonging to the α sublattice ($\alpha\in$ A, B).

It is clear that N denotes the total number of cells, which implies $M = 2N$ where M represents the total number of sites. Thus for an open chain with M atoms, we have $t_M = 0$. In the site basis, the Hamiltonian can be explicitly written as,

$$\mathcal{H} = (c_1^{\dagger}, c_2^{\dagger}, \ldots, c_M^{\dagger})\begin{pmatrix} 0 & t_1 & 0 & . & . & 0 \\ t_1^* & 0 & t_2 & . & . & . \\ 0 & t_2^* & 0 & . & . & . \\ \vdots & \vdots & \vdots & \vdots & \vdots & \vdots \\ 0 & . & . & . & . & 0 \end{pmatrix}\begin{pmatrix} c_1 \\ c_2 \\ \vdots \\ c_M \end{pmatrix}. \tag{4.42}$$

If M is an even number, then $t_{M-1} = t_1$, otherwise, $t_{M-1} = t_2$.

We shall show in the following that a staggered hopping is responsible for opening of a gap in the dispersion, and subject to the fulfilment of a particular condition, the nature of the gap can be topological. To see that, let us study the band structure. We can Fourier transform the electron operators using,

$$c_\alpha(k) = \sum_n e^{ikn} c_{n\alpha} \qquad (\alpha \in \text{A, B}). \tag{4.43}$$

This yields a tight binding Hamiltonian in the sublattice basis, namely $(c_{k\text{A}}, c_{k\text{B}})$ as,

$$\mathcal{H} = \sum_k c^\dagger_{k\alpha} h_{\alpha\beta}(k) c_{k\beta} \tag{4.44}$$

where,

$$h_{\alpha\beta}(\mathbf{k}) = \begin{pmatrix} 0 & t_1 + t_2 e^{-ik} \\ t_1 + t_2 e^{ik} & 0 \end{pmatrix} = \begin{pmatrix} 0 & f(k) \\ f^*(k) & 0 \end{pmatrix}. \tag{4.45}$$

The 2×2 structure of the matrix $h_{\alpha\beta}(k)$ allows us to write,

$$h_{\alpha\beta}(k) = \mathbf{d}(k) \cdot \boldsymbol{\sigma} \tag{4.46}$$

where $\mathbf{d}(k)$ is a vector given by,

$$\mathbf{d}(k) = (d_x(k), d_y(k), d_z(k)) = (t_1 + t_2 \cos k, t_2 \sin k, 0) \tag{4.47}$$

and $\boldsymbol{\sigma} = (\sigma_x \sigma_y, \sigma_z)$ denote the Pauli matrices. The energy dispersion is given by,

$$E(k) = \pm|\mathbf{d}(k)| = \pm\sqrt{(t_1 + t_2 \cos k)^2 + t_2^2 \sin^2 k}\,. \tag{4.48}$$

A little manipulation of the terms inside the square root yields,

$$E(k) = \pm\sqrt{(t_1 - t_2)^2 + 4t_1 t_2 \cos^2 \frac{k}{2}} \tag{4.49}$$

where k is contained in the BZ, that is, $-\pi \leqslant k \leqslant +\pi$. The corresponding normalized eigenvectors are given by,

$$|\psi_\pm\rangle = \frac{1}{\sqrt{2}} \begin{pmatrix} \pm e^{-i\phi(k)} \\ 1 \end{pmatrix} \tag{4.50}$$

where,

$$\phi(k) = \tan^{-1}\left(\frac{t_2 \sin k}{t_1 + t_2 \cos k}\right)$$

We shall explore a few representative cases to make our ongoing discussion clear, namely,

(i) $t_2 = 0$: Extreme dimerized limit (see upper panel of figure 4.6).
(ii) $t_1 > t_2$: Intra-cell hopping is larger than the inter-cell hopping.

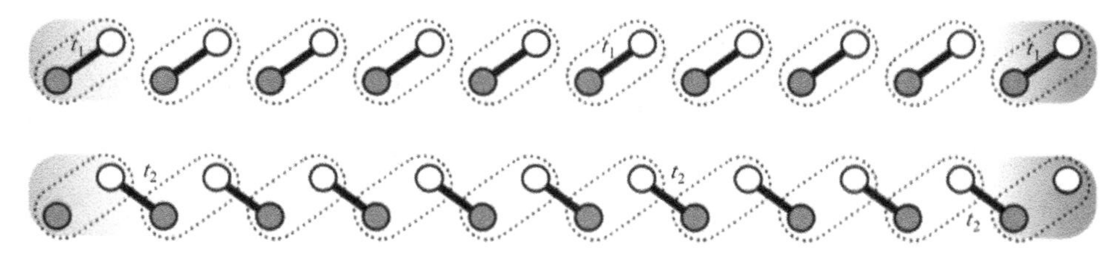

Figure 4.6. The extreme dimerized limit for the trivial (upper panel), and the topological (lower panel) phases of the SSH model are shown.

(iii) $t_1 = t_2$: Intra-cell hopping is same as the inter-cell hopping.
(iv) $t_1 < t_2$: Intra-cell hopping is smaller than the inter-cell hopping.
(v) $t_1 = 0$: Extreme dimerized limit (however, different than (i), see lower panel of figure 4.6).

We plot the band structure and the components of the **d**-vector as a function of the crystal momentum k. The purpose is to define a bulk winding number, which is the topological invariant here. The plot d_x versus d_y for k in the BZ defines a surface (except for the critical case, $t_1 = t_2$). Whether the surface encloses the origin will decide on its topological properties. Further, the unit vector, $\hat{\mathbf{d}}$ defines the direction of the d vector via, $\hat{\mathbf{d}} = \mathbf{d}/|\mathbf{d}|$. At half-filling, the lower band is filled. The two bands are gapped by an amount $2\delta t$, where $\delta t = |t_1 - t_2|$ at $k = \pm\pi$. This is also an insulating phase. However, we shall see that this phase is distinct from case (ii).

4.4.1 Topological properties

The SSH chain hosts both bulk and edge states. The distinction between the bulk and the edge can be understood from the real space analysis. The plot for the energy spectrum as a function of the ratio t_2/t_1 (see figure 4.7) shows that the zero modes start to appear just beyond the critical point. Prior to that, for $0 \leqslant t_2/t_1 \leqslant 1$, the system behaves like a trivial insulator with a bulk band gap. The gap closes at $t_1 = t_2$, and eventually for $t_2 > t_1$, the bulk gap opens again, however, a pair of zero modes appears in the spectrum. These zero modes yield a topological character to the phase. They originate from the two solitary C atoms that reside at the two edges of the chain. The fact that these zero modes indeed arise out of the edges is shown in figure 4.8 via plotting the probability densities, $|\psi_i|^2$ at all sites of the chain. The amplitudes at the left, and the right edges are shown by red and blue colours, respectively, and they vanish everywhere in the bulk of the chain. Further, to emphasize the robustness of the edge modes, we show the inverse participation ratio (IPR) defined by,

$$\text{IPR} = \sum_{i=1}^{L} |\psi_i|^4 \tag{4.51}$$

where IPR = 0 or 1 denotes the extended or the localized phases. However, these extreme values (namely, 0 and 1) can only be obtained in the thermodynamic limit

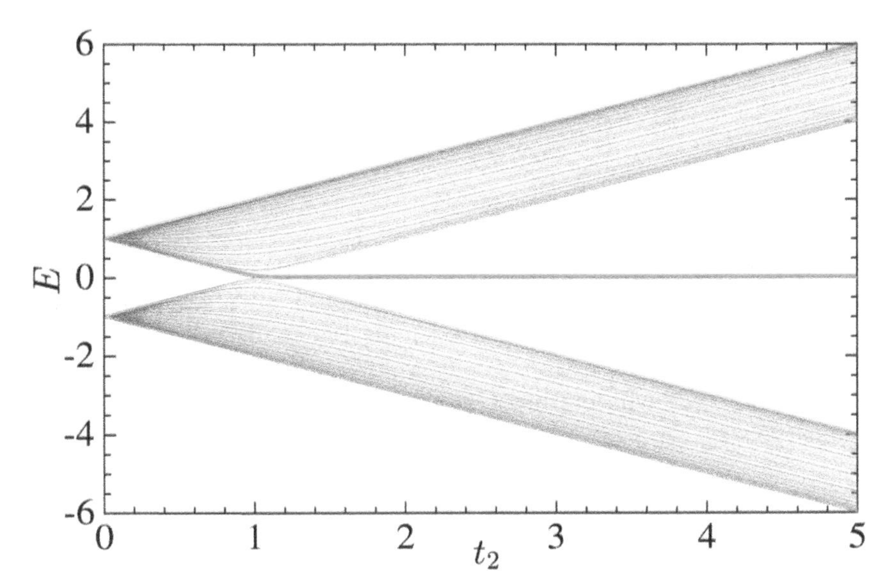

Figure 4.7. The energy is plotted as a function of t_2. A pair of zero modes exist for $t_2 \geqslant 1$ (in unit of t_1). Additionally, the zero mode is shown in red colour, which implies finite value of IPR (see text).

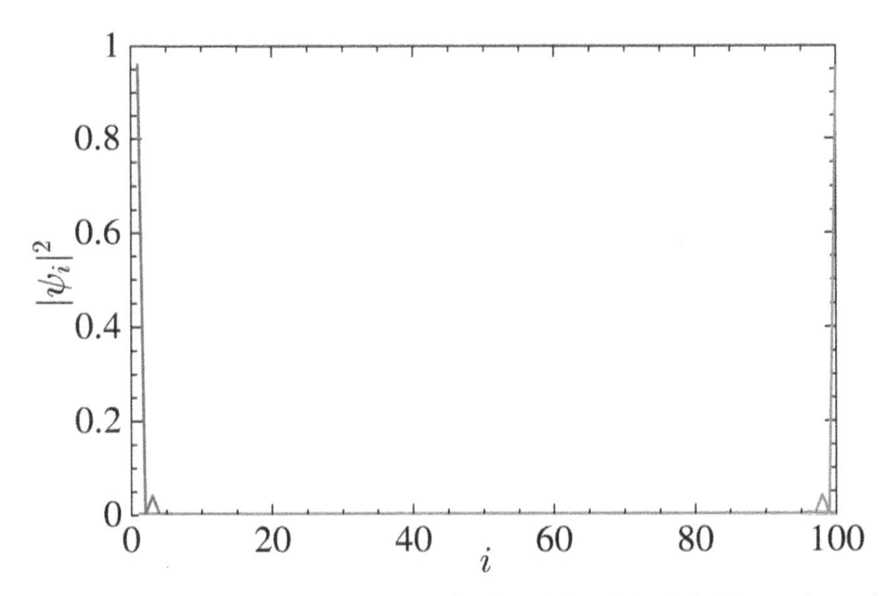

Figure 4.8. The probability amplitude is plotted as a function of sites of the chain. Here we have taken the length, $L = 100$.

($L \to \infty$). Here we denote the edge modes coming from the two edges of the system in red in figure 4.7. In this figure, red denotes a finite value for the IPR. Evidently the zero modes are seen to be localized.

Let us return to the behaviour of the vector $\mathbf{d}(k)$. The components $((t_1 + t_2 \cos k), t_2 \sin k, 0)$ in the BZ defined by $-\pi \leqslant k \leqslant +\pi$ of $\mathbf{d}(k)$ denote the eigenstates with the energy spectrum given by,

$$E(k) = |\mathbf{d}(k)|$$

Corresponding to one of the cases, namely $t_2 > t_1$, the vector $\mathbf{d}(k)$ winds about the origin, while for the other, $t_2 < t_1$, it does not. The origin of the $d_x - d_y$ plane is $\mathbf{d}(k) = 0$ and denotes the gapless (critical) condition. Based on the above information, it is possible to define a winding number, ν which would tell us whether the trajectory of $\mathbf{d}(k)$ winds the origin, as k is varied over the BZ. Thus the winding number is capable of distinguishing the two seemingly equivalent (gapped) scenarios.

Mathematically, the winding number, ν can be written down using the unit $\hat{\mathbf{d}}$ vector defined via,

$$\hat{\mathbf{d}} = \frac{\mathbf{d}(k)}{|\mathbf{d}(k)|}. \tag{4.52}$$

One can now define ν using,

$$\nu = \frac{1}{2\pi} \int_{-\pi}^{+\pi} \left(\hat{\mathbf{d}} \times \frac{d}{dk} \hat{\mathbf{d}} \right)_z dk. \tag{4.53}$$

Let us justify how the above expression on the RHS denotes the winding number. Writing it more explicitly,

$$\begin{aligned}
\nu &= \frac{1}{2\pi} \int_{-\pi}^{\pi} \frac{\mathbf{d}(k)}{|\mathbf{d}(k)|} \times \frac{d}{dk} \frac{\mathbf{d}(k)}{|\mathbf{d}(k)|} dk \\
&= \frac{1}{2\pi} \int_{-\pi}^{\pi} \frac{\mathbf{d}(k)}{|\mathbf{d}(k)|} \times \left(\frac{d}{dk} \frac{\mathbf{d}(k)}{|\mathbf{d}(k)|} - \frac{\mathbf{d}(k) \frac{d}{dk} |\mathbf{d}(k)|}{|\mathbf{d}(k)|^2} \right) dk
\end{aligned} \tag{4.54}$$

since $\mathbf{d}(k) \times \mathbf{d}(k) = 0$

$$\begin{aligned}
\nu &= \frac{1}{2\pi} \int_{-\pi}^{\pi} \frac{\mathbf{d} \times \frac{d}{dk} \mathbf{d}(k)}{|\mathbf{d}(k)|^2} dk \\
&= \frac{1}{2\pi} \int_{-\pi}^{\pi} \frac{\mathbf{d}(k) \times \delta \mathbf{d}(k)}{|\mathbf{d}(k)|^2} dk
\end{aligned} \tag{4.55}$$

$$\frac{d}{dk} \mathbf{d}(k) = \frac{d}{dk} (\mathbf{d}_0 + k \delta \mathbf{d}) \tag{4.56}$$

where in the last line we have used a Taylor expansion of $\frac{d}{dk}\mathbf{d}(k)$. From the definition of the cross product, $\mathbf{d}(k) \times \delta \mathbf{d}(k)$ is the angle (in radian) between $\mathbf{d}(k)$ and $\mathbf{d} + \delta \mathbf{d}$. Thus integrating this over the BZ yields 2π, which when divided by 2π gives 1.

Another useful form for the winding number is given by,

$$\nu = \frac{1}{2\pi i}\int_{-\pi}^{+\pi} dk\frac{d}{dk}\log f(k) \tag{4.57}$$

where, $f(k) = t_1 + t_2e^{-ik}$. Thus,

$$\log f(k) = \log(|f|)e^{i\,\arg(f)}.$$

Consequently, the winding number becomes,

$$\begin{aligned}\nu =& \frac{1}{2\pi i}\int_{-\pi}^{+\pi} \frac{d}{dk}\log(f(k))dk\\ =& \frac{1}{2\pi}\arg(f)_{-\pi}^{+\pi}\\ =& 1 \text{ or } 0\end{aligned} \tag{4.58}$$

depending on whether arg(f) falls in the region of integration, that is, it encloses the origin. The winding number is 1 for the topological phase, and 0 for the trivial phase.

We correlate the band structure with the corresponding winding number calculated above, and show various cases as mentioned above. In the extreme dimerized limit, namely, $t_2 = 0$ and $t_1 = 1$, we get two flat bands at $E = \pm 1$, and the **d**-vector is simply shown by an arrow in figure 4.9. For $t_1 > t_2$, the spectrum is gapped, and corresponds to a trivial insulator because of the absence of winding of the **d**-vector as shown in figure 4.10. Further, the undimerized tight binding chain ($t_1 = t_2$) is shown in figure 4.11 where the gap closes, and the tip of the **d**-vector which executes a circle (shown in red colour) in the $d_x - d_y$ plane just touches the origin at the left, but does not wind it. The fourth case, namely $t_1 < t_2$ again shows a spectral gap, but it is topological in nature as shown by the winding of the **d**-vector in figure 4.12. Finally, the other dimerized case, that is, $t_1 = 0$ and $t_2 = 1$ (see figure 4.13) again shows two flat bands at $E = \pm 1$, however, the trajectory of the **d**-vector shown by the red circle in the right panel of figure 4.13 winds the origin, and hence denotes a topological scenario.

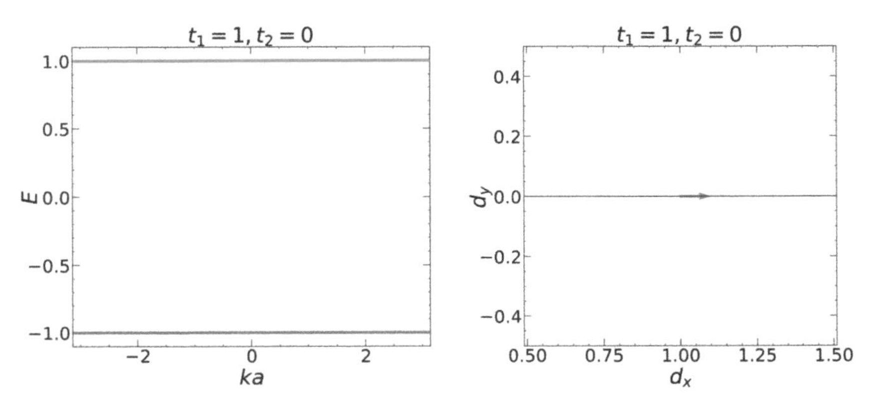

Figure 4.9. The band structure and d-vectors are plotted corresponding to $t_1 = 1$, $t_2 = 0$.

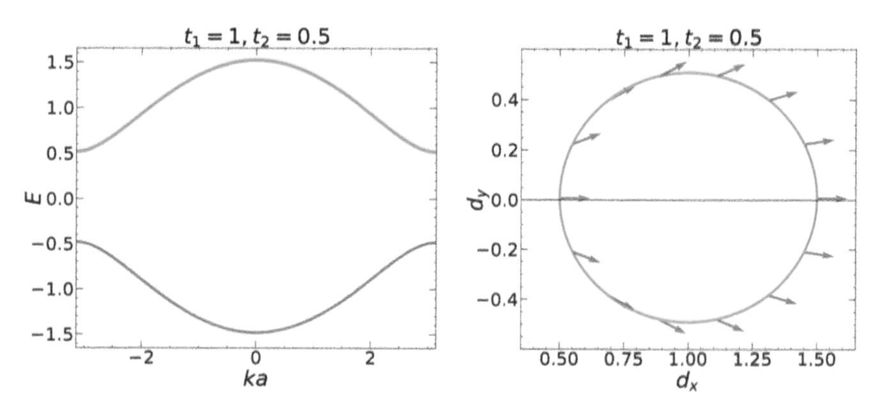

Figure 4.10. The band structure and d-vectors are plotted corresponding to $t_1 = 1$, $t_2 = 0.5$.

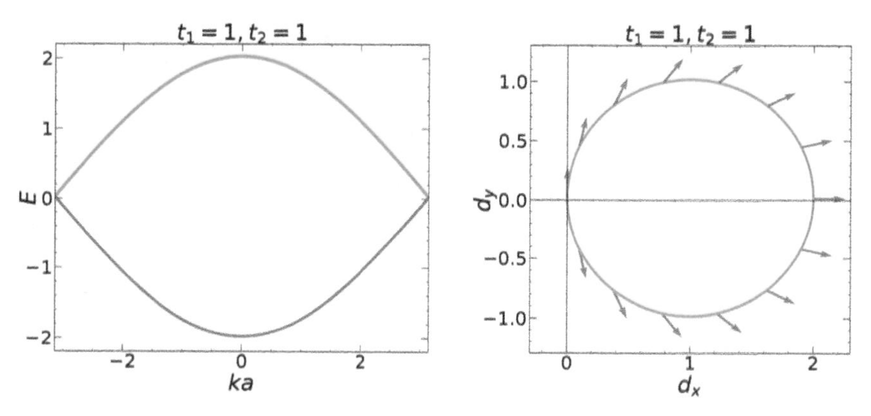

Figure 4.11. The band structure and d-vectors are plotted corresponding to $t_1 = 1$, $t_2 = 1$, that is a simple tight binding chain.

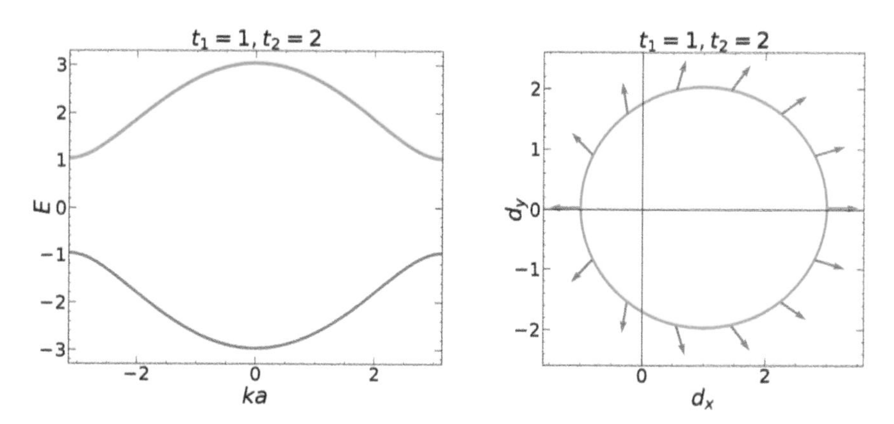

Figure 4.12. The band structure and d-vectors are plotted corresponding to $t_1 = 1$, $t_2 = 2$.

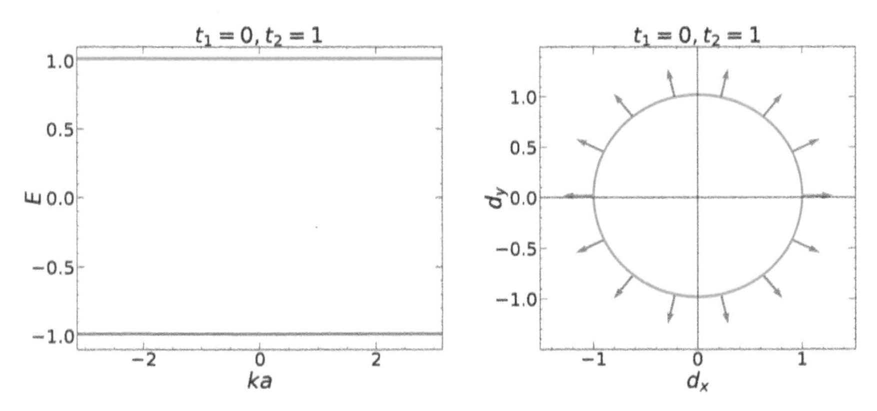

Figure 4.13. The band structure and d-vectors are plotted corresponding to $t_1 = 0$, $t_2 = 1$.

One can also define a Zak phase, Φ_Z [7] (another geometric phase, similar to the Berry phase) defined via[4],

$$\Phi_Z = i \oint \langle \psi | \nabla_k | \psi \rangle dk. \tag{4.59}$$

Using

$$|\psi_\pm\rangle = \frac{1}{\sqrt{2}} \begin{pmatrix} \pm e^{-i\phi(k)} \\ 1 \end{pmatrix} \tag{4.60}$$

$$\phi_Z = \frac{1}{2} \oint \frac{d}{dk} \phi(k) dk = \pm\pi \text{ or } 0 \tag{4.61}$$

which are the values, respectively, for $t_2 > t_1$ and $t_2 < t_1$.

Please note that we have obtained this result without plugging in the explicit form of $\phi(k)$, since the result should be independent of the form of $\phi(k)$. However, if we consider the explicit form of $\phi(k)$, namely,

$$\phi(k) = \tan^{-1}\left(\frac{t_2 \sin k}{t_1 + t_2 \cos k} \right)$$

it throws some subtlety that we need to take care of. If we are in the trivial phase, then the inverse tangent function is present in the first and fourth quadrants because of $t_1 > t_2$. Here the function does not acquire any extra factor, because of which $\phi_Z = 0$. For the topological phase ($t_2 > t_1$), that is, when the inverse tangent function is in the second quadrant, it picks up a phase $\pi - \tan^{-1} x$, while in the third quadrant, the corresponding value is $\pi + \tan^{-1} x$. This can be seen from the sharp change of $\phi(k)$ twice in BZ as seen from figure 4.14. This means that the inverse tangent function acquires an extra phase of 2π. This yields $\phi_Z = -\pi$. Frankly, the

[4] Usually geometric phases that characterize the topological properties of the band structure play a crucial role in the band theory of solids. See reference [7].

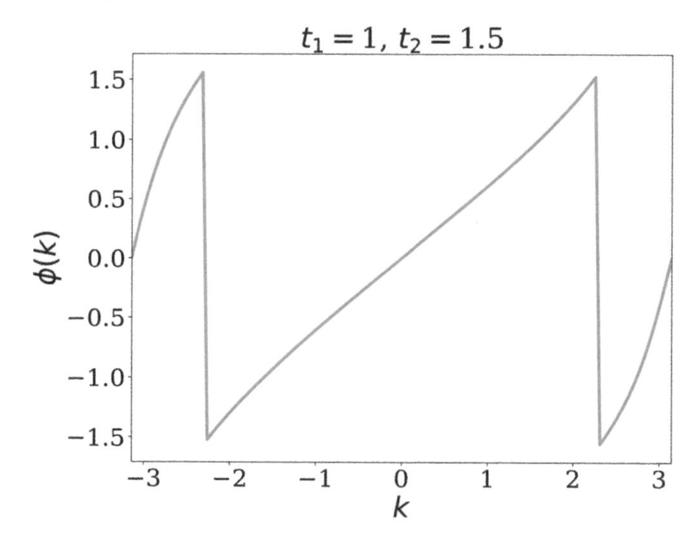

Figure 4.14. ϕ is plotted as a function of k for the topological phase. There are abrupt jumps in the behaviour of ϕ.

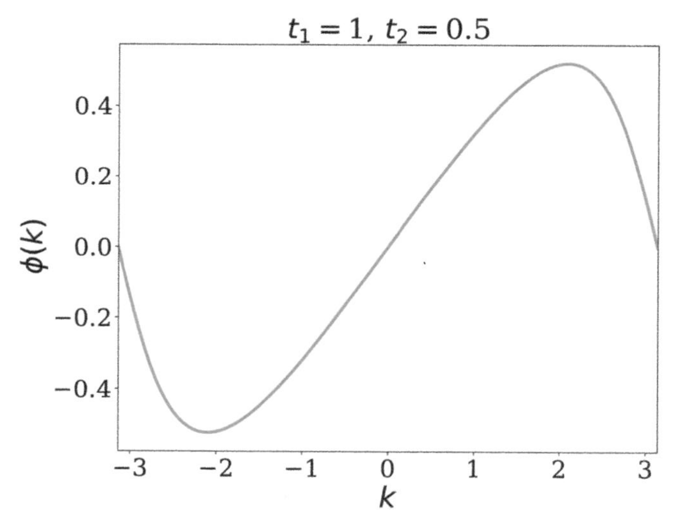

Figure 4.15. ϕ is plotted as a function of k for the trivial phase. It smoothly varies from $-\pi$ to $+\pi$.

negative sign does not mean anything specific. It arises because we have chosen positive sign for the wavefunction. Here the winding number, and the Zak phase are related via $\nu = -\phi_Z/\pi$. Think about $\mathbf{d}(k)$ and ϕ_Z, both of which are obtained from the bulk of the material, yet giving information about the edges of the system. This is traditionally referred to as the *bulk-boundary correspondence* (we shall discuss more about it later). It may also be noted that the behaviour of $\phi(k)$ is smooth over the BZ corresponding to the trivial case (see figure 4.15).

From the preceding discussion it is clear that the two apparently similar insulating phases are topologically different. We have shown that the winding numbers are different (finite in the topological phase and zero in the trivial one) and so are the Zak phases. However, what does it physically entail having different values for the winding number or the Zak phase? Suppose we smoothly deform the Hamiltonian corresponding to the SSH chain via tuning the hopping parameters, t_1 and t_2, we shall get two insulating phases for $t_2 > t_1$, and $t_2 < t_1$ all the while keeping the symmetries preserved and the band gap around $E = 0$ is finite. In tuning from one limit to the other, we have to cross the origin in the $d_x - d_y$ plane (note that $d_z \equiv 0$). This implies that at the intermediate stage, one would obtain the eigensolutions corresponding to the trajectory of $\mathbf{d}(k)$ in the $d_x - d_y$ plane through the origin. Thus a smooth transition from one insulating phase to another is impossible without closing the gap, or satisfying the metallic condition, $t_1 = t_2$. Hence, quite apparently, there is a topological phase transition occurring here.

It is clear that the above discussion will be invalid if there is a z-component of the $\mathbf{d}(k)$ (or a term proportional to σ_z) is present in the system. A simple way to incorporate such a term is through inclusion of an onsite potential. The onsite potential will destroy the zero modes, thereby making it meaningless to talk about the topological properties of the system. However, a disorder in the off-diagonal (hopping) term would retain the zero modes, and hence the system should have a transition from a topological to a trivial phase. The reason for such a distinction between the diagonal and the off-diagonal terms arises because of certain fundamental symmetries that the system possesses. We shall discuss them below.

4.4.2 Chiral symmetry

In standard quantum mechanics, the symmetry of a Hamiltonian, $\mathcal{H}$ is represented by,

$$U\mathcal{H}U^{\dagger} = \mathcal{H}, \quad \text{or} \quad U\mathcal{H} = \mathcal{H}U, \quad \text{or} \quad [\mathcal{H}, U] = 0 \tag{4.62}$$

where U denotes a unitary operator. This implies that U and $\mathcal{H}$ have the same eigenstates and hence can be diagonalized simultaneously. However, in general for topological insulators, such usual unitary symmetries do not have interesting consequences. The reason being, it is mostly possible to make the Hamiltonian block diagonal, thereby reducing the problem to be confined in a single block. In the case of massless Dirac problems, one usually runs out of unitary symmetries, and is left with an irreducible block Hamiltonian which cannot be diagonalized. Thus for the SSH model, there is a different symmetry, that is called the chiral symmetry which is operative here[5]. Here $\mathcal{H}$ obeys,

$$\Gamma\mathcal{H}\Gamma^{\dagger} = -\mathcal{H}, \text{ or } \Gamma\mathcal{H} = \mathcal{H}\Gamma, \quad \text{or} \quad \{\mathcal{H}, \Gamma\} = 0. \tag{4.63}$$

[5] In condensed matter physics, bipartite systems with nearest neighbour hopping, that is, when hopping connects sites with opposite sublattices, obey chiral symmetry.

Here Γ is a unitary operator corresponding to the chiral symmetry. Instead of commuting, it anticommutes with the Hamiltonian. Further, Γ is unitary and Hermitian, implying,

$$\Gamma = \Gamma^{\dagger}, \quad \text{or} \quad \Gamma^{\dagger}\Gamma = \Gamma^2 = \mathbf{1} \tag{4.64}$$

where $\mathbf{1}$ denotes an identity matrix. The above requirement raises a possibility that $\Gamma = e^{i\phi}$ where ϕ is an arbitrary phase. However, this possibility can be eliminated by redefining $\Gamma \to \Gamma e^{-i\phi/2}$. A second requirement is that Γ is a local operator. Thus the matrix elements of Γ survive only within each unit cell, and between the cells they vanish. Hence for the SSH model, the chiral symmetry is equivalent to the sublattice symmetry, which can be expressed through the projectors P_A and P_B corresponding to the A and B sublattices, namely,

$$P_A = \frac{1}{2}(\mathbf{1} + \Gamma) \ ; \qquad P_B = \frac{1}{2}(\mathbf{1} - \Gamma). \tag{4.65}$$

It can be checked that

$$P_A + P_B = \mathbf{1}, \quad \text{and} \quad P_A \cdot P_B = 0.$$

It is also possible to show that,

$$P_A \mathcal{H} P_A = P_B \mathcal{H} P_B = 0. \tag{4.66}$$

The consequence of the chiral symmetry results in a symmetric energy spectrum. That is, corresponding to an energy, E, there is a chiral partner with energy, $-E$. This fact can be seen from the following,

$$\begin{aligned} &\mathcal{H}|\psi\rangle = E|\psi\rangle \\ &\mathcal{H}\Gamma|\psi\rangle = -\Gamma\mathcal{H}|\psi\rangle = -\Gamma E|\psi\rangle = -E\Gamma|\psi\rangle. \end{aligned} \tag{4.67}$$

Of course, the above argument is true for $E \neq 0$. Since the SSH model hosts zero modes, one of them is a partner of the other.

Besides, for all $E \neq 0$, $|\psi\rangle$ and $\Gamma|\psi\rangle$ correspond to distinct and orthogonal eigenstates, which suggests that every non-zero eigenstate of $\mathcal{H}$ derives equal contribution from both the sublattices, that is,

$$\langle\psi|\Gamma|\psi\rangle = \langle\psi|P_A|\psi\rangle - \langle\psi|P_B|\psi\rangle = 0. \tag{4.68}$$

Whereas, for $E = 0$, $\mathcal{H}|\psi\rangle = 0$. Thus,

$$\mathcal{H}P_{A/B}|\psi\rangle = \mathcal{H}[|\psi\rangle + \Gamma|\psi\rangle] = 0. \tag{4.69}$$

Thus the zero energy eigenstates are eigenstates of Γ, and hence are chiral symmetric partners of themselves. It is also owing to the robustness of the chiral symmetry, the zero modes are robust.

One can also define operators for the sublattice symmetry, namely,

$$\Sigma_z = P_A - P_B. \tag{4.70}$$

It can be shown that $\Sigma_z \mathcal{H} \Sigma_z = -\mathcal{H}$ which is similar to the symmetry relation stated earlier, $\Gamma\mathcal{H}\Gamma = -\mathcal{H}$. Thus the chiral symmetry of the SSH model is a re-statement of the sublattice symmetry of the Hamiltonian.

In simple language, the chiral symmetry operator, Γ is actually the z-component of the Pauli matrix, σ_z which yields,

$$\sigma_z \mathcal{H} \sigma_z = -\mathcal{H}. \tag{4.71}$$

A direct multiplication of the three matrices on the LHS can be performed for the proof. We have presented above a crisp description of the SSH model, which despite being simple enough, possesses both trivial and topological phases where the latter shows up via the presence of robust[6] zero energy edge modes, along with a finite value of the winding number. Further, the band structure shows that a phase transition occurs from a trivial to a topological phase (or vice versa) through a gap closing point, where the staggered hopping amplitudes are equal ($t_1 = t_2$).

4.5 Topology in 2D: graphene as a topological insulator

Having studied a prototype model Hamiltonian in 1D, we turn our focus towards 2D, now with the lens on graphene. In particular, we shall explore whether graphene possesses the credibility of becoming a topological insulator. That may happen, provided by some means we are able to open a spectral gap at the Dirac cones. Since a non-zero Berry phase can be a smoking gun for non-trivial properties, let us first look at the Berry phase of graphene.

4.5.1 Berry phase of graphene

For computing the Berry phase, let us consider the low energy Hamiltonian of graphene given by[7],

$$\mathcal{H} = \hbar v_{\mathrm{F}}(\tau_z \sigma_x q_x + \sigma_y q_y) \tag{4.72}$$

where τ_z denotes the valley degree of freedom, that is, $\tau_z = 1$ for K-point, while it is -1 for the K'-point. As usual, σ denotes the sublattice degree of freedom. The Berry connection is obtained as,

$$\mathcal{A} = \langle \psi_- | \nabla | \psi_- \rangle. \tag{4.73}$$

To remind ourselves, the Chern number is defined by integrating the Berry curvature over BZ.

$$n = \frac{1}{2\pi} \oint_{BZ} \mathcal{F} d^2K = C \qquad \text{(a notation we have used earlier)}$$

We rewrite the Dirac Hamiltonian as,

$$h(\mathbf{q}) = \mathbf{q} \cdot \boldsymbol{\sigma} \qquad \text{(the velocity term is dropped).} \tag{4.74}$$

[6] The edge modes are robust as long as the chiral symmetry is intact.

[7] We have discussed the electronic properties of graphene in chapter 3.

In the polar coordinate **q** and $h(\mathbf{q})$ can be represented as,

$$\mathbf{q} = |\mathbf{q}|\begin{pmatrix} \cos\phi \\ \sin\phi \end{pmatrix} = q\begin{pmatrix} \cos\phi \\ \sin\phi \end{pmatrix} \tag{4.75}$$

and

$$h(\mathbf{q}) = q\begin{pmatrix} 0 & \cos\phi - i\sin\phi \\ \cos\phi + i\sin\phi & 0 \end{pmatrix} = q\begin{pmatrix} 0 & e^{-i\phi} \\ e^{i\phi} & 0 \end{pmatrix}. \tag{4.76}$$

The normalized eigenvectors are,

$$\begin{aligned} |\psi_-\rangle &= \frac{1}{\sqrt{2}}\begin{pmatrix} -e^{-i\phi} \\ 1 \end{pmatrix} \text{ and} \\ |\psi_+\rangle &= \frac{1}{\sqrt{2}}\begin{pmatrix} e^{-i\phi} \\ 1 \end{pmatrix}. \end{aligned} \tag{4.77}$$

Next, we calculate the Berry connection $\mathcal{A}$, and remind ourselves only filled bands are to be taken into account. So we shall consider $|\psi_-\rangle$ in the definition of $\mathcal{A}$.

$$\mathcal{A} = i\langle\psi_-|\nabla_q|\psi_-\rangle. \tag{4.78}$$

The gradient operator, ∇_q in polar coordinates is given by,

$$\nabla_q = \left(\frac{\partial}{\partial q}\hat{q} + \frac{1}{q}\frac{\partial}{\partial\phi}\hat{\phi}\right). \tag{4.79}$$

Note that $|\psi_-\rangle$ does not depend upon q. If we now introduce a band index, n, the Chern number corresponding to a band index, n can be written as C_n. The total Chern number is obtained from the contribution from all the bands, namely,

$$\begin{aligned} C &= \sum_n C_n \\ C_n &= \frac{1}{2\pi}\int_S \mathcal{F}_n dS. \end{aligned} \tag{4.80}$$

Here, the S is the surface that encloses the loop. With $\mathcal{A} = \frac{1}{2q}$, $\nabla \times \mathcal{A} = 0$. So $\mathcal{F} = 0$, and hence $C = 0$ which is not a surprise, as for time reversal invariant systems the Chern number should vanish.

The Berry phase around the Dirac points (**K** and **K**$'$) is nothing but the winding number multiplied by π, which is then either +1 or −1. This introduces a measure of the topological *charge* for the Dirac points in the k-space which tells us how the wavefunctions wind around these singular points in k-space differently with respect to each other. The **K** point carries the topological charge +1 (a vortex) and the **K**$'$ point carries a topological charge −1 (an anti-vortex). With the Dirac Fermion sitting at **K** carries a Berry phase, $\Phi_B^{\mathbf{K}} = \pi$, and the Dirac Fermion at **K**$'$ has a Berry phase, $\Phi_B^{\mathbf{K}'} = -\pi$. The overall Berry phase, Φ_B is zero, that is, $\Phi_B = 0$.

4.5.2 Symmetries of graphene

It is fairly well known to readers by now that graphene is represented by a nearest neighbour tight binding model on a honeycomb lattice with a two sublattice basis, namely A and B sublattices. Carbon (C) atoms occupy both the sublattices. The situation is slightly different in boron nitride, in which, despite possessing the same crystal structure, the sublattice symmetry is broken by boron and nitrogen occupying the A and the B sublattices. Thus, graphene is a prototype of a system possessing sublattice symmetry which renders the Hamiltonian block off-diagonal written in the sublattice basis. The low energy physics of this model is denoted by the massless Dirac Hamiltonian that we have seen at length earlier. Here, for the sake of completeness, we recall that the low energy Hamiltonian of graphene at both the Dirac points, namely, $\mathbf{K}$ and $\mathbf{K}'$ is written as,

$$\mathcal{H}_0(\mathbf{k}) = \hbar v_{\mathrm{F}}(k_x\sigma_x\tau_z + k_y\sigma_y) \tag{4.81}$$

where the pseudospins σ_x, σ_y denote sublattice degrees of freedom, and τ_z (again the z-component of the Pauli matrix) distinguishes the valleys at $\mathbf{K}$ and $\mathbf{K}'$. Needless to say, the Hamiltonian is independent of the real spin, which will continue to be a valid description till spin–orbit coupling is included. Now, consider the inversion (or the sublattice) symmetry which, along with switching the two sublattices, also changes the momentum $\mathbf{k}$ to $-\mathbf{k}$ (remember $\mathbf{p} = m\frac{d\mathbf{r}}{dt}$) which again implies that the two valleys are switched under inversion. The corresponding operator that does this operation is given by,

$$\mathcal{P} = \sigma_x\tau_x$$

Under this inversion operator, the Hamiltonian transforms as,

$$\mathcal{P}H(\mathbf{k})\mathcal{P}^{-1} = \hbar v_{\mathrm{F}}\ \sigma_x\ \tau_x\ (k_x\sigma_x\tau_z + k_y\sigma_y)\ \sigma_x\tau_x = \mathcal{H}_0(-\mathbf{k}). \tag{4.82}$$

The above relation can be proved by using product rules of the Pauli matrices, and it ensures inversion symmetry of the Dirac Hamiltonian.

Now we shall discuss time reversal symmetry. In the case of graphene, time reversal symmetry implies changing the momentum vector $\mathbf{k}$ to $-\mathbf{k}$, followed by complex conjugation of the operator (as explained earlier). Under the time reversal symmetry operation, one Dirac point (say, $\mathbf{K}$) goes to another one (say, $\mathbf{K}'$), and thus the two Dirac cones are exchanged. Thus, taking the time reversal operator, $\mathcal{T}$ to be a complex conjugation operator should have been sufficient. However, as discussed earlier, the time reversal symmetry in graphene also implies a transformation from one valley to another, that is $\mathbf{K}$ changing over to $\mathbf{K}'$. This makes us settle for,

$$\mathcal{T} = \tau_x\mathcal{K} \tag{4.83}$$

where τ_x is the x-component of the Pauli matrix. Note that here $\mathcal{T}^2 = 1$ as we are dealing with spinless fermions[8]. To check for the invariance of the Hamiltonian under the operation of $\mathcal{T}$, one needs to prove,

[8] For spin-full systems, $\mathcal{T}^2 = -1$.

$$\mathcal{T}\mathcal{H}_0(\mathbf{k})\mathcal{T}^{-1} = \hbar v_{\rm F}\ \tau_x\ (k_x\sigma_x\tau_z + k_y\sigma_y^*)\tau_x = \mathcal{H}_0(-\mathbf{k}). \tag{4.84}$$

The above relation ensures the invariance of the Dirac Hamiltonian under the time reversal operation.

It may also be mentioned that the above symmetries put together, that is, the product of the sublattice (or inversion), and the time reversal symmetries yield a further discrete symmetry, known as the charge conjugate symmetry, usually denoted by $\mathcal{C}$. It can be checked that $\mathcal{H}_0$ is invariant under the combination of these two symmetries.

To summarize, in the context of graphene, we have seen the emergence of three discrete symmetries, namely, the sublattice symmetry (or, the inversion symmetry, denoted by $\mathcal{P}$), the time reversal symmetry (denoted by $\mathcal{T}$), and finally, a combination of the two, that is, the charge-conjugation symmetry ($\mathcal{C}$). They indeed have different properties, such as, $\mathcal{P}$ is a unitary operator, and anticommutes with the Hamiltonian, $\mathcal{T}$ is an antiunitary operator which commutes with the Hamiltonian, while the charge-conjugation operator $\mathcal{C}$ is anti-unitary (since it is a combination of $\mathcal{P}$ and $\mathcal{T}$) which also anticommutes with the Hamiltonian.

Having discussed the fundamental symmetries of graphene, let us return to its prospects of being a topological insulator. In Haldane's own submission[9] there can be simple efforts to tweak the Hamiltonian to achieve topological properties. Thus the goal is to transform a sheet of graphene into a quantum Hall-like state with conducting edges and an insulating bulk. Further, the edge modes have a chiral character, which means the current is carried in opposite directions at the two edges. There could be two ways of doing this; either break the inversion symmetry, keeping the time reversal symmetry intact, or break the time reversal symmetry, retaining the inversion symmetry. In the following we show that, while the first option does not yield a topological phase, the second one indeed does. Nevertheless, we shall discuss both, which are, respectively, known as the Semenoff insulator (obtained via breaking the inversion symmetry), and a Haldane (or a Chern) insulator (obtained via breaking the time reversal symmetry).

4.5.3 Semenoff insulator

In order to break the inversion symmetry, consider a staggered onsite potential of the form,

$$\mathcal{H}' = \varepsilon_{\rm A}\sum_{\mathbf{r}_{\rm A}} c_{\rm A}^\dagger(\mathbf{r}_{\rm A})c_A(\mathbf{r}_{\rm A}) + \varepsilon_{\rm B}\sum_{\mathbf{r}_{\rm B}} c_{\rm B}^\dagger(\mathbf{r}_{\rm B})c_{\rm B}(\mathbf{r}_{\rm B}) \tag{4.85}$$

where $\varepsilon_{\rm A}$ and $\varepsilon_{\rm B}$ are onsite potentials at the sites A and B, respectively. For $\varepsilon_{\rm A} \neq \varepsilon_{\rm B}$, the inversion symmetry (or the sublattice symmetry) is broken as is the case for hexagonal boron nitride (hBN), where the sites occupied by the C atoms in graphene are occupied by boron (B) and nitrogen (N) at the A and B sublattices sites, thereby causing the onsite energies to be unequal. Thus, including a term that has equal and opposite magnitudes at the two sublattices, the low energy Hamiltonian becomes,

[9] See https://topocondmat.org/w4_haldane/haldane_model.html.

$$\mathcal{H}(\mathbf{q}) = \mathcal{H}_0(\mathbf{q}) + m_I\sigma_z \tag{4.86}$$

where the m_I term makes the massless Dirac particle massive. Here, $m_I = (\varepsilon_A - \varepsilon_B)/2$ can be called the Semenoff mass. $m_I = 0$ for $\varepsilon_A = \varepsilon_B$. Further, σ_z anticommutes with $\mathcal{H}_0(\mathbf{k})$, that is,

$$\{\sigma_z, \mathcal{H}_0\} = 0.$$

The spectrum is given by,

$$E(\mathbf{q}) = \pm\sqrt{\hbar^2 v_F^2 q^2 + m_I^2}. \tag{4.87}$$

In a compact notation, one may write it as,

$$E_\mu(q) = \mu\sqrt{\hbar^2 v_F^2 q^2 + m_I^2} \tag{4.88}$$

where $\mu = \pm 1$ and each sign refers to a valley index. The spectrum is plotted in figure 4.16. Spectral gaps open up of magnitude $2m_I$ at each of the Dirac points. This gap earns the name Semenoff insulator. However, the nature of the gap is a trivial one in the following sense. The gap vanishes as $m_I \to 0$. Besides the wavefunction plotted for a graphene nanoribbon of size $L_x L_y$[10]. There is no trace of the edge states being present. Besides, the Berry phase and the Chern number also vanish as we shall show below, thereby certifying the trivial nature of the energy gap in the spectrum. This eliminates the possibility of any topological properties of the model induced by the inclusion of m_I.

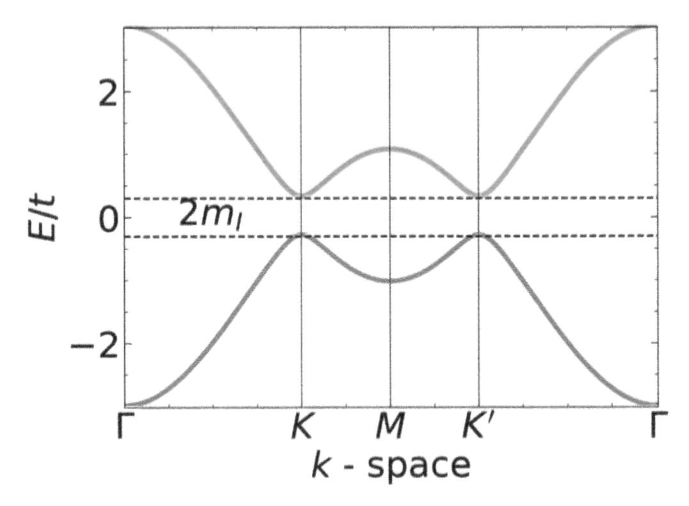

Figure 4.16. The energy band dispersion with a Semenoff mass, m_I. A gap of magnitude $2m_I$ opens up at the Dirac points.

[10] This is called a semi-infinite ribbon. It is finite in y-direction and very large (taken to be infinitely large) along the x-direction ($L_x \gg L_y$). We shall shortly discuss this below.

To gain a bit of detail on the Semenoff insulator, let us complete the mandatory calculations. The eigenfunctions can be written as,

$$\Psi^{\mu}(\mathbf{q}) = \frac{1}{\sqrt{2}}\begin{pmatrix} \sqrt{1 + m_I/E^{\mu}} \\ \mu\sqrt{1 - m_I/E^{\mu}}\, e^{i\theta_q} \end{pmatrix}. \tag{4.89}$$

The corresponding Berry curvature is,

$$\Omega^{\mu} = \frac{v_{\mathrm{F}}^2 m_I}{2\mu\left[v_{\mathrm{F}}^2 q_x^2 + v_{\mathrm{F}}^2 q_y^2 + \beta^2\right]^{\frac{3}{2}}} \tag{4.90}$$

which eventually gives the Berry connection as (see equation (4.73)),

$$\mathcal{A}^{\mu} = \frac{\tau_z}{2}\left(1 + \mu\frac{m_I}{|E^{\mu}|}\right)\frac{\hat{\theta}_q}{q}. \tag{4.91}$$

Finally, the Berry phase is obtained as,

$$\Phi_B = \pi\tau_z\left(1 + \lambda\frac{m_I}{|E^{\mu}|}\right).$$

Thus the Berry phase for a massless Dirac equation is thus renormalized by the Semenoff mass, m_I. One regains the corresponding result for graphene by putting, $m_I = 0$.

A further (and more robust) check on the trivial nature of the spectral gap can be achieved by computing the dispersion for a graphene nanoribbon. A nanoribbon is a system that is infinite along one direction (say, x-direction), and finite along the other direction (y-direction). Usually graphene ribbons are recognized by their edges along the x-axis, for example, with zigzag and armchair edges, and are referred to as zigzag graphene nanoribbon (abbreviated as ZGNR) and armchair graphene nanoribbon (AGNR). There is an important difference between the two. ZGNR is always metallic with gapless edge states, while AGNR is conditionally metallic in the following sense. AGNR has conducting edge states when $N = 3M - 1$, where N is the number of lattice sites in the y-direction, and M is an integer.

We have taken a ZGNR as shown in figure 4.18, with the total number of lattice sites along the y-axis as 256, that is, $N = 256$ (so number of unit cells is 128) along the y-direction and a width given by $(\frac{3N}{2} - 1))a$ (a: lattice constant = 1.42Å), which upon putting $N = 256$ yields $383a$ or 543.86Å. We finally write down the equation of motion, that is, solving the Schrödinger equation, $\mathcal{H}\psi = E\psi$ for the amplitudes at the A and B sublattice sites as below,

$$E_k a_{k,n} = -[t\{1 + e^{(-1)^n ik}\}b_{k,n} + tb_{k,n-1}] + m_I a_{k,n} \tag{4.92}$$

$$E_k b_{k,n} = -[t\{1 + e^{(-1)^{n+1} ik}\}a_{k,n} + ta_{k,n+1}] - m_I b_{k,n}. \tag{4.93}$$

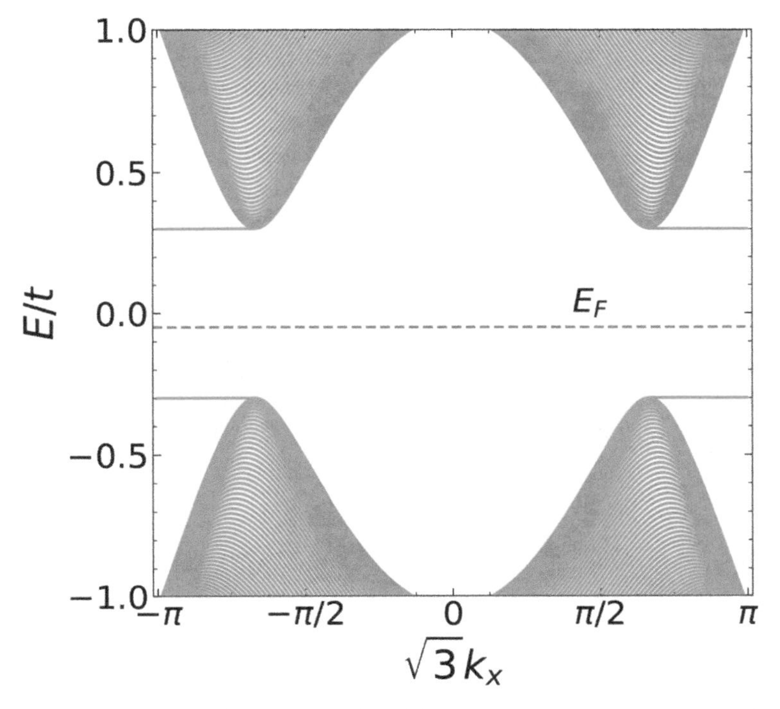

Figure 4.17. The energy dispersion for a Semenoff insulator in a semi-infinite nanoribbon. A (trivial) gap is visible in the spectrum.

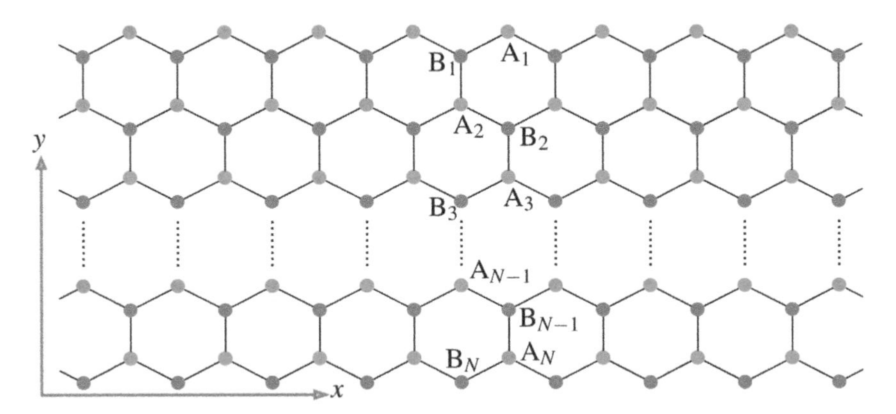

Figure 4.18. A schematic diagram of a semi-infinite nanoribbon is shown. We perform our numeric computation of the edge modes on a geometry such as this.

Along the x-direction, the ribbon is infinite, which is implemented in our numeric computation by assuming the momentum along the x-direction, that is, k_x to be a good quantum number. The above equations are numerically solved. We show the results in figure 4.17 which clearly show the absence of zero modes, which precludes its prospects as a candidate for a topological insulator.

4.5.4 Haldane (Chern) insulator

The second option of breaking the time reversal invariance is more subtle, and yields success in obtaining a topological state. The idea involves including an imaginary second neighbour hopping that assumes opposite signs depending on the direction of hopping. For example, if an anti-clockwise hopping (shown by the blue arrow in figure 4.19) is assumed with a positive sign, then the clockwise hopping (shown by red in figure 4.19) acquires a negative sign. A formal way of writing this term is via,

$$\mathcal{H}'' = t_2 \sum_{\langle\langle ij \rangle\rangle} e^{i\nu_{ij}\phi} c_i^{\dagger} c_j \tag{4.94}$$

where the sum runs over the next nearest neighbour (NNN) sites (double angular bracket $\langle\langle ij \rangle\rangle$ imply NNN sites). ν_{ij} denotes the chiral nature of the hopping term where $\nu_{ij} = -\nu_{ji}$ depending on the direction of the hopping. The convention is $\nu_{ij} = +1$ for clockwise hopping between the NNN sites, while $\nu_{ij} = -1$ for anti-clockwise hopping (see figure 4.19). The phase $e^{i\phi}$ or $e^{-i\phi}$ depending upon the direction of the hopping. Such a complex direction dependent hopping breaks the time reversal invariance, since the time reversal flips the direction of hopping. Only the imaginary part of the phase, ϕ, is interesting. Thus to set the real part to zero, we may choose $\phi = \frac{\pi}{2}$. This is known as the Haldane model [3], which Haldane had prescribed for achieving an anomalous quantum Hall state. It is anomalous in the sense that the Hall effect is realized without an external magnetic field, or equivalently, without the Landau levels. As we have seen earlier, and again shall see shortly, broken time reversal symmetry implies a finite Chern number, a reason why these insulators are known as Chern insulators. The complex phases can be realized by applying staggered magnetic fields pointing at opposite directions at the center of the honeycomb lattice relative to that at the vertices.

Introducing the NNN vectors, $\mathbf{b}_i$, as earlier, where, $\mathbf{b}_1 = \boldsymbol{\delta}_2 - \boldsymbol{\delta}_3$, $\mathbf{b}_2 = \boldsymbol{\delta}_3 - \boldsymbol{\delta}_1$, and $\mathbf{b}_3 = \boldsymbol{\delta}_1 - \boldsymbol{\delta}_2$ where $\boldsymbol{\delta}_i$ denote the vectors connecting NN sites, the Hamiltonian can be written as (the NN tight binding term is also there, but not written here),

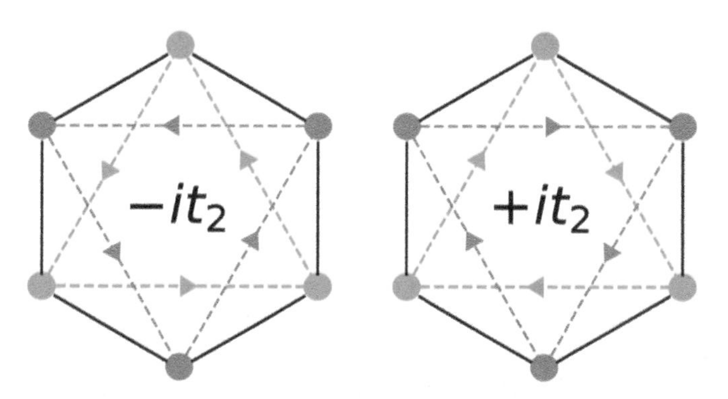

Figure 4.19. Complex next nearest neighbour hopping in the Haldane model.

$$\mathcal{H}'' = t_2 \sum_{i=1}^{3} \left[e^{i\phi} \sum_{\mathbf{r}_A} c_{\mathrm{A}}^{\dagger}(\mathbf{r}_{\mathrm{A}}) c_{\mathrm{A}}(\mathbf{r}_{\mathrm{A}} + \mathbf{b}_i) + e^{-i\phi} \sum_{\mathbf{r}_{\mathrm{B}}} c_{\mathrm{B}}^{\dagger}(\mathbf{r}_{\mathrm{B}}) c_{\mathrm{B}}(\mathbf{r}_{\mathrm{B}} + \mathbf{b}_i) \right]. \tag{4.95}$$

In the momentum space, the full tight binding Hamiltonian reads,

$$\mathcal{H}''(\mathbf{k}) = 2t_2 \left[\cos\phi \sum_{i=1}^{3} \cos(\mathbf{k} \cdot \mathbf{b}_i) \mathbb{1} + \sin\phi \sum_{i=1}^{3} \sin(\mathbf{k} \cdot \mathbf{b}_i) \sigma_z \right]. \tag{4.96}$$

Till this point, the full NNN Hamiltonian is dispersive, that is, it depends upon the $\mathbf{k}$-vector. However, the low energy Hamiltonian, that is, near the $\mathbf{K}$ and $\mathbf{K}'$ points, is independent of $\mathbf{k}$ at the leading order, where the Hamiltonian can be shown to assume the form,

$$\mathcal{H}''_{\pm\mathbf{K}} = m_H \; \tau_z \; \sigma_z \tag{4.97}$$

where we have combined the forms at the two Dirac points by using τ_z where $\tau_z = \pm 1$,

$$m_H = -3\sqrt{3} \; t_2 \sin\phi. \tag{4.98}$$

The above form can be easily obtained by noting that,

$$\sum_{i=1}^{3} \cos(\mathbf{k} \cdot \mathbf{b}_i) = -\frac{3}{2} \quad \text{and} \quad \sum_{i=1}^{3} \sin(\mathbf{k} \cdot \mathbf{b}_i) = \mp \frac{3\sqrt{3}}{2}$$

where $\mathbf{k} \cdot \mathbf{b}_i = \mathbf{K}$. The readers are encouraged to fill up a few steps of algebra.

Therefore, in the leading order $\mathcal{H}''$ is independent of the momentum $\mathbf{k}$. The last term in equation (4.97) breaks the time reversal symmetry. σ_z does not change sign, but τ_z being the valley degree of freedom does. Thus the energy spectrum opens up a gap at the Dirac points for specific values of the complex second neighbour hopping t_2. In fact adding a small t_2 yields a situation similar to adding a small Semenoff mass, m_I. However, when t_2 exceeds a value of $\pm m_H/3\sqrt{3}$, the energy closes at one of the two Dirac points (either $\mathbf{K}$ or $\mathbf{K}'$), and opens up at the other Dirac point for one of the signs mentioned above, say $t_2 = m_H/3\sqrt{3}$. The reverse happens for $t_2 = -m_H/3\sqrt{3}$ where the gap closes at the former Dirac point, while opening at the other. We show this in figure 4.20.

In order to elucidate the topological properties, we repeat the calculations in the same way as that of the Semenoff insulator. The eigenfunctions for the Haldane model can be written as,

$$\Psi^{\mu}(\mathbf{q}) = \frac{1}{\sqrt{2}} \begin{pmatrix} \sqrt{1 + \beta/E^{\mu}} \\ \mu\sqrt{1 - \beta/E^{\mu}} \, e^{i\theta_q} \end{pmatrix} \tag{4.99}$$

where $\mu = \pm 1$, with the energy spectrum is given by,

$$E^{\mu} = \mu\sqrt{v_{\mathrm{F}}^2 q_x^2 + v_{\mathrm{F}}^2 q_y^2 + \beta^2}.$$

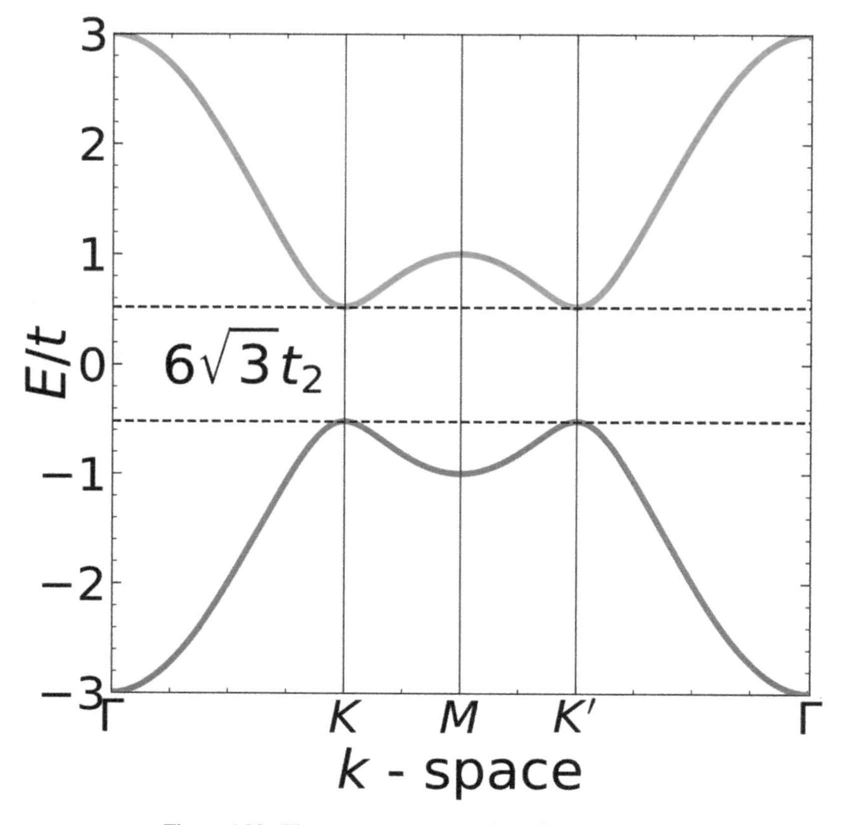

Figure 4.20. The energy spectrum of the Haldane model.

This further yields a Berry curvature which can be shown to have the form,

$$\Omega^{\mu} = \frac{v_F^2 \beta}{2\mu\left[v_F^2 q_x^2 + v_F^2 q_y^2 + \beta^2\right]^{\frac{3}{2}}}. \tag{4.100}$$

The corresponding Berry connection is hence given by,

$$\mathcal{A}^{\mu} = \frac{\tau_z}{2}\left(1 + \mu\frac{\beta}{|E|^{\mu}}\right)\frac{\hat{\theta}_q}{q} \tag{4.101}$$

where $\beta = 3\sqrt{3}\, t_2$. The Berry phase, Φ_B using,

$$\Phi_B = \int \mathcal{A}^{\mu}.\, dq$$

yields,

$$\Phi_B = \pi\tau_z\left(1 + \lambda\frac{\beta}{|\,E^{\mu}\,|}\right). \tag{4.102}$$

Finally, the Chern number can be obtained by integrating the Berry curvature over the BZ,

$$C = \oint \Omega^{\mu}(q)d^2q.$$

For the topological phase, that is, for $|t_2| > m_{\rm H}/3\sqrt{3}$, one obtains a non-zero Chern number. Owing to a non-zero Chern number (C), the model has earned the name 'Chern insulator'. The Chern number is the topological invariant which distinguishes the Semenoff insulator from a Chern insulator. For the Semenoff insulator, $C = 0$. We show the phase diagram in figure 4.21, where the topological phases are shown via the red ($C = 1$) and the blue ($C = -1$) colours, respectively, while the trivial region ($C = 0$) is shown via white colour.

How do we know that this gap is topological in nature, instead of a trivial one as seen for a Semenoff insulator? This is a valid question since the nature of the gaps looks fairly similar in figures 4.17 and 4.20, except that the spectral gaps carry the energy scales proportional to their *masses*, that is, m_I for the Semenoff insulator, and $m_{\rm H}$ for the Chern insulator. In the following we check for the chiral edge modes in a semi-infinite graphene nanoribbon.

In a similar fashion as discussed in the context of a Semenoff insulator, the equations of the motion for the amplitudes at the A and B sublattice sites can now be written as,

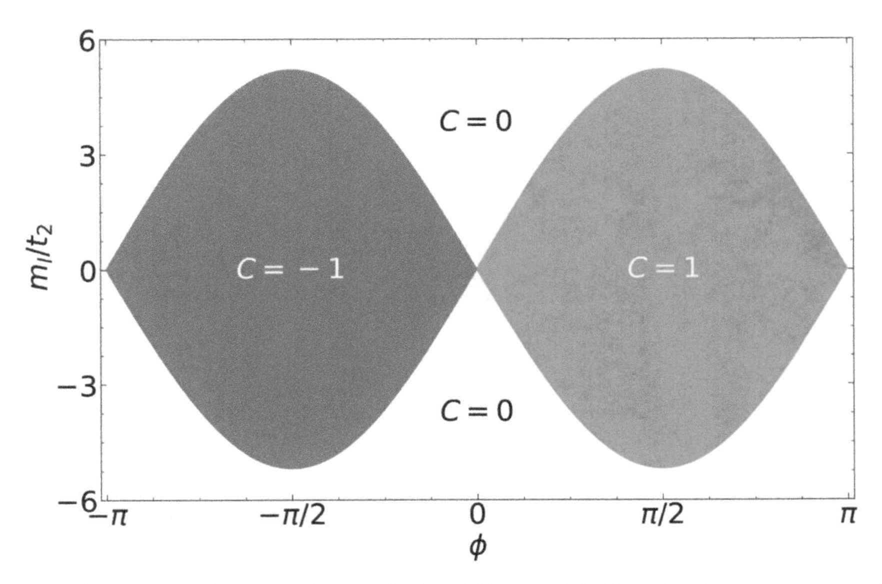

Figure 4.21. The Chern number phase diagram for the Haldane model. The red region corresponds to $C = 1$, while the blue one denotes $C = -1$. The white region outside the lobes refers to a trivial insulator with $C = 0$. Reprinted with permission from [3], Copyright (1988) by the American Physical Society

$$E_k a_{k,n} = -\left[t\{1 + e^{(-1)^n ik}\} b_{k,n} + t b_{k,n-1}\right] - 2t_2\left[\cos(k + \phi) a_{k,n} + e^{(-1)^n \frac{ik}{2}} \cos\left(\frac{k}{2} - \phi\right)\{a_{k,n-1} + a_{k,n+1}\}\right] \quad (4.103)$$

$$E_k b_{k,n} = -\left[t\{1 + e^{(-1)^{n+1} ik}\} a_{k,n} + t a_{k,n+1}\right] - 2t_2\left[\cos(k - \phi) b_{k,n} + e^{(-1)^{n+1} \frac{ik}{2}} \cos\left(\frac{k}{2} + \phi\right)\{a_{k,n-1} + a_{k,n+1}\}\right]. \quad (4.104)$$

In figure 4.22 we show the appearance of the edge modes as t_2 crosses $\pm m_H/3\sqrt{3}$ which are absent at small values of t_2. The appearance of the edge modes implies the

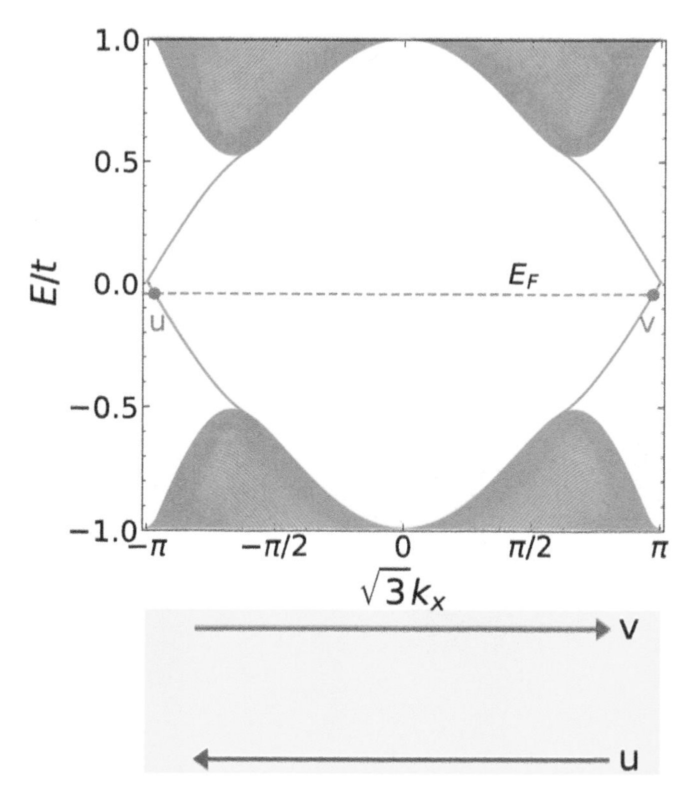

Figure 4.22. The energy dispersion for a Chern insulator in a semi-infinite nanoribbon. The edge states are shown via the red lines that split from the bulk. In the yellow panel below, we show the chiral edge currents that flow in opposite directions along these edge modes.

emergence of the topological phase in the model, and there occurs a phase transition from a topological insulating phase to that of a band insulator. Thus, we get a quantum Hall-like state, with conducting edge modes (and insulating bulk), albeit without an external magnetic field.

4.5.5 Quantum anomalous Hall effect

Finally, we shall present the results for the Hall conductivity. Here, a non-zero Berry curvature yields a finite conductance. To remind ourselves, the full tight binding Hamiltonian (including the NN term which we have excluded earlier in equation (4.96)) is written as [8],

$$\begin{aligned}\mathcal{H} = &-t\left[\cos(\mathbf{k}\cdot\boldsymbol{\delta}_1)+\sum_{i=2}^{3}\cos(\mathbf{k}\cdot\boldsymbol{\delta}_i)\right]\sigma_x - t\left[\sin(\mathbf{k}\cdot\boldsymbol{\delta}_1)+\sum_{i=2}^{3}\sin(\mathbf{k}\cdot\boldsymbol{\delta}_i)\right]\sigma_y \\ &+\left[\Delta - 2t_2\sin\phi\sum_{i=1}^{3}\sin(\mathbf{k}\cdot\boldsymbol{\nu}_i)\right]\sigma_z + \left[2t_2\cos\phi\sum_{i=1}^{3}\cos(\mathbf{k}\cdot\boldsymbol{\nu}_i)\right]I \\ = &\ h_x\sigma_x + h_y\sigma_y + h_z\sigma_z + h_0 I,\end{aligned} \tag{4.105}$$

where h_x, h_y and h_z represent the coefficients of the Pauli matrices σ_i. The low energy expansion of this Hamiltonian is convenient for our purpose. In fact, the computation of the Berry curvature is much easier for the low energy Hamiltonian, than it is for the full tight binding one. Arriving at the low energy Hamiltonian involves expanding the sine and the cosine functions to their leading order in the vicinity of the Dirac points. Applying these simplifications, one arrives at,

$$\mathcal{H} = \mathbf{d}\cdot\boldsymbol{\sigma} \tag{4.106}$$

where the components of the $\mathbf{d}$ differ from those of $\mathbf{h}$, and up to linear in k_x and k_y are given by,

$$d_x(k_x, k_y) = \frac{3}{2}k_x, \quad d_y(k_x, k_y) = \frac{3}{2}k_y, \quad \text{and} \quad d_z(k_x, k_y) = -3\sqrt{3}$$

The above form facilitates computation of the Hall conductivity using the following form for the Berry connection [9],

$$\Omega(E_k) = \frac{\mathbf{d}}{2|\mathbf{d}|^3}\left(\frac{\partial\mathbf{d}}{\partial k_x}\times\frac{\partial\mathbf{d}}{\partial k_y}\right). \tag{4.107}$$

Finally, the Hall conductivity is obtained via,

$$\sigma_{xy} = \frac{e^2}{h}\int\frac{d\mathbf{k}}{(2\pi)^2}f(E_k)\Omega(E_k) \tag{4.108}$$

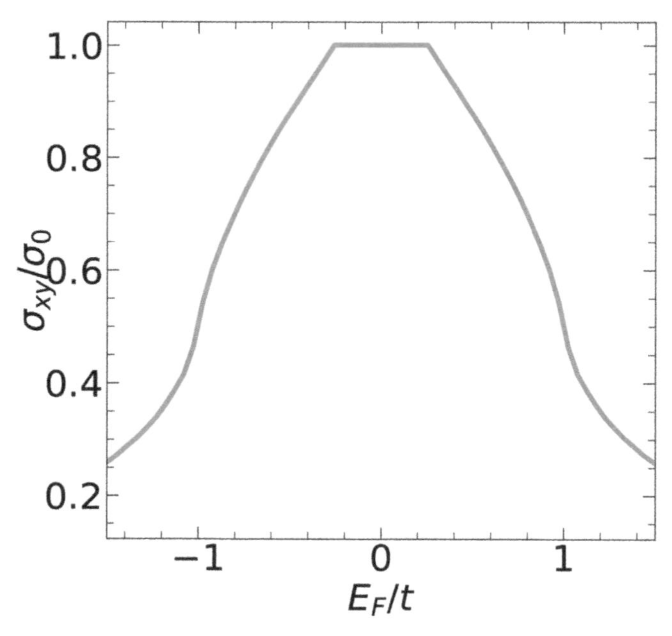

Figure 4.23. The anomalous Hall conductivity is shown as a function of the Fermi energy. There is a distinct plateau in the vicinity of the zero Fermi energy.

where the integral is taken over the BZ, and $f(E_k)$ is the Fermi distribution function. Since our calculations are at zero temperature, we set $f(E_k) = 1$. The Hall conductivity as a function of the Fermi energy is plotted in figure 4.23. A plateau at e^2/h is clearly visible which enunciates the quantization of the Hall conductivity. Further, the presence of only one plateau is confirmed by the value of the Chern number being 1 (or -1), and also that there is only a pair of gapless edge modes. Thus, an anomalous version of the Hall conductivity is indeed distinct from the usual Hall effect (in the presence of an external magnetic field). However, there is experimental realization of systems with higher values of the Chern number, besides being backed up by a library of theoretical proposals.

4.6 Quantum spin Hall insulator

Let us set aside the complex second neighbour hopping due to Haldane for a moment, the Dirac points in graphene are protected by time reversal and the inversion symmetries. The complex second neighbour hopping among sites of the same sublattice breaks the time reversal symmetry as we have seen in the preceding section. C L Kane and E G Mele in 2005 [4] demonstrated that it is possible to restore the time reversal symmetry in the Haldane model if we include (real) spin in the Hamiltonian, thereby making two copies of the Haldane model, one for each spin. Inclusion of the spin opens up the possibility of a spin–orbit coupling, which, however, does not violate any of the fundamental symmetries that we have discussed above. Moreover, the resulting insulating phase is absolutely new, and is referred to as the quantum spin Hall (QSH) phase. It should be clarified that spin–orbit coupling is not an essential ingredient for the realization of the QSH phase. However, to achieve a spin polarized transport in a material, which will aid its usage for spintronic applications, spin–orbit coupling is essential. We shall return to this discussion shortly.

Similar to the quantum Hall phase, the QSH phase is distinct from the trivial insulators by the presence of the conducting states at the edges which are typically protected by the $\mathcal{Z}_2$ topological invariant. However, these edge states are non-chiral, unlike the quantum Hall states. In fact, they are called helical edge states, in the sense that there are two counter propagating edge modes at each edge, one for each spin (see figure 4.24). Such conducting modes are immune to single particle backscattering from defects, disorder or impurities as they are protected by the time reversal symmetry. Thus, as long as there is no time reversal symmetry breaking term, such as a magnetic impurity, etc., the helical edge states are robust, and the QSH phase persists.

It was initially thought that graphene would host a QSH-like phase, however, it is almost impossible to realize such a phase owing to an extremely weak spin–orbit coupling [13]. However, a theoretical proposal of a QSH phase happened soon after when Bernevig, Hughes and Zhang [14] predicted that quantum wells made of CdTe/HgTe/CdTe host a QSH phase for a certain critical width of the HgTe layer where the band inversion occurs. The corresponding Hamiltonian is called the BHZ model (after Bernevig, Hughes, and Zhang). Quite fortunately, immediately afterwards Molenkamp and co-workers [15] experimentally achieved such a scenario where an

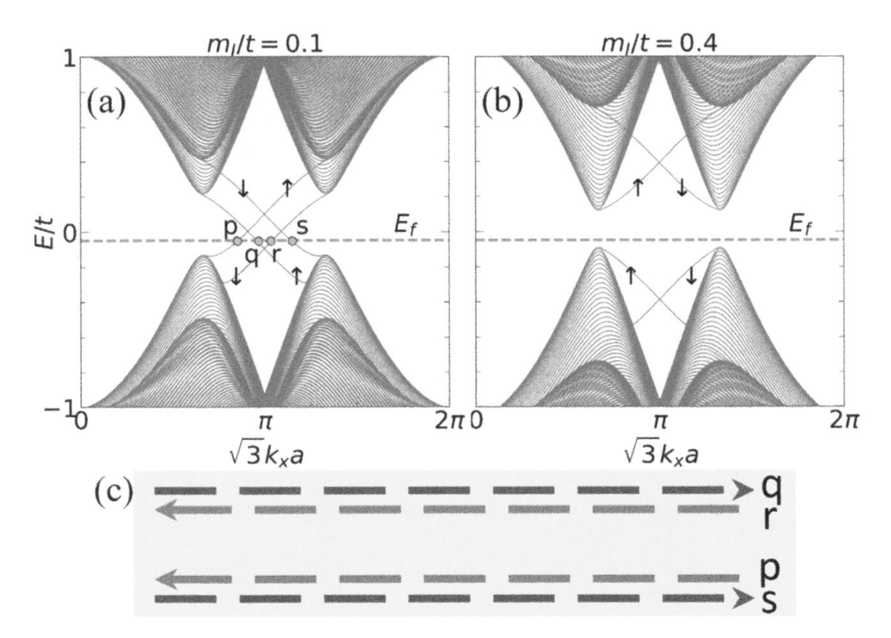

Figure 4.24. The helical edge modes for the Kane–Mele model on a nanoribbon showing (a) topological, and (b) trivial phases. The yellow panel in (c) shows the spin polarized helical modes carrying current.

inverted band structure occurs, followed by the realization of the helical (instead of chiral) edge states.

In the following we shall describe the Kane–Mele model for graphene which serves as a toy model for a QSH phase that hosts counter propagating edge modes one for each spin at each of the edges. Further, these modes are found to be robust in the presence of a special kind of spin–orbit coupling, known as the Rashba spin–orbit coupling (RSOC). Here, as we shall show below, owing to the restoration of the time reversal symmetry, the Chern number is zero. However, the helical edge modes are still protected by bulk $\mathbb{Z}_2$ topological invariant, which is a consequence of Kramer's theorem applicable for band properties of fermions in a time reversal invariant system.

4.6.1 Kane–Mele model

The Dirac points have been shown to be protected by the time reversal and the inversion symmetries. Breaking any one of them opens a gap at the Dirac points by splitting the degeneracy. Throughout our discussion on graphene thus far, the spin of the electron never played a role[11]. Kane and Mele included the spin, thereby writing two Haldane Hamiltonians, one for spin-↑ fermions, and the other for spin-↓ fermions. This results in,

$$\begin{aligned}\mathcal{H} &= -t\sum_{\langle ij\rangle,\alpha} c^{\dagger}_{i\alpha}c_{j\alpha} + it_2\sum_{\langle\langle ij\rangle\rangle,\alpha\beta} \nu_{ij}c^{\dagger}_{i\alpha}(S_z)_{\alpha\beta}c_{j\beta} \\ &= \mathcal{H}_0 + \mathcal{H}_{\rm KM}\end{aligned} \tag{4.109}$$

where $\mathcal{H}_0$ is the usual NN tight binding term, and $\mathcal{H}_{\rm KM}$ is like the Haldane term, now summed over the (real) spins. Traditionally, this term is called as the *intrinsic spin–orbit coupling.* S_z is the z-component of the spin of the electrons, and (α, β) denote the spin indices. S_z is indeed the z-component of the Pauli matrices $\boldsymbol{\sigma}$. However, to distinguish it from the sublattice and the valley indices we have written it with S_z. The NNN hopping term (the second term) describes a spin–orbit coupling that couples the chirality of the electrons, described by ν_{ij} with the z-component of spin ($S_z = \pm 1$). It is as if the orbital angular momentum vector $\mathbf{L}$ is associated with the chirality in a familiar $\mathbf{L}\cdot\mathbf{S}$ term, thus justifying its identification as the spin–orbit coupling.

The second term, even though it resembles the Haldane term, respects all the symmetries of graphene. Time reversal flips the direction of hopping, that is reversing the motion, but simultaneously it also flips the spin, thereby yielding another negative sign. This term respects all the symmetries of graphene. It may also be noted that the term does not involve spin flip, and hence the two bands of the Haldane model (one for each spin) behave distinctly. In a mathematical sense, it means that the Hamiltonian retains a block diagonal form, and hence is easy to deal with. The bands corresponding to the up-spin electrons are identical to the Haldane model (Chern insulators) discussed earlier. That is, they correspond to the phase of

[11] The Pauli matrices denote sublattice and valley degrees of freedom.

the complex NNN hopping to be $\phi = \pi/2$, and hence have opposite masses at the **K** and the **K**′ (remember the term $m_H\tau_z\sigma_z$ in the low energy limit of the Haldane model). Further, it has a Chern number $C_\uparrow = +1$. Please note that we have brought in a spin index to the Chern number. For the down-spin electrons for which $\phi = -\pi/2$, there will be an extra negative sign, which implies reversed signs for m_H at the **K** and the **K**′ points as compared to the situation for the up-spin. This yields $C_\downarrow = -1$. Thus the total Chern number, $\sum_\sigma C_\sigma = 0$, which is a consequence of the time reversal symmetry.

A simple way of seeing the Kane–Mele model being two copies of the Haldane model is that

$$[\mathcal{H}_{KM}, S_z] = 0 \tag{4.110}$$

which signifies that $\mathcal{H}_{KM}$ decouples into Hamiltonian one for each spin. The situation is akin to a Haldane flux $\phi = \frac{\pi}{2}$ for one type of spin, and $\phi = -\frac{\pi}{2}$ for the other. The low energy Kane-Male Hamiltonian can be shown to have a form (readers are encouraged to complete the derivation), written as,

$$\mathcal{H}_{KM} = m_H\sigma_z\tau_z S_z$$

where the amplitude $m_H = -3\sqrt{3}\ t_2$ is the Haldane mass as stated earlier.

Let us convince ourselves that time reversal is indeed a valid symmetry operation for the Kane–Mele model. For spinor particles we have seen that the time reversal operator $\mathcal{T}$ is written as,

$$\mathcal{T} = i\sigma_y K$$

where K is complex conjugation operator. Here we write it as,

$$\mathcal{T} = iS_y K.$$

However, since the time reversal transformation are one valley to another, an operator that can be implemented by incorporating a τ_x, we can write,

$$\mathcal{T} = \tau_x\ iS_y K.$$

It is fairly trivial to see that $\mathcal{H}_{KM}$ is even under time reversal. Please recall that the time reversal inflicts complex conjugation, flips the real spin, reverses the valley degree of freedom, and in addition, reverses the direction of the momentum $\mathbf{k} \to -\mathbf{k}$. While the last one is not relevant, since the low energy Hamiltonian is independent of $\mathbf{k}$, the first two yield under the time reversal,

$$\tau: \qquad \tau_z \to -\tau_z, \qquad S_z \to -S_z.$$

However, σ_z does not change sign as it denotes the sublattice degree of freedom. Hence, two negative signs cancel and we get $\mathcal{H}_{KM}$ to be even under $\mathcal{T}$.

To remind ourselves on the other fundamental symmetry, the inversion symmetry $\mathcal{P}$, which yields

$$\mathcal{P}: \qquad \sigma_z \to -\sigma_z \qquad \tau_z \to -\tau_z, \qquad S_z \to S_z.$$

Hence $\mathcal{H}_{KM}$ respects all symmetries of graphene as claimed earlier.

Let us look at the topological phase transition in a little more detail. For this, it is instructive to look at only one spin at a time, for example, $S_z = +1$ (that is, up-spin). The Hamiltonian including a Semenoff mass becomes,

$$\mathcal{H}_{KM}(\mathbf{k}) = \hbar v_{\mathrm{F}}\left(k_x\sigma_x\tau_z + k_y\sigma_y\right) + (m_I + m_H\tau_z)\sigma_z \tag{4.111}$$

Explicitly writing the above Hamiltonian for the two valleys,

$$\mathcal{H}_{KM}^{\mathbf{K}}(\mathbf{k}) = \hbar v_{\mathrm{F}}(k_x\sigma_x + k_y\sigma_y) + (m_I + m_H)\sigma_z \tag{4.112}$$

$$\mathcal{H}_{KM}^{\mathbf{K}'}(\mathbf{k}) = \hbar v_{\mathrm{F}}(-k_x\sigma_x + k_y\sigma_y) + (m_I - m_H)\sigma_z. \tag{4.113}$$

Now consider two possibilities: (i) $m_I > m_H$ and (ii) $m_H > m_I$. In the first case, consider the extreme limits (for convenience), that is $m_I \gg m_H$ where we have a trivial band insulator. Now consider the other case where $m_H > m_I$: Nothing happens to $\mathcal{H}_{KM}^{\mathbf{K}}(\mathbf{k})$ in equation (4.112), but for $\mathcal{H}_{KM}^{\mathbf{K}'}(\mathbf{k})$ in equation (4.113), the gap closes and reopens. Thus, the insulating phase with $m_H > m_I$ is distinct from that of a band insulator by a *gap closing* phase transition, which by definition is a topological phase transition.

The situation for $S_z = -1$ is identical, except that the sign of m_H changes, which results in a similar phase transition at the other Dirac point, that is at the **K** point. Now defining,

$$\tilde{m} = m_I - m_H \tag{4.114}$$

yields, corresponding to,

$$\begin{aligned} &\tilde{m} < 0, \quad C = 1 \\ \text{and} \quad &\tilde{m} > 0, \quad C = 0. \end{aligned} \tag{4.115}$$

4.7 Bulk-boundary correspondence

Bulk boundary correspondence (BBC) yields a guide to the phenomenology of topological insulators. The topological invariants computed from the bulk properties corresponding to a particular phase of the system uniquely reflect the conducting edge modes. Let us try to answer the question that we have posed above, that is, how is $\tilde{m}<0$ fundamentally different from $\tilde{m}>0$? Again consider a semi-infinite nanoribbon, that is, infinite in x-direction and finite in y-direction. The Schrödinger equation with the Hamiltonian given earlier can now be solved for a semi-infinite system as we have discussed earlier.

Before we discuss the numerical solution for a nanoribbon, let us explore an analytic solution. We can assume that the Hamiltonian has an edge at $y = 0$, so that the system exists for $y < 0$ and a vacuum for $y > 0$. In addition, let us assume a particular value of k_x, namely, $k_x = 0$ (remember k_x is a good quantum number owing to translational invariance in the x-direction). Hence, we can write down the Hamiltonian,

$$\mathcal{H}(y) = -iv_F\sigma_y\frac{\partial}{\partial y} + (m_I - m_H)\sigma_z \tag{4.116}$$

where, $\hbar = 1$ and $m_I - m_H = \tilde{m}(y)$. The RHS of equation (4.116) resembles a y-dependent potential energy in a 1D free Hamiltonian. Further, let us insist on,

$$\begin{aligned} \tilde{m}(y) < 0, &\qquad \text{for } y < 0 \\ \tilde{m}(y) > 0, &\qquad \text{for } y > 0. \end{aligned} \tag{4.117}$$

Thus, as if there is a physical boundary between topological and trivial states. Let's look at the zero energy solution.

Now make an ansatz for the y-dependent wavefunction (like variational wavefunction)

$$\psi(y) = i\sigma_y\ e^{f(y)}\ \phi \tag{4.118}$$

where ϕ is a 2-component spinor. Putting equation (4.118) in equation (4.116)

$$\left(iv_F\frac{df}{dy} + \tilde{m}(y)\sigma_x\right)\phi = 0 \qquad \text{(using} \quad \sigma_y\sigma_z = i\sigma_x). \tag{4.119}$$

The formal solution for $f(y)$ is obtained as,

$$f(y) = -\frac{1}{v_F}\int_0^y dy'\ \tilde{m}(y') \tag{4.120}$$

where ϕ is assumed to be eigenstate of σ_x with eigenvalue +1.

Also, the effect of $i\sigma_y = e^{i\frac{\pi}{2}\sigma_y}$ is to rotate by π around the y-axis.

$$\psi(y) = \exp\left(-\frac{1}{v_F}\int_0^y \tilde{m}(y')\ dy'\right)|\sigma_x = -1\rangle. \tag{4.121}$$

The $\exp(-\frac{1}{v_F}\int_0^y \tilde{m}(y')\ dy')$ factor allows it fall off at the inside of the sample. So $\psi(y)$ the edge is maximum at the edges. Also it is an eigenstate of σ_x as it has to mix the two sublattices by hopping along the boundary. For the other Dirac point, the state traverses in the other direction. At larger energies, $\epsilon(k_x) = -v_F k_x$, so that,

$$v_F(\text{or } v) = \frac{\partial\varepsilon(k_x)}{\partial x} = -v. \tag{4.122}$$

For $\tilde{m} \to -\tilde{m}$, we have an electron traversing in the opposite direction at the other cone.

Finally, we show the numeric computation of the edge modes in Kane–Mele nanoribbon by solving the following sets of equations.

$$
\begin{aligned}
E_k a_{k,n} = & [t\{1 + e^{(-1)^{n+1}ik}\}b_{k,n} + tb_{k,n+1}]s_0 + m_I a_{k,n} s_0 \\
& + 2t_2\left[a_{k,n}\sin k + e^{(-1)^{n+1}\frac{ik}{2}}\sin\frac{k}{2}\{a_{k,n-1} + a_{k,n+1}\}\right]s_z \\
& + i\lambda_R\left[\left\{-\frac{1}{2}(1 + e^{(-1)^{n+1}ik})b_{k,n} + b_{k,n+1}\right\}s_y\right. \\
& \left. - \left\{(-1)^n\frac{\sqrt{3}}{2}(1 - e^{(-1)^{n+1}ik})b_{k,n}\right\}s_x\right]
\end{aligned}
\tag{4.123}
$$

$$
\begin{aligned}
E_k b_{k,n} = & [t\{1 + e^{(-1)^n ik}\}a_{k,n} + ta_{k,n-1}]s_0 - m_I b_{k,n} s_0 \\
& + 2t_2\left[b_{k,n}\cos k + e^{(-1)^n\frac{ik}{2}}\cos\frac{k}{2}\{a_{k,n-1} + a_{k,n+1}\}\right]s_z \\
& + i\lambda_R\left[\left\{\frac{1}{2}(1 + e^{(-1)^n ik})b_{k,n} + b_{k,n-1}\right\}s_y\right. \\
& \left. - \left\{(-1)^{n+1}\frac{\sqrt{3}}{2}(1 - e^{(-1)^n ik})a_{k,n}\right\}s_x\right].
\end{aligned}
\tag{4.124}
$$

Figure 4.24(a) clearly shows existence of spin filtered edge modes in the topological phase, while they are absent in figure 4.24(b). The presence of the helical modes carrying spin resolved currents at each edge are shown in the yellow panel below (in figure 4.24(c)). Each of those conducting modes denotes a channel for each spin. Thus the model supports spin polarized conduction via the edge modes, while the bulk remains gapped. It must be kept in mind that the Chern number is identically equal to zero in this case owing to the time reversal symmetry being intact. However, the topological invariant here is the $\mathcal{Z}_2$ index which is non-zero.

4.8 Spin Hall conductivity

We have seen either that even though the individual conducting edge states have a non-zero Chern number, the total Chern number, C still vanishes owing to the time reversal symmetry being present. Thus the charge Hall conductivity vanishes, that is, $\sigma_{xy} = 0$. However, the spin Hall conductivity survives.

In order to calculate the spin Hall conductivity, let us rewind the Corbino disc argument due to Laughlin. When a quantum of flux Φ_0 is added to the inner edge of the disc, an electron is transferred from an inner to an outer edge of the disc. Say it happens for the up-spin leading to a e^2/h (charge) Hall conductivity. For the down-spin sector, in the presence of Φ_0, an electron is transferred backwards, that is, from the outer edge to the inner one. Including both the spins, the total Hall conductivity is zero, as demanded by the time reversal invariance. However, in the process, a net

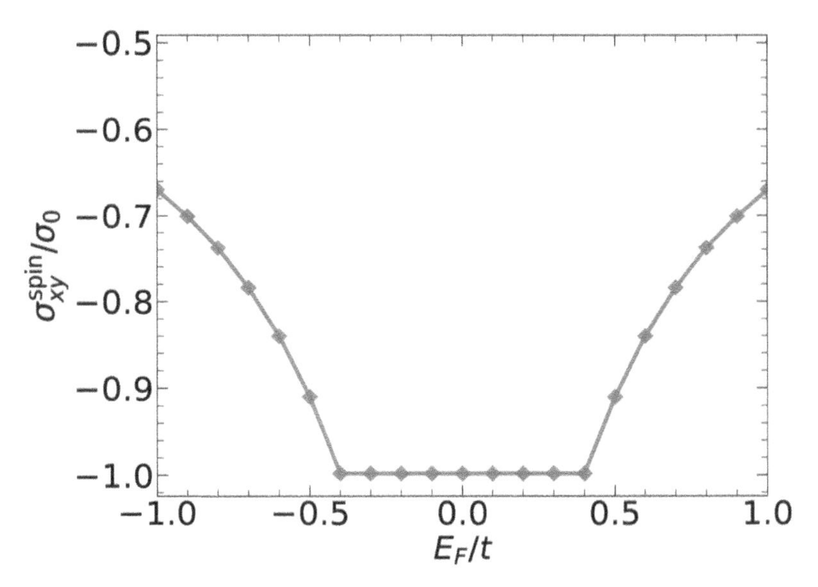

Figure 4.25. The spin Hall conductivity as a function of the Fermi energy is plotted. A quantization plateau in the vicinity of the zero Fermi energy is clearly visible.

spin is transferred from the inner to the outer edge. The corresponding spin Hall conductance is given by,

$$G_s = \frac{\hbar}{2e}\left(\frac{e^2}{h} + \frac{e^2}{h}\right) = \frac{e}{2\pi}. \tag{4.125}$$

Thus G_s is quantized in unit of $e/2\pi$. The spin Hall conductance as a function of the Fermi energy is presented for graphene in figure 4.25. There a plateau in the vicinity of the zero Fermi energy (zero bias), a signature of the quantized nature of the spin Hall conductivity, is visible.

There is a subtle point about the intrinsic spin–orbit coupling that deserves special mention, namely, the second term in equation (4.110) KM Hamiltonian commutes with the Hamiltonian, that is $[\mathcal{H}_{KM}, S_z] = 0$. However, usually SOC terms are spin non-conserving which means they can mix different spins. Thus in addition to the intrinsic SOC, other types of SOC can also be present. One such SOC is the Rashba spin–orbit coupling (RSOC), in which we shall see different spin components.

4.8.1 Rashba spin–orbit coupling

In solids, free (or nearly free) electrons do not feel the strong attraction of the nucleus of their host atoms. However, the electrons may still experience an electric field or a potential gradient due to internal effects. As we know, if the electrons experience a strong electric field of potential gradient, then there is the possibility of emerging of a spin–orbit coupling. So if a potential gradient exists across the interface due to the structural inversion asymmetry, there will be a spin–orbit coupling, and named after its discoverer, E I Rashba, that is, Rashba spin–orbit coupling (RSOC) [10]. The importance of the RSOC lies in the fact that asymmetry in the confinement potential can be varied by electrostatic means, allowing one to tune the RSOC strength by an external gate voltage. The strength of the RSOC also depends on the crystal structure in quantum wells, and is largest for narrow gap III–V semiconductors, such as InAs and InGaAs, etc. In the following subsection we shall describe the RSOC in a continuum model. Later on, we shall extend our discussion on graphene.

RSOC couples the wave vector with the spin degrees of freedom of the electrons. Further, it leads to orientation of spins which point perpendicular to the direction of the electron propagation wave vector. The free particle Hamiltonian including RSOC is described by,

$$\begin{aligned}\mathcal{H}_R &= -\boldsymbol{\mu}.\,\mathbf{B} = -\boldsymbol{\mu}.\,\frac{\mathbf{v}\times\mathbf{E}}{c^2}\\ &= \frac{eE}{mc^2}\mathbf{S}\cdot(\mathbf{v}\times\hat{z}) = \frac{eE\hbar^2}{8\pi^2m^2c^2}\boldsymbol{\sigma}.\,(\mathbf{k}\times\hat{z})\\ &= \alpha_R(\hat{z}\times\mathbf{k})\cdot\boldsymbol{\sigma}\end{aligned} \tag{4.126}$$

where $\alpha_R = \frac{eE\hbar^2}{8m^2\pi^2c^2}E$ is the strength of the RSOC, $\boldsymbol{\sigma}$ is a vector of Pauli spin matrices, $\mathbf{E} = -\nabla V$ is electric field along $\hat{z}$ direction. α_R can be tuned using an external gate voltage. In the absence of any Zeeman coupling, assuming elastic scattering and for $\hat{n} = \hat{z}$ (as per convention), the total Hamiltonian for the electron is given by,

$$\mathcal{H} = \frac{p^2}{2m} + \alpha(\mathbf{p}\times\boldsymbol{\sigma})\cdot\hat{z} = \frac{p^2}{2m} + \alpha(\sigma_x p_y - \sigma_y p_x). \tag{4.127}$$

This Hamiltonian yields the following energy spectrum,

$$E(k) = \frac{\hbar^2k^2}{2m} \pm \alpha\ \hbar|k| \tag{4.128}$$

where $|k|$ is the modulus of electron momentum with the plus and the minus signs denoting two possible spin directions. The associated wavefunctions are given by,

$$\Psi_{\pm}(x, y) = e^{i(k_x x + k_y y)}\frac{1}{\sqrt{2}}\begin{pmatrix}1\\ \pm\, ie^{-i\theta}\end{pmatrix} \tag{4.129}$$

where $\theta = tan^{-1}(k_y/k_x)$. It is easily understood that the spin states are always perpendicular to the direction of motion (equation (4.129)). If an electron moves along the x-direction, the spinor part of the eigenvector becomes $(1, \pm i)$, that is, the spin up and the spin down are locked in y-direction. By contrast, if the electron moves along the y-direction, the eigenvectors become $(1, \pm 1)$, that is, the spin up and the spin down states are constrained in the x direction (see figure 4.26). In figures 4.26(c)–(e), the energy spectrum as a function of momentum, k_y (keeping k_x constant) for a 2DEG are plotted corresponding to the following situations. Figure 4.26(c) is related to a free electron in 2DEG where the spin degeneracy is present. Figure 4.26(d) represents the energy spectrum for an electron in the presence

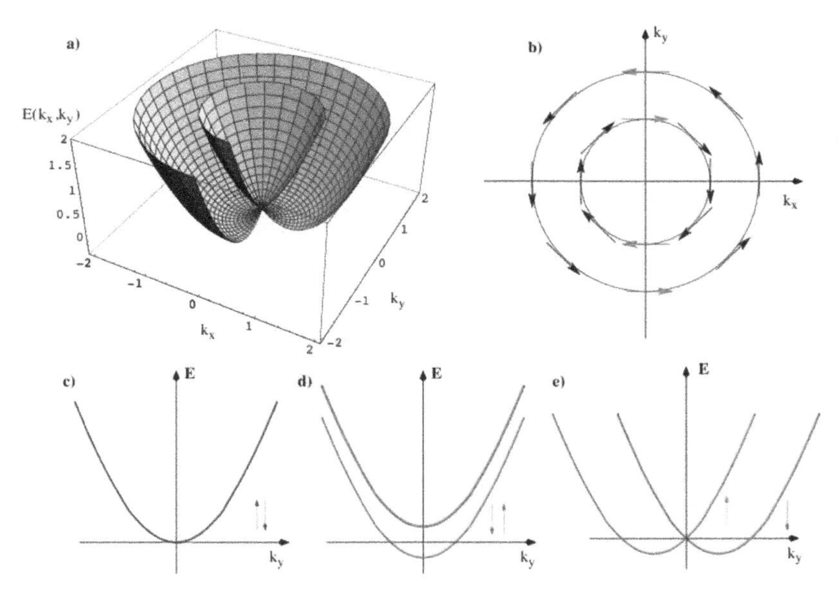

Figure 4.26. (a) 3D energy spectrum of the Hamiltonian $\mathcal{H}$ (equation (4.127)). (b) The Fermi energy contours for the Hamiltonian $\mathcal{H}$. (c) Energy spectrum for a free electron. (d) Energy spectrum for an electron in presence of a magnetic field (Zeeman splitting). (e) Energy spectrum for an electron in presence of Rashba spin–orbit coupling. Copyright Carlo Zucchetti.

of a magnetic field **B**, the spin degeneracy is lifted by the Zeeman splitting, and the gap separating spin up and spin down bands is equal to $g\mu_B B$ where g is the Bohr magneton. Figure 4.26(e) presents a 1D view of the energy spectrum for an electron in the presence of RSOC. The spin degeneracy is lifted up except for $k_y = 0$. In this situation, the degeneracy is removed without opening of any gap. At $k_y = 0$, the spin spectra are degenerate.

4.8.2 Rashba spin–orbit coupling in graphene

Writing the low energy Hamiltonian in the vicinity of the Dirac points for graphene

$$\mathcal{H}_R = \lambda_R(S_x q_y - S_y q_x) \tag{4.130}$$

which mixes up and down spins, which is why $[\mathcal{H}_R, S_z] \neq 0$. We prefer to call the strength as λ_R here, instead of α_R which was earlier used by us. The corresponding spectrum is given by,

$$E_{\gamma\delta}(q) = \mu\sqrt{q^2 + (m_H + \delta\lambda_R)^2} + \delta\lambda_R \tag{4.131}$$

where the indices $\mu = \pm 1$ and $\delta = \pm 1$ yield the conduction and the valence band spectra at the **K** and the **K**$'$ points. As we have already realized that the spectrum is gapped in the presence of the intrinsic spin–orbit coupling (namely, the Haldane term) itself, and now when λ_R is included, it will start competing with m_H when

$\delta = -1$. Also, with increasing λ_R, the energy gap decreases. At $m_H = \lambda_R$, the spectrum consists of a Dirac cone with two gapped parabolic bands.

For the sake of completeness, we reiterate that the strength of RSOC is too weak to yield any observable effects. For example, $\lambda_R \sim 10^{-3}$ K in graphene, while the kinetic energy is much larger. There are techniques to enhance RSOC by using heavier adatoms, using an external gate voltage or bending the graphene layer. The basic idea is to create a strong gradient of the electric potential. We shall not discuss this any further, and suggest more specialized reviews on the subject.

4.8.3 $\mathbb{Z}_2$ invariant

Since we shall include the Rashba SOC term in the KM Hamiltonian (see equation (4.109)), and that it respects all symmetries of graphene (for example, Chern number equal to zero), a new topological invariant has to emerge. *A priori*, it is the $\mathbb{Z}_2$ invariant that we are talking about, however, we refer to the topological classification by Altland and Zirnbauer in references [11, 12].

Let us discuss the topological invariant relevant here, namely, the $\mathbb{Z}_2$ index that characterizes the topological properties of the system. We shall only talk about an inversion symmetric system. For the calculation of the $\mathbb{Z}_2$ index, one may consider the Bloch wavefunctions, $u_i(\mathbf{k}_i)$ of the occupied bands corresponding to a pair of points $\mathbf{k}_1$ and $\mathbf{k}_2$ in the Brillouin zone. These two points denote the locations of the band extrema (minima for the conduction band and maximum for the valence band) in the BZ. The wavefunction at one of these points can be obtained by time reversing the wavefunction corresponding to the other one, that is, $|\, u_i(\mathbf{k}_1)\rangle = \mathcal{T}|\, u_i(\mathbf{k}_2)\rangle$, and vice versa where $\mathcal{T}$ denotes the time reversal operator. Since the Hamiltonian is time reversal invariant, so we can decompose the Hamiltonian, $\mathcal{H}(\mathbf{k})$ and its corresponding occupied band wavefunctions, $|u_i(\mathbf{k})\rangle$ into even and odd subspaces. The even subspace has the property that $\mathcal{T}|\, u_i(\mathbf{k})\rangle$ is equivalent to $|\, u_i(\mathbf{k})\rangle$ up to a $U(2)$ rotation. Whereas, the wavefunctions corresponding to the odd subspace has the property that the space spanned by $\mathcal{T}|\, u_i(\mathbf{k})\rangle$ is orthogonal to that of $|\, u_i(\mathbf{k})\rangle$. Now the $\mathbb{Z}_2$ invariant can be calculated by considering the momenta which belong to the odd subspace. We compute the expectation value of the time reversal operator between $|u_i(\mathbf{k})\rangle$ and $|u_j(\mathbf{k})\rangle$, namely, $\langle u_i(\mathbf{k})|\mathcal{T}|u_j(\mathbf{k})\rangle$. This yields a matrix which is antisymmetric. Hence we have,

$$\langle u_i(\mathbf{k})|\mathcal{T}|u_j(\mathbf{k})\rangle = \epsilon_{ij}P(\mathbf{k}), \tag{4.132}$$

where ϵ_{ij} is the Levi-Civita symbol and $P(\mathbf{k})$ is the Pfaffian of the matrix defined as,

$$P(\mathbf{k}) = \mathrm{Pf}\big[\langle u_i(\mathbf{k})|\Theta|u_j(\mathbf{k})\rangle\big]. \tag{4.133}$$

For a 2×2 antisymmetric matrix A_{ij}, the Pfaffian picks up the off-diagonal component. Now the absolute value of this Pfaffian is unity in the even subspace, while it is zero in the odd subspace. Hence, we dissect the BZ into two halves, such that the points $\mathbf{k}_1$ and $\mathbf{k}_2$ lie in different halves. Thus the $\mathbb{Z}_2$ index can be computed via performing the integral,

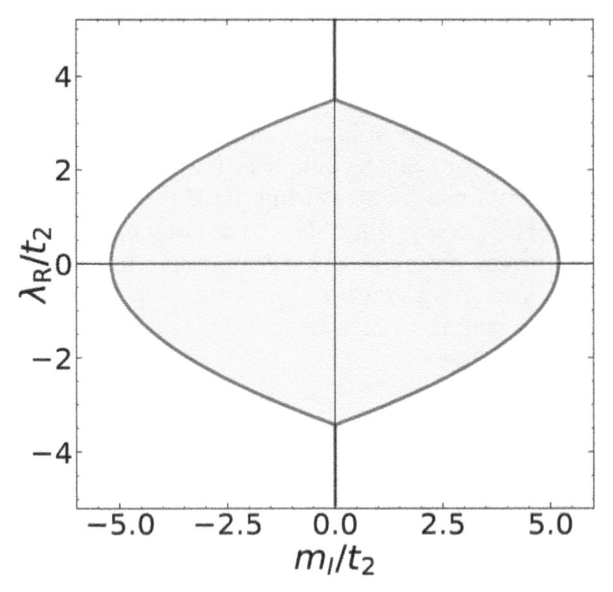

Figure 4.27. The phase diagram for the Kane–Mele model. The light blue region corresponds to $\mathbb{Z}_2 = 1$, which denotes a quantum spin Hall insulator. The white region outside the lobe refers to a trivial insulator with $\mathbb{Z}_2 = 0$.

$$\mathbb{Z}_2 = \frac{1}{2\pi i} \oint_{\mathrm{c}} \mathrm{d}\mathbf{k} \cdot \nabla \log \left(P(\mathbf{k}) + i\delta\right) \tag{4.134}$$

where δ is the convergence factor and the contour C is the circumference of the halved BZ discussed above.

The variation of the $\mathbb{Z}_2$ index is shown in the parameter plane defined by the Rashba coupling, λ_R and the Semenoff mass, m_I (both scaled by the NNN hopping t_2), which is shown to have a value 1 in the light blue region, and vanishes outside in figure 4.27. The region with non-zero $\mathbb{Z}_2$ invariant will host spin filtered chiral edge modes, and will denote a quantum spin Hall insulator, and the region outside, denotes a trivial insulator. Figure 4.27 denotes the phase diagram of a quantum spin Hall insulator, in the same spirit as figure 4.21 denotes the phase diagram for a Chern insulator. To remind readers, the Chern number vanishes here due to the present of the time reversal symmetry, thereby ruling out the possibility of a quantum Hall-like state. However, a new topological phase emerges, known as the QSH phase.

4.9 Spin Hall effect

To shed light on possible applications, we give a brief description of the spin hall effect (SHE), and cursorily on the subject of spintronics. The SHE is the generation of the spin current perpendicular to the applied charge current. It leads to accumulation of spins of opposite kinds at the edges of the sample. The spin selection can be facilitated by a strong spin–orbit coupling (SOC). Strong SOC may be intrinsic to doped semiconductors. The proposal has triggered intense investigation of the phenomenon, and the lead has been taken by the first observation of SHE in n doped semiconductors [16, 17], and 2D hole gases [18]. Both the experiments directly measure the spin accumulation induced at the edges of the sample through different optical techniques. However, more quantitative and accurate estimates are obtained by measuring the Hall angle[12]. An excellent review on the family of the SHEs, comprising SHE (discussed briefly above), inverse SHE in which a pure spin current generates a charge current, and even an anomalous Hall effect (AHE) in which a charge current generates a polarized transverse charge current in a ferromagnetic material, can be found in the review by J Sinova *et al* [19].

4.9.1 Spin current

Measuring the spin current is central to the study of SHE and hence to the emerging field of spintronics. The spin current has to be contrasted with the charge current. The charge current density, $\mathbf{j}_{el}(\mathbf{r}, t)$ is given by,

$$\mathbf{j}_{el}(\mathbf{r}, t) = \mathrm{Re}\,[\psi^{\dagger}(\mathbf{r}, t)(e\mathbf{v})\psi(\mathbf{r}, t)] \tag{4.135}$$

which further obeys a continuity equation of the form,

$$\frac{d\rho^{\mathrm{el}}}{dt} + \nabla.\,\mathbf{j}_{el} = 0 \tag{4.136}$$

where $\mathbf{v}$ is the velocity of the electrons (charges), and $\rho^{\mathrm{el}}(\mathbf{r}, t) = e\psi^{\dagger}(\mathbf{r}, t)\psi(\mathbf{r}, t)$ is the charge density. The continuity equation in equation (4.135) is the consequence of the invariance of charge. However, in the case of spin current, there is an ambiguity that arises from the fact that the spin is not an invariant quantity in spin transport owing to the presence of the spin–orbit coupling [20]. Usually the spin current is defined as $\langle \mathbf{v}.\,\mathbf{s}\rangle$ which is a non-conserved quantity. However, in the classical sense, just as the charge current density, the spin current density, $\mathbf{j}_s$ (the subscript s refers to the spin) can be written as,

$$\mathbf{j}_s = \mathrm{Re}\,[\psi^{\dagger}(\mathbf{r}, t)(\mathbf{v}.\,\mathbf{s})\psi(\mathbf{r}, t)]. \tag{4.137}$$

For a generic Hamiltonian with spin–orbit coupling, it can be shown that,

$$\mathrm{Re}(\psi^{\dagger}v_{\alpha}s_{\beta}\psi) = \mathrm{Re}(\psi^{\dagger}s_{\alpha}v_{\beta}\psi)$$

where α and β refer to the components in an orthogonal coordinate system. From $\mathbf{j}_s(\mathbf{r}, t)$ one can get the total spin current using,

$$\mathrm{I}_{s\alpha}(t) = \int dA\hat{\boldsymbol{\alpha}}.\,\mathbf{j}_s(\mathbf{r}, t) = \int dA\left[\psi^{\dagger}(\mathbf{r}, t)\frac{1}{2}(\mathbf{v}.\,\mathbf{s} + \mathbf{s}.\,\mathbf{v})\psi(\mathbf{r}, t)\right] \tag{4.138}$$

where dA is the elemental area and $\hat{\boldsymbol{\alpha}}$ denotes a certain direction (that is, $\hat{\boldsymbol{\alpha}} \in (\hat{x}, \hat{y}, \hat{z})$).

[12] Hall angle is the angle that the resultant of the applied electric field vector (E_x) and the Hall field (E_{H}) makes with E_x.

This clearly tells us that the spin current density operator, $\mathbf{j}_s$ is an anticommutator of $\{s_\alpha, v_\beta\}$ multiplied by a factor of 1/2. In terms of the Pauli matrices,

$$j_s^{\alpha\beta} = \frac{1}{4}\{s_\alpha, v_\beta\} \qquad (\hbar = 1). \tag{4.139}$$

Quite strikingly, unlike the charge current which is odd under time reversal, the spin current is invariant under time reversal operation. However, Ohm's law ($\mathbf{j} = \sigma\mathbf{E}$) holds for both $\mathbf{j}_{el}$ and $\mathbf{j}_s$. Since the electric field, $\mathbf{E}$ is even under time reversal, the charge conductivity, σ_{el} is odd, while the spin conductivity, σ_s is even.

For concreteness, let us specialize in a particular case, where we choose $\alpha = z$ and $\beta = y$. The y-component of the velocity is obtained from the Hamilton's equation of motion,

$$v_y = \frac{\partial \mathcal{H}}{\partial p_y}. \tag{4.140}$$

Considering a Hamiltonian for a 2DEG with Rashba spin–orbit coupling,

$$\mathcal{H} = \frac{p^2}{2m} - \lambda_R \quad \boldsymbol{\sigma}.\,(\hat{z} \times \boldsymbol{p}) \qquad \hbar = 1 \tag{4.141}$$

which yields,

$$v_y = \frac{p_y}{m} + \lambda_R \sigma_x. \tag{4.142}$$

Finally, the spin current density assumes a form,

$$j_s^{yz} = \frac{1}{2m}\sigma_z p_y. \tag{4.143}$$

Just as an electric current induces a magnetic field, a pure spin current induces an electric field. The magnetic moment due to the spin of the electron generates a current [21]. This can be understood as follows. Since the motion of a single magnetic moment is equivalent to an electric dipole which creates an electric field in its vicinity, there will be an electric field due to the motion of a magnetic moment. An estimation of the electric field can be made as follows. Consider two equal and opposite magnetic charges $\pm q_m$ separated by a small distance d moving in opposite directions. Such '*moving*' magnetic dipoles whose magnetic moments are given by, $\mathbf{m} = (q_m d)\hat{r}$ ($\hat{r}$ denotes the polarization direction) constitute a spin current. Each member of the group, that is a single magnetic moment will generate a magnetic field. This magnetic field in turn generates an electric field, which is given by,

$$\mathbf{E} \sim \frac{\mu}{4\pi}\int dV j_s \times \frac{1}{R^3}\left(\hat{r} - \frac{3\mathbf{R}(\mathbf{R}\cdot\hat{r})}{R^2}\right) \tag{4.144}$$

where dV is an elemental volume. This electric field is quite tiny in magnitude, yet can produce measurable effects [22].

Just as a current carrying wire experiences a force in a magnetic field ($\sim\mathbf{j} \times \mathbf{B}$), a spin current experiences a force $\sim\mathbf{j}_s \times \mathbf{E}$. Despite being small, it is able to control the motion of a spin, including zitterbewegung (jittery motion) of the Dirac electrons. Thus in semiconductors, where the spin–orbit coupling can be fairly strong, or can even be enhanced by external means, electrons with opposite spins are deflected along the opposite edges of the sample. Thus a spin unpolarized (paramagnetic) system can yield a pure spin current perpendicular to the direction of the electric field.

Over the last decade and a half, studies concerning the spin current and its application to spintronics in terms of efficiently generating, manipulating and detecting the spin accumulation phenomena have received a plethora of attention. Some progress has also occurred from the device fabrication perspective, via techniques, such as, spin injection, etc. A major advantage in dealing with the spin current lies in the non-dissipative (or much less dissipation) nature which arises owing to the time reversal invariance of the spin current. This property is in direct contrast with that of the charge current. A simple way to understand the role of time reversal invariance in the phenomenon of dissipation that can be understood with the aid of a damped harmonic oscillator, whose Hamiltonian may be written as,

$$\mathcal{H} = \frac{p^2}{2m} + \frac{1}{2}kx^2 + \alpha\dot{x}$$

where the $\alpha\dot{x}$ denotes the damping and breaks the time reversal symmetry. Without this term, the time reversal invariance holds and the scenario is non-dissipative. Thus a time reversal invariant system presents a non-dissipative scenario, which is precisely the main advantage for the spin transport phenomena.

4.9.2 Summary and outlook

Quantum Hall states are the first examples of topological insulators which demonstrate completely contrasting electronic behaviour between the bulk and the edges of the sample. The bulk of the system is insulating, while there exist conducting states at the edges. Moreover, the Hall conductivity is quantized in units of a universal constant, e^2/h. It became clear later on that the quantization is actually related to a topological invariant known as the Chern number. The geometric interpretation of this invariant is provided by the Gauss–Bonnet theorem, which relates the integral of the Gaussian curvature over a closed surface to a constant which simply counts the number of 'genus' (or holes) of the object. In solid state physics, the closed surface is the Brillouin zone and the Gaussian curvature is analogous to a quantity known as the Berry curvature. This brings us to the study of Berry-o-logy where the topological invariants are defined in terms of the Berry phase (a geometric phase picked by a particle during a complete revolution), Berry connection (analogous to the vector potential in electrodynamics), and the Berry curvature (similar to a magnetic field). The machinery is applied to a simple case, in both one and two dimensions, such as, an SSH model and graphene. The topology in the SSH model is induced by a dimerized hopping with two atoms per unit cell, and is stabilized by the chiral symmetry of the Hamiltonian. As long as the chiral symmetry is intact, and when the intra-cell hopping amplitude is larger than the inter-cell one, the model displays localized zero modes at the edges. The topological phase is further characterized by a finite value of the winding number. Further, the symmetry aspects, such as the inversion symmetry (parity), time reversal symmetry, etc are discussed with a view to explore topological properties of graphene. The bulk-boundary correspondence and the existence of the edge states in a nanoribbon geometry are investigated which serve as an acid test for the topological state. Eventually, following Haldane's conjecture, a topological state emerges by breaking the time reversal symmetry where the system acquires a topological gap at the Dirac points. The presence of such a non-trivial gap is confirmed via the presence of chiral edge states in a ribbon. Further progress is reported in terms of proposal of a scenario in which the broken time reversal symmetry is revoked using two copies of the Haldane model for each kind of spin of the carriers. This is called the Kane–Mele model which respects all the symmetries that graphene has. Yet there is an important difference which can be brought about by adding the Rashba spin–orbit coupling, which may be weak, but inherent to the 2D systems. The Rashba term leaves the time reversal symmetry intact. Thus the Kane–Mele model in the presence of the Rashba spin–orbit contribution yields yet another distinct topological state of matter, namely the quantum spin Hall phase. Owing to the time reversal symmetry being intact, the Chern number vanishes, although the system is characterized by a new topological invariant, known as the $\mathcal{Z}_2$ invariant which yields the spin Hall conductivity to be non-zero. The prospects of manipulating the spin degree of freedom give birth to an emerging field known as spintronics. With the fact that the spin current obeys time reversal symmetry, one gets a non-dissipative transport and this holds prospects of transmitting information with no (or very little) decay. Thus the spin transport mechanism does not have any associated Joule heating phenomena.

References

[1] Perelman G 2006 *Science* **314** 1848
[2] Thouless D J, Kohmoto M, Nightingale M P and den Nijs M 1982 *Phys. Rev. Lett.* **49** 405
[3] Haldane F D M 1988 *Phys. Rev. Lett.* **61** 2015
[4] Kane C L and Mele E J 2005 *Phys. Rev. Lett.* **95** 226801
Kane C L and Mele E J 2005 *Phys. Rev. Lett.* **95** 146802
[5] Kittel C 1986 *Introduction to Solid State Physics* (New York: Wiley)
[6] Berry M V 1984 *Proc. R. Soc.* A **392** 45–57
[7] Zak J 1988 *Phys. Rev. Lett.* **62** 2747
[8] Mondal S and Basu S 2021 *J. Phys.: Condens. Matter.J. Phys.: Condens. Matter.* **33** 225504
[9] Xiao D, Chang M-C and Niu Q 2010 *Rev. Mod. Phys.* **82** 1959
[10] Bychkov Y A and Rashba E I 1984 *Sov. Phys. - JETP Lett.* **39** 78
[11] Altland A and Zirnbauer M R 1997 *Phys. Rev.* B **55** 1142
[12] Ryu S *et al* 2010 *New J. Phys.* **12** 065010
[13] Min H, Hill J E, Sinitsyn N A, Saha B R, Kleinman L and McDonald A H 2006 *Phys. Rev.* B **74** 165310

[14] Bernevig B A, Hughes T L and Zhang S C 2006 *Science* **314** 1757

[15] König M *et al* 2007 *Science* **318** 766

[16] Kato Y K, Myers R C, Gossard A C and Awschalom D D 2004 *Science* **304** 1910

[17] Kato Y K, Myers R C, Gossard A C and Awschalom D D 2004 *Phys. Rev. Lett.* **93** 176601

[18] Wunderlich J, Kaestner B, Sinova J and Jungwirth T 2005 *Phys. Rev. Lett.* **94** 047204

[19] Sinova J, Valenzuela S O, Wunderlich J, Back C H and Jungwirth T 2015 *Rev. Mod. Phys.* **87** 1213

[20] Sun Q-F and Xie X C 2005 *Phys. Rev.* B **72** 245305

[21] Fisher G P 1971 *Am. J. Phys.* **39** 1528

[22] Shen S-Q 2008 *AAPPS Bull.* **18** 29

IOP Publishing

Condensed Matter Physics: A Modern Perspective

Saurabh Basu

Chapter 5

Green's functions

5.1 Introduction

In this chapter, we shall discuss the method of second quantization, a technique that formulates the study of many-particle systems. The formulation relies on the algebra of the creation and the annihilation operators. The two remarkable features of this technique are: (i) it provides a compact way of representing the many-body state, and (ii) the commutation and the anticommutation relations of the operators, respectively, for bosons and fermions, encode the symmetry properties of the wavefunction, and importantly, Pauli's exclusion principle for the fermions. In fact, the formalism of second quantization is considered as the first cornerstone for the development of quantum field theory. Here, we shall provide a minimal amount of mathematical details to familiarize the readers with the technique that will enable them to apply these in writing model Hamiltonians.

We begin with the notations that are traditionally used for developing the formalism. Hence we introduce the Fock basis, which simplifies the description of a many-particle state. We discuss the development of the technique of writing one and two-particle operators (quadratic and quartic in terms of the creation and the annihilation operators) which, respectively, denote the kinetic and the potential (interaction) energies. Hence we introduce different representations in quantum mechanics, such as the Schrödinger, Heisenberg and the interaction representations which aid us in writing down the Green's function that enriches our understanding of interacting many-body systems. The zero temperature Green's function, and hence the finite temperature Green's function are introduced afterwards, along with discussing the Feynman diagrams which are useful in understanding scattering processes in many-particle systems.

5.2 Second quantization

Consider a normalized set of wavefunctions $|\lambda\rangle$ of a certain single-particle Hamiltonian, $\mathcal{H}$, such that,

doi:10.1088/978-0-7503-3031-2ch5

$$\mathcal{H}|\lambda\rangle = \epsilon_\lambda|\lambda\rangle \tag{5.1}$$

where ϵ_λ denotes the eigenvalues of $\mathcal{H}$. Extending beyond the single-particle case, the normalized two-particle wavefunction for fermions and bosons, denoted, respectively, by ψ_F and ψ_B are given by the anti-symmetrized and symmetrized products as shown below.

$$\begin{aligned}\psi_F(x_1, x_2) &= \frac{1}{\sqrt{2}}\left[\phi_1(x_1)\phi_2(x_2) - \phi_1(x_2)\phi_2(x_1)\right]\\ \psi_B(x_1, x_2) &= \frac{1}{\sqrt{2}}\left[\phi_1(x_1)\phi_1(x_2) + \phi_1(x_2)\phi_2(x_1)\right]\end{aligned} \tag{5.2}$$

x_1 and x_2 denote the positions of the particles. In Dirac's notation, the above can be written in a compact fashion as,

$$|\lambda_1, \lambda_2\rangle_{F(B)} = \frac{1}{\sqrt{2}}[|\lambda_1\rangle \otimes |\lambda_2\rangle + \mathrm{P}|\lambda_2\rangle \otimes |\lambda_1\rangle]$$

where $|\lambda\rangle$s denote the single-particle states and $\mathrm{P} = -1$ for fermions and $\mathrm{P} = +1$ for bosons.

For a system of three particles, we can write the wavefunction for the fermions as,

$$\begin{aligned}\psi_F(x_1, x_2, x_3) = \frac{1}{\sqrt{6}}[&\psi_1(x_1)\psi_2(x_2)\psi_3(x_3) + \psi_1(x_2)\psi_2(x_3)\psi_3(x_1) + \psi_1(x_3)\psi_2(x_1)\psi_3(x_2)\\ &- \psi_1(x_2)\psi_2(x_1)\psi_3(x_3) - \psi_1(x_3)\psi_2(x_2)\psi_3(x_1) - \psi_1(x_1)\psi_2(x_3)\psi_3(x_2)]\end{aligned}$$

where an even number of swaps yields $\rightarrow +1$, while an odd number of swaps yields a factor $\rightarrow -1$. However, for bosons, such swaps do not yield any sign. For example, the bosonic wavefunction is given by,

$$\begin{aligned}\psi_B(x_1, x_2, x_3) = \frac{1}{\sqrt{6}}[&\psi_1(x_1)\psi_2(x_2)\psi_3(x_3) + \psi_1(x_2)\psi_2(x_3)\psi_3(x_1) + \psi_1(x_3)\psi_2(x_1)\psi_3(x_2)\\ &+ \psi_1(x_2)\psi_2(x_1)\psi_3(x_3) + \psi_1(x_3)\psi_2(x_2)\psi_3(x_1) + \psi_1(x_1)\psi_2(x_3)\psi_3(x_2)].\end{aligned}$$

Generalizing the above state for a system of N indistinguishable particles, one can write the N-particle state as [1],

$$|\lambda_1, \lambda_2 \ldots \lambda_N\rangle = \frac{1}{\sqrt{N! \prod_{\lambda=0}^{\infty} n_\lambda!}} \sum_k P^{(1-\mathrm{sgn}\ k)/2} |\lambda_{k_1}\rangle \otimes |\lambda_{k_2}\rangle \otimes \ldots |\lambda_{k_n}\rangle \tag{5.3}$$

where n_λ represents the total number of particles in state λ (for fermions Pauli's exclusion principle demands $n_\lambda \leqslant 1$; however, no such restriction exists for bosons). For an even number of permutations k takes a value $+1$, and for odd, k takes a value -1. The summation runs over all $N!$ permutations of the set of quantum numbers (that characterize the system) $\{\lambda_1 \ldots \lambda_N\}$. sgn k denotes the sign of the number of permutations k done via swapping the positions of the particles. Thus sgn k denotes a value 1 (-1) if the number of transpositions of two elements which brings the permutation $(k_1, k_2 \ldots k_n)$ back to its original configuration $(1, 2, \ldots N)$ is even (odd).

The pre-factor $\frac{1}{\sqrt{N! \prod_{\lambda=0}^{\infty} n_\lambda!}}$ normalizes the many-body wavefunction. In the fermionic case, this is known as the Slater determinant. Further, $\text{sgn}\, k = +1$ for an even number of permutations and $\text{sgn}\, k = -1$ for an odd number of permutations. It is also important to realize that the quantum numbers $\{\lambda_i\}$ defining the state $|\lambda_1\lambda_2\ldots\lambda_N\rangle$ are ordered. Suppose we label our quantum states by integers, then $\lambda_i = 1, 2, 3\ldots$. Simply speaking, any three-component initially non-ordered state, e.g. $|2, 1, 3\rangle$ can be brought into an ordered form $|1, 2, 3\rangle$ at the cost of, at most, a sign change.

5.2.1 Fock basis

We are now ready to formulate a second quantized formulation for a many-body system. Suppose a quantum state is represented by $|1, 1, 1, 2, 2, 2, 3, 3, 4, 5, 6\ldots\rangle$ [2]. This implies that 3 quantum states containing 1 and 2 particles each, and 2 quantum states containing 3 particles each, etc. Obviously this notation contains a description which has a lot of redundancy and can be avoided in the following manner. A more efficient encoding of the state can be written as, $|3, 3, 2, 1, 1, 1\ldots\rangle$. The number i in this new representation signals how many particles occupy a particular state which are arranged in ascending order of occupancy. For fermions, the occupation number takes values either 0 or 1. Thus $|n_1, n_2, \ldots\rangle$ defines an occupation number represented with $\sum_i n_i = N$. This space is known as Fock space (after Russian scientist V A Fock [3]) and denoted by $\mathcal{F}$. The many-body state is written as,

$$|\Psi\rangle = \sum_{n_1, n_2, \ldots} C_{n_1, n_2, \ldots} |n_1, n_2, \ldots\rangle \quad \text{with} \quad \sum n_i = N$$

where $C_{n_1, n_2, \ldots}$ denotes the coefficient. In a grand canonical sense, we can relax the constraint of a fixed number of particles, thus allowing a Hilbert space large enough to accommodate a state with an infinite number of particles. The entire Fock space can be represented by a direct sum,

$$\mathcal{F} \equiv \bigoplus_{N=0}^{\infty} \mathcal{F}^N \qquad \bigoplus : \text{direct sum.} \tag{5.4}$$

The direct sum includes $\mathcal{F}^0$ which is a vacuum and is denoted by $|0\rangle$. It is important to include this vacuum state in the family of basis states which will be clear in a while.

To obtain the basis of $\mathcal{F}$, we shall use our previous basis $\{|n_1, n_2, \ldots\rangle\}$ and drop the constraint $\sum_i n_i = N$. The occupation number representation is helpful but does not take into account the exchange of particles, and hence the formidable sum over the permutation group appearing in equation 5.3. Thus we have to symmetrize $\mathcal{O}(10^{23})$ number of particles. The formalism of second quantization introduced below elegantly removes this obstacle.

Specifically, we define a creation operator for our purpose which is denoted by,

$$a_i^\dagger |n_1 \ldots n_i \ldots\rangle = \sqrt{(n_i + 1)}\ \mathrm{P}^{s_i} |n_1 \ldots n_i + 1 \ldots\rangle \quad \text{with } s_i = \sum_{j=1}^{i-1} n_j, \tag{5.5}$$

where, again P = +1 for bosons and P = −1 for fermions. In a fermionic system, this n_i is termed as modulo 2[1].

Now P^{s_i} makes arrangement for the symmetrization of the many-particle state. Suppose we operate the many-particle state by $a_3^\dagger$.

$$a_3^\dagger|n_1, n_2, n_3, \ldots\rangle = \sqrt{n_3 + 1}\ P^{(n_1+n_2)}|n_1, n_2, n_3 + 1, \ldots\rangle.$$

Thus P^{s_i} symmetrizes the many-particle wavefunction. Let us take an example where we consider a fermionic state comprising $|n_1, n_2, n_3\rangle$. Now operate with $a_3^\dagger$,

$$a_3^\dagger|1\ 0\ 0\rangle = -|1\ 0\ 1\rangle.$$

Hence operate with $a_2^\dagger$,

$$a_2^\dagger a_3^\dagger|1\ 0\ 0\rangle = -\sqrt{1}\ P^1|1\ 1\ 1\rangle = |1\ 1\ 1\rangle$$
$$a_2^\dagger|1\ 0\ 0\rangle = \sqrt{1}\ P^1|1\ 1\ 0\rangle = -|1\ 1\ 0\rangle.$$

Next operate in reverse order[2],

$$a_3^\dagger a_2^\dagger|1\ 0\ 0\rangle = -a_3^\dagger|1\ 1\ 0\rangle = -\sqrt{1}\ P^{1+1}|1\ 1\ 1\rangle = -|1\ 1\ 1\rangle.$$

So one gets,

$$\underbrace{\{a_3^\dagger + a_3^\dagger a_2^\dagger\}}_{=0}|1\ 0\ 0\rangle = 0$$

where we have used the anticommutation relations, $\{a_3^\dagger, a_2^\dagger\} = 0$, $\{a_3, a_2\} = 0$, and $\{a_2, a_2^\dagger\} = 1$. In principle, an n-particle state can be written as,

$$|n_1, n_2, \ldots\rangle = \prod_i \frac{1}{\sqrt{n_i!}} (a_i^\dagger)^{n_i}\ |0\rangle. \tag{5.6}$$

This claims that the complicated permutation entanglement encoded in equation (5.3) can be generated in a straightforward application of a set of linear operators to a single reference state. Physically, an N-fold application of the operator $a^\dagger$ to the empty vacuum state (which is why we need the vacuum state in the Fock space) generates an N-particle state. For the fermionic case, $n_i = 0, 1$ encode complicated correlations between the different $a_i^\dagger$. We shall show this in the following.

Consider two operators $a_i^\dagger$ and $a_j^\dagger$ for $i \neq j$. We demand a commutation relation of the type,

$$(a_i^\dagger a_j^\dagger - P a_j^\dagger a_i^\dagger) = 0.$$

[1] Modulo operation computes the remainder after division of one number by another. Modulo 2 means dividing a number by 2 leaves a remainder which is 0 or 1. For example, 5 modulo 2 is 1.

[2] The original state $|1\ 0\ 0\rangle$ is operated, respectively, by $a_2^\dagger$ and $a_3^\dagger$.

This can be written as,

$$[a_i^\dagger, a_j^\dagger]_P = 0$$

where, $[A, B]_P = AB - PBA$, and $P = \pm 1$ denote commutator (+1) and anticommutator (−1). Also $a_i^{\dagger 2} = 0$ (nilpotent) for fermions, that is, double creation (or double annihilation) on any of the two fermionic state ($|1\rangle$ or $|0\rangle$) leads to a null state. If the indices 1 and 2 are interchanged, $a_2^\dagger a_1^\dagger |0\rangle$, as per the anticommutation relation, we should get a negative sign. Summarizing, we have found that the creation operators obey the commutation relation,

$$[a_i^\dagger, a_j^\dagger]_P = 0; \quad \forall_{i,j}.$$

Just as a creation operator transforms an n-component Fock space, $\mathcal{F}^N$ to $\mathcal{F}^{N+1}$, annihilation operators transform $\mathcal{F}^N$ to $\mathcal{F}^{N-1}$. Finally, in a compact notation, the commutation relations for both fermions and bosons can be written as,

$$[a_i, a_j^\dagger]_P = \delta_{ij} \qquad [a_i, a_j]_P = 0 \qquad [a_i^\dagger, a_j^\dagger]_P = 0.$$

The upshot of the above discussion is that the complicacies of the Fock space associated with the symmetrization of the wavefunction are taken care of by the simple commutation relations. This yields a remarkable simplification of a rather complicated problem.

For the sake of completeness, let us list out the commutation relations. For bosons, they are:

$$[b_n, b_m^\dagger] = \delta_{nm} \quad [b_n, b_m] = 0 = [b_n^\dagger, b_m^\dagger].$$

Similarly for fermions,

$$[c_n, c_m^\dagger]_- = \{c_n, c_m^\dagger\} = c_n c_m^\dagger + c_m^\dagger c_n = \delta_{nm}; \qquad \{c_n, c_m\} = \{c_n^\dagger, c_m^\dagger\} = 0$$
$$c_m|1\rangle_m = |0\rangle_m; \quad c_m^\dagger|1\rangle_m = 0; \; c_m|0\rangle_m = 0; \quad c_m^\dagger|0\rangle_m = |1\rangle_m.$$

Hence,

$$c_n c_m = \begin{cases} |1\rangle_m \\ |0\rangle_m \end{cases} = 0; \quad c_n^\dagger c_m^\dagger = \frac{|1\rangle_m}{|0\rangle_m} = 0.$$

With spin included, the commutation relations assume the form,

$$\{c_{n\sigma}, c_{m\sigma'}^\dagger\} = \delta_{nm}\delta_{\sigma\sigma'}; \quad \{c_{n\sigma}, c_{m\sigma'}\} = 0; \quad \{c_{n\sigma}^\dagger, c_{m\sigma'}^\dagger\} = 0.$$

This simply reiterates Pauli's exclusion principle. Another way to see this result is to consider the number operator for the state $N_m = c_m^\dagger c_m$. The square of it is $N_m^2 = c_m^\dagger c_m c_m^\dagger c_m$. Using the anticommutation relation, $c_m c_m^\dagger = 1 - c_m^\dagger c_m$.

$$N_m^2 = c_m^\dagger(1 - c_m^\dagger c_m)c_m; = c_m^\dagger c_m - c_m^\dagger c_m^\dagger c_m c_m = N_m$$
$$N^2 = \sum_{n=0}^{1} |n\rangle\langle n|n^2 = |0\rangle\langle 0|0^2 + |1\rangle\langle 1|1^2.$$

The only numbers which are equal to their own squares are 0 and 1. Thus $N_m = 0$ or 1. The anticommutator relations are built into the properties of a fermion, that is, no more than one particle is there in state $|m\rangle$.

5.2.2 Representation of a one-body operator in second quantized notation

Single-particle or one-body operators $\hat{O}_1$ acting on N-particle Hilbert space $\mathcal{F}^N$ generally take the form,

$$\hat{O}_1 = \sum_{i=1}^{N} \hat{O}_i \tag{5.7}$$

where $\hat{O}_i$ is a single-particle operator acting on the ith particle. A typical example is a kinetic energy operator,

$$\hat{T} = \sum_i \frac{\hat{p}_i^2}{2m}$$

where $\hat{p}_i$ is the momentum operator acting on the ith particle. Other examples include a one-body onsite potential, $\sum_i \hat{V}(x_i)$ or the spin operator $\sum_i s_i$, etc. Now we want to represent any such operator $\hat{O}_1$ in the representation comprising of the creation and the annihilation operators. It is convenient to represent the occupation number operator as,

$$\hat{n}_\lambda = a_\lambda^\dagger a_\lambda.$$

Such that,

$$\hat{n}_{\lambda_j}|n_{\lambda_1}, n_{\lambda_2}\ldots\rangle = n_{\lambda_j}|n_{\lambda_1}, n_{\lambda_2}\ldots\rangle$$

which simply counts the number of particles in state λ (λ is the relevant quantum number that describes a quantum state).

Let us now consider a one-body operator, $\hat{O}_1$ which is diagonal in the basis $|\lambda\rangle$ with,

$$\hat{O}_1 = \sum O_{\lambda_i}|\lambda_i\rangle\langle\lambda_i| \quad \text{and} \quad O_{\lambda_i} = \langle\lambda_i|\hat{O}_1|\lambda_i\rangle.$$

With this nomenclature, one finds for the matrix elements of $\hat{O}_1$,

$$\langle n_{\lambda_1'}, n_{\lambda_2'}\ldots|\hat{O}_1|n_{\lambda_1}, n_{\lambda_2}\ldots\rangle = \sum_i O_{\lambda_i} n_{\lambda_i}\langle n_{\lambda_1'}, n_{\lambda_2'}\ldots|n_{\lambda_1}, n_{\lambda_2}\ldots\rangle.$$

Since this equality holds for any set of states, one can infer the second quantized representation of the operator $\hat{O}_1$ as,

$$\hat{O}_1 = \sum_{\lambda=0}^{\infty} O_\lambda n_\lambda = \sum_{\lambda=0}^{\infty} \langle\lambda|\hat{O}|\lambda\rangle a_\lambda^\dagger a_\lambda$$

where a_λs are the single-particle operators[3] .

The above result is straightforward in the following sense that a one-body operator engages a single- particle state at a time, the others act just as spectators. In the representation of the number operators (diagonal representation), one simply counts the number of particles in a state λ and multiplies it by the corresponding eigenvalue of the one-body operator. One can generalize this from a diagonal basis to any basis to obtain a general result,

$$\hat{O}_1 = \sum_{\mu\nu} \langle\mu|\hat{O}|\nu\rangle a_\mu^\dagger a_\nu$$

where μ and ν are two quantum numbers that characterize the state.

As a specific example, take the spin operator. The matrix elements $(s_i)_{\alpha\alpha'}$ are the components of the spin $=\frac{1}{2}(\sigma_i)_{\alpha\alpha'}$, where α, α' are the two spin indices and σ_i are the Pauli matrices, namely,

$$\sigma_x = \begin{pmatrix} 0 & 1 \\ 1 & 0 \end{pmatrix} \quad \sigma_y = \begin{pmatrix} 0 & -i \\ i & 0 \end{pmatrix} \quad \sigma_z = \begin{pmatrix} 1 & 0 \\ 0 & -1 \end{pmatrix}.$$

The spin operator of a many-body system assumes the form,

$$\hat{S} = \sum a_{\lambda\alpha'}^\dagger \; S_{\alpha\alpha'} \; a_{\lambda\alpha}$$

where λ may denote an additional set of quantum numbers, for example, the lattice sites, etc.

When second quantized in the position representation, we can show that the one-body Hamiltonian for a free particle is given as a sum of kinetic and an onsite potential energy as,

$$\hat{\mathcal{H}} = \int d^d\mathbf{r} \; a^\dagger(\mathbf{r})\left[\frac{\hat{p}^2}{2m} + V(\mathbf{r})\right]a(\mathbf{r})$$

where $\mathbf{r}$ is a continuous coordinate variable, and hence the integral is used. The momentum operator is given by, $\hat{p} = -i\hbar\partial_{\mathbf{r}}$. Further, the local density operator at a site $\mathbf{r}$ is given by $\hat{\rho}(\mathbf{r}) = a^\dagger(\mathbf{r}) \; a(\mathbf{r})$.

5.2.3 Representation of a two-body operator

Two-body operators, $\hat{O}_2$ are needed to describe pair-wise interaction between the particles. Although pair interaction potentials are straightforwardly included in classical many-body theories, however, including them in many-body quantum mechanics is cumbersome owing to the indistinguishability of the particles. To begin with, let us consider particles subject to the symmetric two-body potential, that is,

$$V(\mathbf{r}_m, \mathbf{r}_n) = V(\mathbf{r}_n, \mathbf{r}_m)$$

[3] In a general sense, we are using the single-particle operators by a and $a^\dagger$, while for writing the Hamiltonians we have used c operators in most of the cases in this text.

between any two particles at positions $\mathbf{r}_m$ and $\mathbf{r}_n$. Our aim is to find an operator $\hat{V}$ in the second quantized form whose action on a many-body state yields the inter-particle interaction[4],

$$\hat{V}|\mathbf{r}_1, \mathbf{r}_2 \ldots \mathbf{r}_N\rangle = \sum_{n<m}^{N} V(\mathbf{r}, \mathbf{r}')|\mathbf{r}_1, \mathbf{r}_2, \ldots, \mathbf{r}, \mathbf{r}', \ldots \mathbf{r}_N\rangle$$
$$= \frac{1}{2}\sum_{n \neq m}^{N} V(\mathbf{r}, \mathbf{r}')|\mathbf{r}_1, \mathbf{r}_2, \ldots \mathbf{r}, \mathbf{r}', \ldots \mathbf{r}_N\rangle$$

where the interaction term $\hat{V}(\mathbf{r}, \mathbf{r}')$ is written as,

$$\hat{V}(\mathbf{r}, \mathbf{r}') = \frac{1}{2}\int d^d\mathbf{r} \int d^d\mathbf{r}' \; a^\dagger(\mathbf{r}) \; a^\dagger(\mathbf{r}') \; V(\mathbf{r}, \mathbf{r}') \; a(\mathbf{r}') \; a(\mathbf{r}).$$

The interaction term thus contains two creation and two annihilation operators. To remind ourselves, for the fermions,

$$[a_i, a_j^\dagger]_+ = \delta_{ij} = a_i a_j^\dagger + a_j^\dagger a_i = \delta_{ij}; \quad a_i^\dagger a_i = 1 - a_i a_i^\dagger; \quad a_j^\dagger a_i = \delta_{ij} - a_i a_j^\dagger.$$

The above term is interpreted as describing two-particle scattering events. We can write down the two-body scattering term as,

$$\hat{O}_2 = \sum_{\lambda\lambda'\mu\mu'} O_{\mu,\mu',\lambda,\lambda'} \; a_\mu^\dagger \; a_{\mu'}^\dagger \; a_\lambda \; a_{\lambda'} \qquad \text{where } O_{\mu,\mu',\lambda,\lambda'} = \langle \mu, \mu'|\hat{O}_2|\lambda, \lambda'\rangle.$$

Thus one particle in state μ scatters to state λ another one in μ' scatters to λ'. Each of the indices μ, μ', λ, λ' can assume all of their possible values. It could happen that, for $\mu = \lambda$, that is, the first particle does not scatter and only the second particle scatters to a different state $\mu' \to \lambda'$. See figure 5.1.

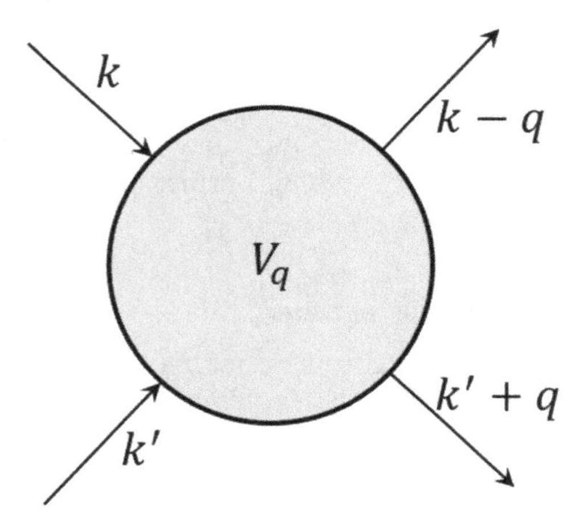

Figure 5.1. A typical interaction process where two particles of momenta $\mathbf{k}$ and $\mathbf{k}'$ scattered to $\mathbf{k} - q$ and $\mathbf{k} + q$ is shown schematically. $V_\mathbf{q}$ is the interaction vertex. We shall formally learn the diagrammatic technique in this chapter.

[4] Here we prefer to use a real space representation, instead of the occupation number representation.

An example of such a two-body interaction is a spin–spin interaction in a magnetic system, such as,

$$\hat{V} = \frac{1}{2}\int d^d\mathbf{r}\int d^d\mathbf{r}' \sum_{\alpha\alpha'\beta\beta'} J(\mathbf{r}, \mathbf{r}')\, \mathbf{S}_{\alpha\beta}\cdot\mathbf{S}_{\alpha'\beta'}$$

$J(\mathbf{r}, \mathbf{r}')$ denotes the exchange interaction. In principle one can proceed to write down an n-body interaction in second quantized from. However, interaction beyond two particles is rare (and mostly unsolvable), so we shall stick to two-particle interactions only.

5.2.4 Applications of the second quantized method

Let us take a prototype electronic system whose Hamiltonian can be written as,

$$\mathcal{H} = \mathcal{H}_0 + \hat{V}_{\text{e-e}}$$

$$\text{where}\quad \mathcal{H}_0 = \int d^d\mathbf{r}\; a_\sigma^\dagger(\mathbf{r})\left[\frac{\mathbf{p}^2}{2m} + V(\mathbf{r})\right]a_\sigma(\mathbf{r})$$

$$\text{and}\quad \hat{V}_{\text{e-e}} = \frac{1}{2}\int d^d\mathbf{r}\int d^d\mathbf{r}'\; V(\mathbf{r}-\mathbf{r}')\; a_\sigma^\dagger(\mathbf{r})\; a_{\sigma'}^\dagger(\mathbf{r}')\; a_{\sigma'}(\mathbf{r}')\; a_\sigma(\mathbf{r})$$

where $\sigma(\sigma') = \uparrow(\downarrow)$ denote the spin indices. This Hamiltonian correctly describes a wide variety of electronic systems, such as, metals, magnets, insulators, etc. To enumerate the physical properties of the model, it is helpful to consider a long-range (Coulomb) interaction in momentum space, namely,

$$\hat{V}_{\text{e-e}} = \frac{1}{2L^d}\sum_{k,k',q,\sigma,\sigma'} \hat{V}_{\text{e-e}}(\mathbf{q})\; a_{k-q,\sigma}^\dagger\; a_{k'+q,\sigma'}^\dagger\; a_{k',\sigma'}\; a_{k,\sigma}.$$

In the simplest form, that is, in its spin independent form, $V_{\text{e-e}} = \frac{e^2}{q^2}$[5] is the Fourier transform of the Coulomb interaction potential $\frac{e^2}{|r|}$. The interaction is diagrammatically represented by the diagram in figure 5.1. So two electrons with momentum $\mathbf{k}-\mathbf{q}$ and $\mathbf{k}'+\mathbf{q}$ are, respectively, scattered to $\mathbf{k}$ and $\mathbf{k}'$, so that the interaction vertex carries the momentum $\mathbf{q}$. In all of the condensed matter physics, one typically considers low excitation energies.

5.3 Green's function

Green's function was originally formulated to solve linear differential equations by G Green who developed a theory of partial differential equations with appropriate boundary conditions. He published his own work relatively recently in 2008 in the form of an essay [4].

[5] The reader should learn the Fourier transform of the Coulomb potential, namely, $V(r) = \frac{e^2}{r}$. The standard trick is to use a screening term, that is, $V(r) = \frac{e^2}{r}e^{-\lambda r}$ (λ is the screening parameter), and at the end of the Fourier transform put $\lambda \to 0$. Also see discussion in chapter 1.

In the classical version of the Green's function, the differential equation that one has to solve is,

$$Df(x) = g(x) \tag{5.8}$$

where D is a linear differential operator, and $f(x)$, $g(x)$ being functions of x. $f(x)$ can be obtained in terms of $g(x)$ with the aid of the Green's function $G(x, x')$ using,

$$f(x) = \int G(x, x')g(x')dx' \tag{5.9}$$

where the Green's function obeys,

$$DG(x, x') = \delta(x - x'). \tag{5.10}$$

A familiar example in electrodynamics is Poisson's equation for the electrostatic potential, that is,

$$\nabla^2 V(\mathbf{r}) = -\frac{\rho(\mathbf{r})}{\epsilon_0} \tag{5.11}$$

where $\rho(\mathbf{r})$ is the charge density, and the potential $V(\mathbf{r})$ obeys the boundary condition, $V(\mathbf{r}) \to 0$ for $|\mathbf{r}| \to \infty$. $V(\mathbf{r})$ can be obtained using the Green's function via,

$$V(\mathbf{r}) = -\frac{1}{\epsilon_0}\int G(\mathbf{r}, \mathbf{r}')\rho(\mathbf{r}')d\mathbf{r}' \tag{5.12}$$

where $G(\mathbf{r}, \mathbf{r}')$ satisfies,

$$\nabla^2 G(\mathbf{r}, \mathbf{r}') = \delta(\mathbf{r} - \mathbf{r}'). \tag{5.13}$$

Thus the solution of the differential equation (equation (5.11)) is converted into an integral equation, that is, equation (5.12)).

For a moment let us shift our focus to finding out the Green's function for a known problem in electrostatics, that is, say a unit point charge at $\mathbf{r}'$ for which the electric potential is,

$$V(|\mathbf{r}|) = \frac{1}{4\pi\epsilon_0|\mathbf{r} - \mathbf{r}'|} \tag{5.14}$$

where the source point is $\mathbf{r}'$ and the field is measured at the point $\mathbf{r}$. The corresponding Green's function is given by,

$$G(\mathbf{r}, \mathbf{r}') - \frac{1}{|\mathbf{r} - \mathbf{r}'|} \tag{5.15}$$

It is easy to see that this form of the Green's function obeys the boundary condition for $V(|\mathbf{r}|)$ via,

$$G(\mathbf{r}, \mathbf{r}') \to 0; \quad \text{as} \quad |\mathbf{r} - \mathbf{r}'| \to \infty.$$

There are several excellent texts on the subject and readers are encouraged to look at some of them [5].

5.3.1 Green's function for a single particle

In quantum mechanics the behaviour of particles is governed by the time-dependent Schrödinger equation,

$$-\frac{\hbar^2}{2m}\nabla^2\psi(\mathbf{r}, t) + V(\mathbf{r}, t)\psi(\mathbf{r}, t) = i\hbar\frac{\partial\psi(\mathbf{r}, t)}{\partial t} \tag{5.16}$$

where, with all generality, we have assumed the potential to be dependent on both space and time variables. One can equivalently solve the Green's function for the following '*inhomogeneous*' problem, namely,

$$\left[i\hbar\frac{\partial}{\partial t} + \frac{\hbar^2}{2m}\nabla^2\right]G(\mathbf{r}, t, \mathbf{r}', t') = \delta(\mathbf{r} - \mathbf{r}')\delta(t - t'). \tag{5.17}$$

The wavefunction is obtained using,

$$\psi(\mathbf{r}, t) = \int G(\mathbf{r}, t, \mathbf{r}', t')\psi(\mathbf{r}', t')d^3\mathbf{r}'. \tag{5.18}$$

Thus the Green's function propagates the wavefunction from one space–time point $(\mathbf{r}, t)$ to another $(\mathbf{r}', t')$. Hence, it is the inner product of the wavefunctions at these two points, that is,

$$G(\mathbf{r}, t, \mathbf{r}', t') = \langle\psi(\mathbf{r}, t)|\psi(\mathbf{r}', t')\rangle \tag{5.19}$$

where $\psi(\mathbf{r}, t) = e^{-i\mathcal{H}t}\psi(\mathbf{r})$ ($\hbar = 1$, $\mathcal{H}$ being the Hamiltonian of the system). In the Dirac notation, the same can be written as,

$$G(\mathbf{r}, t, \mathbf{r}', t') = \langle\mathbf{r}|\mathbf{r}'\rangle e^{-i\mathcal{H}t}. \tag{5.20}$$

Introducing the completeness of basis states,

$$\begin{aligned} G(\mathbf{r}, t, \mathbf{r}', t') &= \sum_{\mathbf{n}}\langle\mathbf{r}|\mathbf{n}\rangle\langle\mathbf{n}|\mathbf{r}'\rangle e^{-i\mathcal{H}t} \\ &= \sum_{\mathbf{n}}\phi_{\mathbf{n}}^*(\mathbf{r})\phi_{\mathbf{n}}(\mathbf{r}')e^{-iE_n(t-t')} \end{aligned} \tag{5.21}$$

where $\phi_{\mathbf{n}}(\mathbf{r}) = \langle\mathbf{r}|\mathbf{n}\rangle$ denotes the basis states. Writing the Green's function in terms of energy, E yields,

$$G(\mathbf{r}, \mathbf{r}', E) = i\sum_{\mathbf{n}}\frac{\phi_{\mathbf{n}}^*(\mathbf{r})\phi_{\mathbf{n}}(\mathbf{r}')}{E - E_{\mathbf{n}}}. \tag{5.22}$$

Thus the Green's function has poles at the eigenenergies of the system.

5.3.2 Green's function for a many-particle system

In an interacting many-particle system, the Green's function or the propagator not only yields the energy spectrum, but also carries information about the nature of inter-particle interactions. In a typical thought experiment, imagine a particle is created (or introduced) at an initial time t_0 in the ground state of the many-particle system. Eventually the particle is annihilated (or retracted) from the system at later time t, thereby facilitating it to come back to the ground state. The main purpose is to probe the effect of the interaction present in the system via the eigenenergies and the eigenfunctions. We shall calculate the Green's function that will track the propagation of the particle by computing the correlation of the creation ($c^\dagger$) of the particle at time t_0 and annihilation (c) at time t taken between the many-particle ground state.

5.3.3 Representations in quantum mechanics

Before we proceed with the calculation of the Green's function for an interacting many-body system, there are a few simple concepts that should be repeated for readers. These are about representations in quantum mechanics which are usually taught at the undergraduate level to discuss how a system evolves with time via the time evolution of wavefunction and the operators. In this regard, one learns the so called Schrödinger and the Heisenberg representations; however, a third one, known as the interaction representation is most helpful for our subsequent discussions on the Green's function.

Our aim is to solve a many-particle Hamiltonian which can be written in the form $\mathcal{H} = \mathcal{H}_0 + \mathcal{H}'$. More often than not, even for the single-particle problem, the Hamiltonian cannot be solved exactly. Very importantly (and in general true), $\mathcal{H}_0$ and $\mathcal{H}'$ do not commute. Assuming that $\mathcal{H}$ is not an explicit function of time, the behaviour of the wavefunction and the operators can be written in the following manner.

We begin with the Schrödinger representation. The Schrödinger equation can be written as,

$$i\hbar \frac{\partial \psi(t)}{\partial t} = \mathcal{H}\psi(t). \tag{5.23}$$

So that the solution for the wavefunction can be written as,

$$\psi(t) = e^{-iEt}\psi(0). \tag{5.24}$$

Here the wavefunction is time dependent through its usual time dependence, while the Hamiltonian and other operators are time independent. We have put $\hbar = 1$ and will carry over the same in subsequent discussions.

In the Heisenberg representation, the wavefunctions are independent of time, while the operators are time dependent. One can write the time evolution of the operators as,

$$\hat{O}(t) = e^{i\mathcal{H}t}\hat{O}(0)e^{-i\mathcal{H}t} \tag{5.25}$$

and the corresponding equation of motion (EOM) is,

$$i\frac{d\hat{O}(t)}{dt} = [\hat{O}(t), \mathcal{H}]. \tag{5.26}$$

However, the physical observables, namely, the expectation values of operators, etc will remain invariant. This can be seen via computing the matrix elements of an operator $\hat{O}(0)$ between any two arbitrary states. Let us first calculate the $\langle \hat{O} \rangle$ in the Schrödinger representation, which can be written as,

$$\langle \psi_1(t)|\hat{O}(0)|\psi_2(t)\rangle = \langle \psi_1(0)|e^{i\mathcal{H}t}\hat{O}(0)e^{-i\mathcal{H}t}|\psi_2(0)\rangle. \tag{5.27}$$

Also in the Heisenberg representation, we should obtain the same result, that is,

$$\langle \psi_1(0)|\hat{O}(t)|\psi_2(0)\rangle = \langle \psi_1(0)|e^{i\mathcal{H}t}\hat{O}(0)e^{-i\mathcal{H}t}|\psi_2(0)\rangle. \tag{5.28}$$

Now let us look at the above scenario in the interaction representation where both the operators and the wavefunction depend on time. The operators have a time dependence via,

$$\hat{O}(t) = e^{i\mathcal{H}_0 t}\hat{O}e^{-i\mathcal{H}_0 t}. \tag{5.29}$$

While for the wavefunction, the time dependence can be denoted by,

$$\psi(t) = e^{i\mathcal{H}_0 t}e^{-i\mathcal{H}t}\psi(0). \tag{5.30}$$

As said earlier, it is assumed that $[\mathcal{H}_0, \mathcal{H}'] \neq 0$. Remember the exponentials $e^{i\mathcal{H}_0 t}$ and $e^{-i\mathcal{H}t}$ cannot be combined because, $e^A e^B = e^{A+B}$ only if $[A, B] = 0$, where A and B can be $\mathcal{H}_0$ and $\mathcal{H}$.

We shall now show the time dependence of the operators. Of course, the interaction representation preserves the expectation values of operators. In the following, we show that the time dependence of the wavefunction is governed by the interaction term, $\mathcal{H}'$.

$$\begin{aligned}\frac{\partial\psi(t)}{\partial t} &= ie^{i\mathcal{H}_0 t}(\mathcal{H}_0 - \mathcal{H})e^{-i\mathcal{H}t}\psi(0)\\ &= -\, ie^{i\mathcal{H}_0 t}\mathcal{H}'e^{-i\mathcal{H}t}\psi(0)\\ &= -\, ie^{i\mathcal{H}_0 t}\mathcal{H}'e^{-i\mathcal{H}_0 t}(e^{i\mathcal{H}_0 t}e^{-i\mathcal{H}t}\psi(0)).\end{aligned} \tag{5.31}$$

Thus,

$$\frac{\partial\psi(t)}{\partial t} = -i\mathcal{H}'(t)\psi(t).$$

Now let us introduce an operator,

$$U(t) = e^{i\mathcal{H}_0 t}e^{-iHt} \quad \text{with} \quad U(0) = 1. \tag{5.32}$$

The EOM for $U(t)$ is,

$$\begin{aligned}\frac{\partial U(t)}{\partial t} &= -ie^{i\mathcal{H}_0 t}(\mathcal{H}_0 - \mathcal{H})e^{-i\mathcal{H}t} = -ie^{i\mathcal{H}_0 t}\mathcal{H}'(e^{-i\mathcal{H}_0 t}e^{i\mathcal{H}_0 t})e^{-i\mathcal{H}t} \\ &= -i\mathcal{H}'(t)U(t).\end{aligned} \tag{5.33}$$

Thus the EOMs for $\psi(t)$ and $U(t)$ are identical. The solutions have to be identical as well. The solution for $U(t)$ can be written as,

$$\begin{aligned}U(t) - U(0) &= -i\int_0^t \mathcal{H}'(t_1)U(t_1)dt_1 \\ U(t) &= 1 - i\int_0^t \mathcal{H}'(t_1)U(t_1)dt_1.\end{aligned} \tag{5.34}$$

Again the same information is required on the RHS, that is inside the integrand, which necessitates for an iterative solution of the form,

$$\begin{aligned}U(t) &= 1 - i\int_0^t H'(t_1)dt_1 + (-i)^2\int_0^t dt_1\int_0^{t_1} \mathcal{H}'(t_1)\mathcal{H}'(t_2)dt_2 + \cdots \\ &= \sum_{n=0}^{\infty}(-i)^n\int_0^t dt_1\int_0^{t_1} dt_2\cdots\int_0^{t_{n-1}} dt_n\mathcal{H}'(t_1)\mathcal{H}'(t_2)\cdots\mathcal{H}'(t_n).\end{aligned} \tag{5.35}$$

It is now convenient to introduce the time ordering operator T, which puts the term with the earliest time on the left. That is,

$$T[\mathcal{H}'(t_1)\mathcal{H}'(t_2)\mathcal{H}'(t_3)] = \mathcal{H}'(t_3)\mathcal{H}'(t_1)\mathcal{H}'(t_2). \tag{5.36}$$

Here, $t_3 > t_1 > t_2$. The following step function is useful in this respect, which can be expressed as,

$$\begin{aligned}\Theta(x) &= 1 \text{ if } x > 0 \\ &= 0 \text{ if } x < 0 \\ \text{Additionally,}\quad \Theta(x) &= \frac{1}{2} \text{ if } x = 0.\end{aligned} \tag{5.37}$$

For two operators, we can use the Θ function, such that,

$$T[H'(t_1)H'(t_2)] = \Theta(t_1 - t_2)\mathcal{H}'(t_1)\mathcal{H}'(t_2) + \Theta(t_2 - t_1)\mathcal{H}'(t_2)\mathcal{H}'(t_1). \tag{5.38}$$

If $\mathcal{H}'(t_1)$ and $\mathcal{H}'(t_2)$ commute with each other then the ordering is unimportant, however, that is not true in general. Now consider the integral using the time ordering operator,

$$\begin{aligned}\frac{1}{2!}\int_0^t dt_1\int_0^t T[\mathcal{H}(t_1)\mathcal{H}'(t_2)]dt_2 &= \frac{1}{2!}\int_0^t dt_1\int_0^{t_1} \mathcal{H}'(t_1)\mathcal{H}'(t_2)dt_2 \\ &+ \frac{1}{2!}\int_0^t dt_2\int_0^{t_2} \mathcal{H}'(t_2)\mathcal{H}'(t_1)dt_1.\end{aligned} \tag{5.39}$$

The second term is identical to the first if we redefine $t_1 \rightarrow t_2$. Thus,

$$\frac{1}{2!}\int_0^t dt_1 \int_0^t T[\mathcal{H}'(t_1)\mathcal{H}'(t_2)]dt_2 = \int_0^t dt_1 \int_0^{t_1} \mathcal{H}'(t_1)\mathcal{H}'(t_2)dt_2. \tag{5.40}$$

Similarly, it is easy to show that,

$$\begin{aligned}\frac{1}{3!}\int_0^t dt_1 \int_0^t dt_2 \int_0^t T[\mathcal{H}'(t_1)\mathcal{H}'(t_2)\mathcal{H}'(t_3)]dt_3 = \\ \int_0^t dt_1 \int_0^{t_1} dt_2 \int_0^{t_2} \mathcal{H}'(t_1)\mathcal{H}'(t_2)\mathcal{H}'(t_3)dt_3.\end{aligned} \tag{5.41}$$

Thus the expression for $U(t)$ is,

$$\begin{aligned}U(t) = 1 + \sum_{n=1}^{\infty}\frac{(-i)^n}{n!}\int_0^t dt_1 \cdots \\ \int_0^t dt_n T[\mathcal{H}'(t_1)\cdots\mathcal{H}'(t_n)] = Texp[-i\int_0^t dt_1\mathcal{H}'(t_1)].\end{aligned} \tag{5.42}$$

Let us introduce another operator, namely the S-matrix. We have discussed that the wavefunction in the interaction representation is given by,

$$\psi(t) = e^{i\mathcal{H}_0 t}e^{-i\mathcal{H}t}\psi(0). \tag{5.43}$$

Now define an S-matrix, namely, $S(t, t')$ which connects $\psi(t)$ and $\psi(t')$ by,

$$\psi(t) = S(t, t')\psi(t') = S(t, t')U(t')\psi(0) = U(t)\psi(0). \tag{5.44}$$

Thus,

$$S(t, t') = U(t)U^{\dagger}(t').$$

Some properties of the S-Matrix can be enumerated as,

(i) $S(t, t) = 1 = U(t)U^{\dagger}(t) = (e^{i\mathcal{H}_0 t}e^{-i\mathcal{H}t})(e^{i\mathcal{H}t}e^{-i\mathcal{H}_0 t})$
(ii) $S^{\dagger}(t, t') = U(t')U^{\dagger}(t) = S(t', t)$
(iii) $S(t, t')S(t', t'') = S(t, t'')$

The last one, that is, (iii), can be proved as follows.

$$\psi(t) = S(t, t')\psi(t') = S(t, t')S(t'', t')\psi(t'') = S(t, t'')\psi(t'').$$

The EOM for S is,

$$\frac{dS(t, t')}{dt} = \frac{dU(t)U^{\dagger}(t')}{dt} = -i\mathcal{H}'(t)S(t, t') \tag{5.45}$$

whose solution can be written as,

$$S(t, t') = T\text{exp}\left[-i\int_{t'}^t dt_1\mathcal{H}'(t_1)\right]. \tag{5.46}$$

There is still an unsolved issue in the formalism, namely, the interacting many-body state at $t = 0$, that is, $\psi(0)$ is not known. However, the eigensolutions of $\mathcal{H}_0$ are known. Let the ground state of $\mathcal{H}_0$ be ϕ_0. Somehow we have to establish a connection between $\psi(0)$ and ϕ_0. The connection will be provided by the Gell-Mann–Low theorem, which we simply state here as,

$$\psi(0) = S(0, -\infty)\phi_0. \tag{5.47}$$

Thus the interacting many-body ground state must have evolved from the non-interacting one in the distant past. Also, it is known that $\psi(t) = S(t, 0)\psi(0)$. Now if we operate $\psi(t)$ by $S(0, t)$, then,

$$\psi(0) = S(0, t)\psi(t). \tag{5.48}$$

For $t \to -\infty$

$$\psi(0) = S(0, -\infty)\psi(-\infty) = S(0, -\infty)|\phi_0\rangle \tag{5.49}$$

$\psi(-\infty)$ before the interaction is switched is ϕ_0.

There is an additional property that will be needed for the discussion of the Green's function. At $t \to \infty$, that is,

$$\psi(\infty) = S(\infty, 0)\psi(0). \tag{5.50}$$

It can be assumed that the effect of interaction is switched off at $t = \infty$. So $\psi(\infty)$ and ϕ_0 may be the same or differ by a phase factor of χ.

$$\phi_0 e^{i\chi} = \psi(\infty) = S(\infty, 0)\psi(0) = S(\infty, -\infty)\phi_0. \tag{5.51}$$

Thus, the phase factor is given by the matrix elements of $S(\infty, -\infty)$ between the non-interacting ground state, that is,

$$e^{i\chi} = \langle\phi_0|S(\infty, -\infty)|\phi_0\rangle.$$

5.3.4 Electron Green's function at zero temperature

At zero temperature, the Green's function is defined as,

$$G(\lambda, t - t') = -i\langle\psi_0|T[c_\lambda(t)c_\lambda^\dagger(t')]|\psi_0\rangle. \tag{5.52}$$

Here $t > t'$ and λ denotes a relevant quantum number. For example, in the free-electron case, $\lambda = (\mathbf{k}, \sigma)$ (where the eigenstates are labelled, $|\psi_0\rangle$ is the many-body ground state, that is, eigenstate of $\mathcal{H} = \mathcal{H}_0 + \mathcal{H}'$). Of course, the interacting ground state is not known, however, we know the eigenstates of $\mathcal{H}_0$. Now writing $c_\lambda(t)$ as,

$$c_\lambda(t) = e^{i\mathcal{H}t}c_\lambda(0)e^{-i\mathcal{H}t} \quad \text{(in Heisenberg representation)}. \tag{5.53}$$

For $t' > t$, the Green's function is defined as,

$$G(\lambda, t' - t) = i\langle\psi_0|c_\lambda^\dagger(t')c_\lambda(t)|\psi_0\rangle. \tag{5.54}$$

The sign in front is changed because the fermion operators are swapped.

As we have discussed earlier, the interacting many-body ground state can be obtained from the non-interacting ground state as, $|\psi_0\rangle = S(0, -\infty)|\phi_0\rangle$. In the interaction representation, we can write,

$$\begin{aligned} c_\lambda(t) &= e^{i\mathcal{H}_0 t} c_\lambda(0) e^{-i\mathcal{H}_0 t} \\ \text{Inverting the relation,} \quad c_\lambda(0) &= e^{-i\mathcal{H}_0 t} c_\lambda(t) e^{i\mathcal{H}_0 t}. \end{aligned} \tag{5.55}$$

These operators have to be converted to Heisenberg representation, which can be written as,

$$c_\lambda(t) = e^{i\mathcal{H}t} e^{-iH\mathcal{H}_0 t} c_\lambda(t) e^{i\mathcal{H}_0 t} e^{-i\mathcal{H}t} = U^\dagger(t) c_\lambda(t) U(t)\, e^{i\mathcal{H}t}\, e^{-\mathcal{H}t} e^{i\mathcal{H}_0 t} c_\lambda(t)\, e^{i\mathcal{H}_0 t} e^{i\mathcal{H}t} \tag{5.56}$$

$$\begin{aligned} G(\lambda, t - t') = &- i\Theta(t - t')\langle\phi_0|S(-\infty, 0)S(0, t)c_\lambda(t)S(t, 0) \\ &S(0, t')c_\lambda^\dagger(t')S(t', 0)S(0, -\infty)|\phi_0\rangle + \\ &i\Theta(t' - t)\langle\phi_0|S(-\infty, 0)S(0, t')c_\lambda^\dagger(t')S(t', 0)S(0, t) \\ &c_\lambda(t)S(t, 0)S(0, -\infty)|\phi_0\rangle. \end{aligned} \tag{5.57}$$

The extreme left-hand bracket is replaced by,

$$\begin{aligned} \langle\phi_0|S(-\infty, 0) &= e^{-i\chi}\langle\phi_0|S(\infty, -\infty)S(-\infty, 0) \\ &= \frac{\langle\phi_0|S(\infty, 0)}{\langle\phi_0|S(\infty, -\infty)|\phi_0\rangle}. \end{aligned} \tag{5.58}$$

Finally, the electron Green's function becomes,

$$\begin{aligned} G(\lambda, t - t') = &\frac{-i}{\langle\phi_0|S(\infty, -\infty)|\phi_0\rangle} \\ &[\Theta(t - t')\langle\phi_0|S(\infty, t)c_\lambda(t)S(t, t')c_\lambda^\dagger(t')S(t', -\infty)|\phi_0\rangle \\ &- \Theta(t' - t)\langle\phi_0|S(\infty, t')c_\lambda^\dagger(t')S(t', t)c_\lambda(t)S(t, -\infty)|\phi_0\rangle]. \end{aligned} \tag{5.59}$$

The first term can be simplified by writing,

$$\begin{aligned} &\Theta(t - t')\langle\phi_0|S(\infty, t)c_\lambda(t)S(t, t')c_\lambda^\dagger(t')S(t', -\infty)|\phi_0\rangle \\ &= \Theta(t - t')\langle\phi_0|Tc_\lambda(t)c_\lambda^\dagger(t')S(\infty, -\infty)|\phi_0\rangle. \end{aligned} \tag{5.60}$$

This is because $S(\infty, -\infty)$ contains operators which act in the three time intervals, namely, (∞, t), (t, t') and $(t', -\infty)$. The T operator automatically sorts these segments so that they act in the proper sequence, which are, respectively, to the left of $c_\lambda(t)$, between $c_\lambda(t)$ and $c_\lambda^\dagger(t')$ and to the right of $c_\lambda^\dagger(t')$. The same can be done for the second term, yielding a form for the total Green's function as,

$$G(\lambda, t - t') = -i\frac{\langle\phi_0|Tc_\lambda(t)c_\lambda^\dagger(t')S(\infty, -\infty|\phi_0\rangle}{\langle\phi_0|TS(\infty, -\infty)|\phi_0\rangle}. \tag{5.61}$$

In fact, it does not matter whether one puts $S(\infty, -\infty)$ in the numerator, because the time ordering operator finally puts things in order.

5.3.5 Example: a degenerate electron gas

Let us take a simple case, for example, consider a metal at zero temperature. Consider the single-particle energies and the chemical potential of the system given by ϵ_k and μ, respectively. All the single-particle states with $\epsilon_k < \mu$ are occupied, and all others, with $\epsilon_k > \mu$ are empty. It is customary to measure all energies relative to μ[6], and hence the energy scale becomes,

$$\xi_k = \epsilon_k - \mu. \tag{5.62}$$

For a uniform spherical Fermi surface with a Fermi wave-vector, k_F,

$$\langle \phi_0 | c_k^\dagger c_k | \phi_0 \rangle = \langle \eta_k \rangle = \Theta(k_F - k) = \lim_{\beta \to \infty} \frac{1}{e^{\beta \xi_k} + 1} \tag{5.63}$$

$$\langle \phi_0 | c_k c_k^\dagger | \phi_0 \rangle = 1 - \langle \eta_k \rangle = \Theta(k - k_F). \tag{5.64}$$

The unperturbed Green's function or the non-interacting (free) propagator corresponding to $\mathcal{H}' = 0$ is written as,

$$G(\lambda, t - t') = -i \langle \phi_0 | T c_\lambda(t) c_\lambda^\dagger(t') | \phi_0 \rangle. \tag{5.65}$$

Expanding,

$$\begin{aligned} G^{(0)}(\lambda, t - t') &= -i\Theta(t - t') \langle \phi_0 | c_\lambda(t) c_\lambda^\dagger(t') | \phi_0 \rangle \\ &\quad + i\Theta(t' - t) \langle \phi_0 | c_\lambda^\dagger(t') c_\lambda(t) | \phi_0 \rangle \\ &= -i[\Theta(t - t')\Theta(\xi_k) - \Theta(t' - t)\Theta(-\xi_k)] e^{-i\xi_k(t-t')}. \end{aligned} \tag{5.66}$$

The Fourier transform of $G^{(0)}$ can be written as,

$$\begin{aligned} G^{(0)}(\mathbf{k}, \omega) &= -i\left[\Theta(\xi_k)\int_0^\infty dt \ e^{it(\omega - \xi_k + i\eta)} - \Theta(-\xi_k)\int_{-\infty}^0 dt \ e^{it(\omega - \xi_k - i\eta)}\right] \\ &= \frac{\Theta(\xi_k)}{\omega - \xi_k + i\eta} + \frac{\Theta(-\xi_k)}{\omega - \xi_k - i\eta}. \end{aligned} \tag{5.67}$$

A simpler way to write $G^{(0)}(\mathbf{k}, \omega)$ is given by,

$$G^{(0)}(\mathbf{k}, \omega) = \frac{1}{\omega - \xi_k + i\delta_k} \tag{5.68}$$

[6] It is a worthwhile reminder that the chemical potential, μ and the Fermi energy, ϵ_F are identical at zero temperature, while they differ at finite temperatures. In fact, the Fermi surface loses its definition at finite temperature.

where,

$$\delta_k = \text{sgn}(\xi_k) = \text{sgn}(\epsilon_k - \mu)$$
$$= + \text{ when } \epsilon_k > \mu \text{ first term in equation (5.67)} \tag{5.69}$$
$$= - \text{ when } \epsilon_k < \mu \text{ second term in equation (5.67).}$$

Thus, the free fermion Green's function can be written in a compact form as,

$$G_{\alpha\beta}^{(0)}(\mathbf{k}, \omega) = \delta_{\alpha\beta}\left[\frac{\Theta(k - k_\text{F})}{\omega - \xi_k + i\eta} + \frac{\Theta(k_\text{F} - k)}{\omega - \xi_k - i\eta}\right]. \tag{5.70}$$

5.4 Retarded and advanced Green's functions

From the discussion we had so far, it is clear that the Green's function aids in propagating a many-particle wavefunction in time. However, there is a subtle point that we have not mentioned yet. It is that the Green's function can be useful in propagating a state both ahead and backward in time. In fact, both have different physical implications associated with them, which we wish to highlight in the subsequent discussion.

A quick reminder of the story related to time evolution appears in the following. To evolve a state with time, we need to solve a time-dependent Schrödinger equation,

$$\left(i\frac{\partial}{\partial t} - \mathcal{H}\right)\psi(\mathbf{r}, t) = 0 \tag{5.71}$$

which has an integral solution that connects the states at two different space–time points via,

$$\psi(\mathbf{r}, t) = \int d\mathbf{r}' G(\mathbf{r}, t, \mathbf{r}', t')\psi(\mathbf{r}', t'). \tag{5.72}$$

Now we can define a retarded propagator, G^R for $t > t'$, that is, the particle propagates from an initial time t' (from a coordinate point $\mathbf{r}'$) to a later time t (whose coordinate is $\mathbf{r}$). Thus,

$$G^\text{R}(\mathbf{r}, t, \mathbf{r}', t') = \Theta(t - t')G(\mathbf{r}, t, \mathbf{r}', t') \tag{5.73}$$

which satisfies the Schrödinger equation,

$$\left(i\frac{\partial}{\partial t} - \mathcal{H}\right)G^\text{R}(\mathbf{r}, t, \mathbf{r}', t') = \delta(\mathbf{r} - \mathbf{r}')\delta(t - t'). \tag{5.74}$$

In a many-body system, the retarded propagator holds the same meaning, except that now it takes a form,

$$G^\text{R}(\mathbf{r}, t, \mathbf{r}', t') = -i\Theta(t - t')\left\langle\left\{c(\mathbf{r}, t), c^\dagger(\mathbf{r}', t')\right\}\right\rangle \tag{5.75}$$

where the expectation value is taken with respect to the many-body ground state, and $\{..\}$ denotes the anticommutation relation for fermions. For bosons, it should be replaced by a commutation relation.

In a similar spirit, an advanced Green's function is defined as,

$$G^{\mathrm{A}}(\mathbf{r}, t, \mathbf{r}', t') = i\Theta(t' - t)\left\langle\left\{c(\mathbf{r}, t), c^{\dagger}(\mathbf{r}', t')\right\}\right\rangle. \tag{5.76}$$

In addition, in certain situations one introduces $G^{<}$ (called the lesser Green's function) and $G^{>}$ (greater Green's function) that are defined by,

$$\begin{aligned} G^{<}(\mathbf{r}, t, \mathbf{r}', t') &= -i\langle c^{\dagger}(\mathbf{r}', t'), c(\mathbf{r}, t)\rangle \\ G^{>}(\mathbf{r}, t, \mathbf{r}', t') &= -i\langle c(\mathbf{r}, t), c^{\dagger}(\mathbf{r}', t')\rangle. \end{aligned} \tag{5.77}$$

In fact, combinations of $G^{<}$ and $G^{>}$ yield G^{R} and G^{A}.

5.4.1 Spectral representation

We have been expressing the Green's function in the time domain, however expressing it in the frequency domain turns out to be more insightful. To do so, the Green's function should depend upon $t - t'$, instead of individually on t and t' (or equivalently, there should be time translational symmetry). For our convenience of demonstrating this, we take a free-particle Hamiltonian with onsite energies $\epsilon(\mathbf{r})$ as,

$$\mathcal{H} = \sum_{\mathbf{r}} \epsilon(\mathbf{r}) c_{\mathbf{r}}^{\dagger} c_{\mathbf{r}}. \tag{5.78}$$

The time evolution of the $c_{\mathbf{r}}$ operators obeys Heisenberg's EOM and can be written as,

$$\begin{aligned} \dot{c}_{\mathbf{r}}(t) &= -i[c_{\mathbf{r}}, \mathcal{H}] = -i\epsilon(\mathbf{r})c_{\mathbf{r}} \\ \dot{c}_{\mathbf{r}}^{\dagger}(t) &= -i[c_{\mathbf{r}}^{\dagger}, \mathcal{H}] = i\epsilon(\mathbf{r})c_{\mathbf{r}}^{\dagger}. \end{aligned} \tag{5.79}$$

The retarded and the advanced Green's functions can be written as,

$$\begin{aligned} G^{\mathrm{R}}(\mathbf{r}, \mathbf{r}', t - t') &= -i\Theta(t - t')e^{-i\epsilon(\mathbf{r})(t-t')}\delta_{\mathbf{r},\mathbf{r}'} \\ G^{\mathrm{A}}(\mathbf{r}, \mathbf{r}', t - t') &= i\Theta(t' - t)e^{i\epsilon(\mathbf{r})(t-t')}\delta_{\mathbf{r},\mathbf{r}'} \end{aligned} \tag{5.80}$$

where we have used anticommutation relation for the fermions, namely,

$$\left\langle\left\{c_{\mathbf{r}}, c_{\mathbf{r}'}^{\dagger}\right\}\right\rangle = \delta_{\mathbf{r},\mathbf{r}'}$$

The Green's function is thus diagonal in the frequency domain, which is a feature of the non-interacting system. That is, it does not involve different states (here in the position basis).

Writing the integral representation of the Θ function,

$$\Theta(t - t') = -\frac{1}{2\pi i}\int_{-\infty}^{\infty} d\omega \frac{e^{-i\omega(t-t')}}{\omega + i\eta} \tag{5.81}$$

where, with $\eta = 0^{+}$ we displace the pole at $\omega = 0$ in the downward direction by an infinitesimal amount. This yields the retarded Green's function to acquire a form,

$$G^{\mathrm{R}}(\mathbf{r}, \mathbf{r}', t - t') = \frac{1}{2\pi}\int_{-\infty}^{\infty} d\omega \frac{e^{-i(\omega+\epsilon(\mathbf{r}))(t-t')}}{\omega + i\eta}. \tag{5.82}$$

Redefining $(\omega + \epsilon(\mathbf{r})) \to \omega'$, or simply ω ($\delta_{\mathbf{r},\mathbf{r}'}$ is implied),

$$G^{\mathrm{R}}(\mathbf{r}, \mathbf{r}', t - t') = \frac{1}{2\pi}\int_{-\infty}^{\infty} d\omega \frac{e^{-i\omega(t-t')}}{\omega - \epsilon(\mathbf{r}) + i\eta}. \tag{5.83}$$

Using,

$$G^{\mathrm{R}}(\mathbf{r}, \mathbf{r}', t - t') = \frac{1}{2\pi}\int d\omega G^{\mathrm{R}}(\mathbf{r}, \mathbf{r}', \omega) \tag{5.84}$$

helps us to identify,

$$G^{\mathrm{R}}(\mathbf{r}, \mathbf{r}', \omega) = \frac{1}{\omega - \epsilon_n + i\eta} \tag{5.85}$$

where ϵ_n denotes the eigenspectrum. Similarly for the advanced Green's function, one obtains,

$$G^{\mathrm{A}}(\mathbf{r}, \mathbf{r}', \omega) = \frac{1}{\omega - \epsilon_n - i\eta}. \tag{5.86}$$

A few comments are in order.

(i) The poles of the Green's function correspond to the eigenspectrum of the system.

(ii) The imaginary part of G yields the local density of states,

$$\rho(\mathbf{r}, \omega) = -\frac{1}{\pi}\,\mathrm{Im}\, G^{\mathrm{R}}(\mathbf{r}, \mathbf{r}', \omega) = \frac{1}{\pi}\,\mathrm{Im}\, G^{\mathrm{A}}(\mathbf{r}, \mathbf{r}', \omega)$$

where we have used,

$$\frac{1}{x \pm i\eta} = \mathcal{P}\left(\frac{1}{x}\right) \mp i\pi\delta(x).$$

5.4.2 Wick's theorem and Feynman diagrams

Here we shall learn the techniques of including interaction terms in the Green's function. This will eventually help us in expressing the scattering events in terms of the Feynman diagrams. However, that will require us to split the interacting Green's function (G) into non-interacting Green's functions (G^0), and interaction vertices, (V) which are in general functions of energy and momentum. Such a process is done via Wick's theorem which we shall discuss below. Since electronic systems are mainly of importance to us, we primarily discuss electronic Green's function. However, in real materials, particularly at finite temperatures, lattice excitations, or phonons become increasingly important. We cite reference [2] to readers for this.

The interacting electronic Green's function is evaluated by expanding the S-matrix, that is, $S(\infty, -\infty)$ in the form of a series,

$$G(\mathbf{k}, t - t') = \Sigma_{n=0}^{\infty}\frac{(-i)^{n+1}}{n!}\int_{-\infty}^{\infty} dt_1 \cdots \int_{-\infty}^{\infty} dt_n \frac{\langle\phi_0|Tc_k(t)\mathcal{H}'(t_1)\mathcal{H}'(t_2)\mathcal{H}'(t_3)\cdots\mathcal{H}'(t_n)c_k^\dagger(t')|\phi_0\rangle}{\langle\phi_0|S(\infty, -\infty)|\phi_0\rangle}. \tag{5.87}$$

Let us for the moment ignore the denominator which is merely a normalization constant. The immediate aim is to compute the time-ordered brackets, such as terms like,

$$\langle\phi_0|Tc_k(t)\mathcal{H}'(t_1)\mathcal{H}'(t_2)\mathcal{H}'(t_3)c_k^\dagger(t')|\phi_0\rangle. \tag{5.88}$$

To have a concrete discussion, let us consider a specific case of Coulomb interaction among the particles, such that the interaction, $\mathcal{H}'(t)$ is denoted by,

$$\mathcal{H}'(t) = \frac{1}{2}\sum_{k,\, k',\, q}\sum_{\sigma,\, \sigma'}\frac{4\pi e^2}{q^2}c_{k+q,\,\sigma}^\dagger c_{k'-q,\,\sigma'}^\dagger c_{k',\sigma'}c_{k,\sigma}e^{it(\xi_{k+q}+\xi_{k'-q}-\xi_{k'}-\xi_k)}. \tag{5.89}$$

For this particular case, for the time-ordered product above, we have 7 creation and 7 annihilation operators with 6 creation and 6 annihilation operators contributed by $\mathcal{H}'$ at different times, and one each coming from the definition of the Green's function. It is a difficult task to take into account the time ordering and compute pairing of the creation and the annihilation operators so as to be able to write each in terms of non-interacting Green's functions. Since, in a general sense at the nth level, one always encounters $n + 1$ creation and $n + 1$ annihilation operators, that is,

$$\langle\phi_0|Tc(t)c_1(t_1)c_1^\dagger(t_1)\cdots c_n(t_n)c_n^\dagger(t_n)c^\dagger(t')|\phi_0\rangle. \tag{5.90}$$

The angular brackets containing an unequal number of creation and annihilation operators vanish.

To remind ourselves, the job of a creation operator is to create a particle in a certain state $|n\rangle$, and the destruction operator destroys the state $|m\rangle$, such that, for $m = n$, the system has to return to the ground state[7]. For example,

$$\langle\phi_0|Tc_\alpha(t)c_\beta^\dagger(t')|\phi_0\rangle = 0 \text{ (unless } \alpha = \beta). \tag{5.91}$$

Similarly,

$$\langle\phi_0|Tc_\alpha(t)c_\beta^\dagger(t_1)c_\gamma(t_2)c_\delta^\dagger(t')|\phi_0\rangle$$

will yield zero unless $\alpha = \beta$, and $\gamma = \delta$ or $\alpha = \delta$, and $\gamma = \beta$. Thus in the expression of equation (5.88), there are many such pairings and several orderings need to be accounted for. However, only a limited number of them are physically meaningful and interesting. Wick's theorem does this sorting of the physically interesting terms

[7] Imbalanced situation in terms of particle densities will yield zero expectation values.

in a tractable manner. Wick's theorem is really just an observation that the time ordering can be taken care of in a simple way. Also, while making possible pairing between creation and annihilation operators, each pair should be time ordered.

For example, note that there is a time-ordering operator in each of the pairing brackets. For $n = 3$, there will be 6 possible pairing brackets. Thus for n operators (half of them being creation and the other half annihilation operators), there are $n!$ pairing brackets. The above is written so as to get a number of non-interacting Green's functions which yield the Green's function for the interacting system that we are looking for. Thus,

$$\begin{aligned}(-i)^2\langle\phi_0|TC_\alpha(t)C^\dagger_\beta(t_1)C_\gamma(t_2)C^\dagger_\delta(t')|\phi_0\rangle &= \delta_{\alpha\beta}\delta_{\gamma\delta}G^{(0)}(\alpha, t - t_1)G^{(0)}(\gamma, t_2 - t') \\ &\quad - \delta_{\alpha\delta}\delta_{\beta\gamma}G^{(0)}(\alpha, t - t')G^{(0)}(\gamma, t_2 - t_1).\end{aligned} \tag{5.92}$$

In summary, Wick's theorem states that a time-ordered bracket may be evaluated by expanding it into all possible pairings. Each of these pairings is either a Green's function or the number operator n. It should be kept in mind that the Wick's theorem is valid when $\mathcal{H}_0$ is bilinear in creation and annihilation operators.

Let us summarize our discussion so far.

(i) It should be kept in mind that all the operators are written in the interaction representation.

(ii) The Green's function consists of time-ordered products of creation and annihilation operators and interaction terms $\mathcal{H}'$ occurring at distinct times. For example,

$$G(\mathbf{k}, t - t') \sim \langle\phi_0|Tc_k(t)\mathcal{H}'(t_1)\mathcal{H}'(t_2)c^\dagger_k(t')|\phi_0\rangle.$$

(iii) Each order of $\mathcal{H}'$ included in the expansion of G yields the order of the Green's function. For example, three $\mathcal{H}'$ included gives a third order Green's function. In principle, an infinite number of $\mathcal{H}'$ inside the time-ordered product yields the fully interacting Green's function.

(iv) Each of the interaction terms, $\mathcal{H}'$ consists of two creation and annihilation operators, that is,

$$\mathcal{H}' \sim c^\dagger cc^\dagger c.$$

(v) The expectation value of the time-ordered product is taken with respect to the non-interacting ground state, which differs from the many-body ground state before that interaction is switched on (at $t = -\infty$), at the most by a phase factor.

(vi) The interacting Green's function is finally written in terms of product of the non-interacting Green's functions and the number operators.

(vii) Such products are most conveniently computed using Feynman diagrams.

Let us discuss the Feynman diagrams for the first order Green's function written above. However, before we do that, we shall lay down the rules for drawing the diagrams. We enumerate them below.

(i) A solid line with an arrow pointing from t' (t is smaller or later time) to t (t earlier or larger time) for the free-electron Green's function with a momentum corresponding to the propagation.
(ii) A dashed line with no arrow denotes the phonon Green's function. There is no arrow since the phonons can move in either direction in time. We shall not discuss this here.
(iii) The electron density $n(\xi_p)$ is shown via a loop to imply that it starts and ends at the same time.
(iv) Any interaction term V_q is represented by a wiggly line. For example, $\mathcal{H}' = G^{(0)}(p)V_q(q)G^{(0)}(p-q)G^{(0)}(p)$.
(v) At the vertex, there must be conservation of momentum and energy. Thus, it is necessary to integrate over all the intermediate energies (frequencies) and sum over all intermediate momenta.
(vi) Multiply by a factor $(-i)^n \times (-1)^F$ where F denotes the number of closed fermion loops.

To express the full Green's function in terms of Feynman diagrams, let us look at the first order Green's function. This means that we shall compute the Green's function up to first order in $\mathcal{H}'(t)$. We shall see that the number of diagrams increases very rapidly as one considers second order in the interaction term. Thus, for the purpose of illustration, we restrict ourselves only up to first order. The essential part of the Green's function[8] can be written as,

$$V_q\langle Tc_k(t)c_{k_1}^\dagger(t_1)c_{k_2}^\dagger(t_1)c_{k_2+q}(t_1)c_{k_1-q}(t_1)c_k^\dagger(t')\rangle.$$

Before we consider several combinations of the expectation of one creation and one annihilation operator, we need to number them. Schematically (leaving out the V_q at the moment),

$$\langle T \overset{(\mathrm{I})}{c_k(t)}\,\underbrace{\overset{(\mathrm{II})}{c_{k_1}^\dagger(t_1)}\,\overset{(\mathrm{III})}{c_{k_2}^\dagger(t_1)}\,\overset{(\mathrm{IV})}{c_{k_2+q}(t_1)}\,\overset{(\mathrm{V})}{c_{k_1-q}(t_1)}}_{\mathcal{H}'(t_1)}\,\overset{(\mathrm{VI})}{c_k^\dagger(t')}\rangle.$$

There are six possible combinations with three each of creation and annihilation operators, namely,

$$\left.\begin{array}{l}
\text{(a) (I,II),(III,IV),(V,VI)}\\
\text{(b) (I,III),(II,IV),(V,VI)}\\
\text{(c) (I,III),(II,V),(IV,VI)}\\
\text{(d) (I,II),(III,V),(IV,VI)}\\
\text{(e) (I,VI),(II,IV),(III,V)}\\
\text{(f) (I,VI),(II,V),(III,IV)}
\end{array}\right\} \quad 6 = 3! \quad \text{combinations.} \tag{5.93}$$

[8] There are factors of i, sum over internal momenta, and integration over the internal time (or energy) variables.

They are explicitly written as,

$$\text{(a)}\ \underbrace{\langle Tc_k(t)c^\dagger_{k_1}(t_1)\rangle}_{\delta_{k,k_1}G^{(0)}(k,t-t_1)}\underbrace{\langle Tc_{k_2+q}(t_1)c^\dagger_{k_2}(t_1)\rangle}_{\delta_{q=0}\ \eta(\xi_{k_2})}\underbrace{\langle Tc_{k_1-q}(t_1)c^\dagger_k(t')\rangle}_{\delta_{q=0}G^{(0)}(k_1,t_1-t')\delta_{k,k_1}}$$

$$\text{(b)}\ \underbrace{\langle Tc_k(t)c^\dagger_{k_2}(t_1)\rangle}_{\delta_{k,k_2}G^{(0)}(k,t-t_1)}\underbrace{\langle Tc_{k_2+q}(t_1)c^\dagger_{k_1}(t_1)\rangle}_{\delta_{k_1=k_2+q}\ G^{(0)}(k,t_1)}\underbrace{\langle Tc_{k_1-q}(t_1)c^\dagger_k(t')\rangle}_{\delta_{k_1-q,k}G^{(0)}(k,t_1-t')}$$

$$\text{(c)}\ \underbrace{\langle Tc_k(t)c^\dagger_{k_2}(t_1)\rangle}_{\delta_{k,k_2}\ G^{(0)}(k,t-t_1)}\underbrace{\langle TC_{k_1-q}(t_1)c^\dagger_{k_1}(t_1)\rangle}_{\delta_{q=0}\ n(\xi_{k_1})}\underbrace{\langle Tc_{k_2+q}(t)c^\dagger_k(t')\rangle}_{\delta_{k,k_2+q}G^{(0)}(k,t-t')}$$

$$\text{(d)}\ \underbrace{\langle Tc_k(t)c^\dagger_{k_1}(t_1)\rangle}_{\delta_{k,k_1}G^{(0)}(k,t-t_1)}\underbrace{\langle Tc_{k_1-q}(t_1)c^\dagger_{k_2}(t_1)\rangle}_{\delta_{k_1-q,k_2}\ G^{(0)}(k-q,t_1)}\underbrace{\langle Tc_{k_2+q}(t_1)c^\dagger_k(t')\rangle}_{\delta_{k,k_2+q}G^{(0)}(k,t_1-t')}$$

$$\text{(e)}\ \underbrace{\langle Tc_k(t)c^\dagger_k(t')\rangle}_{G^{(0)}(k,t-t')}\underbrace{\langle Tc_{k_2+q}(t_1)c^\dagger_{k_1}(t_1)\rangle}_{\delta_{k_1,k_2+q}\ n(\xi_{k_1})}\underbrace{\langle Tc_{k_1-q}(t_1)c^\dagger_{k_2}(t_1)\rangle}_{\delta_{k_2,k_1-q}\ n(\xi_{k_2})}$$

$$\text{(f)}\ \underbrace{\langle Tc_k(t)c^\dagger_k(t')\rangle}_{G^{(0)}(k,t-t')}\underbrace{\langle Tc_{k_2+q}(t_1)c^\dagger_{k_2}(t_1)\rangle}_{\delta_{q=0}\ n(\xi_{k_2})}\underbrace{\langle Tc_{k_1-q}(t_1)c^\dagger_{k_2}(t_1)\rangle}_{\delta_{k_1-q,k_2}\ n(\xi_{k_1})}.$$

In addition to these terms, there is a V_q that goes with each one of them, shown by wavy lines in the diagrams in figure 5.2. The six terms above are denoted by six diagrams in figure 5.2. As per the rules laid down above, the solid lines denote non-interacting Green's functions, $G^{(0)}$, and the wavy line stands for the interaction term, V_q. The closed loops (bubble) denote the density term.

In a similar manner, a second order Green's function will contain a total of 5 creation and 5 annihilation operators. Hence the number of combinations that arises will be 5! number of terms, that is, 120 terms A third order term will have 7! (= 5040) number of terms.

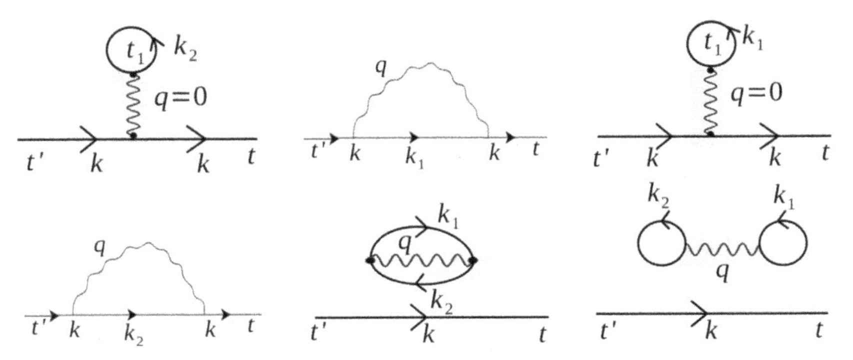

Figure 5.2. First order Feynman diagrams are shown here. The bold straight lines represent electron Green's functions, wavy lines denote Coulomb interactions, while the circles denote electron densities.

5.5 Self-energy: Dyson equation

While the perturbation theory is fine in some cases, it turns out that in many cases, one needs to sum over whole classes of diagrams to get a reasonable answer. We know that the energy spectrum of the interacting system is given by the poles of the Green's function. It is not clear whether a desired result can be obtained by performing a finite order perturbation theory. Besides, in terms of computational logistics, in going from zeroth order to first order, the number of diagrams increased from one to six. At the second order, it is 120 and at the third order, the number of diagrams is 5040, and so on. Thus computation of diagrams up to a finite order is itself a tedious task. Fortunately, there is a simple way to sum up infinite classes of diagrams.

For the purpose of illustration, again consider the two first order diagrams[9] in figure 5.3. Figure 5.3 (left) is known as the Hartree term, and (right) denotes the Fock term. We shall see that the Hartree term gives no contribution for Coulomb interaction. At the first order perturbation theory, the energy change is computed using,

$$\langle \mathcal{H}' \rangle = \langle \Psi | \mathcal{H}' | \Psi \rangle = \langle \phi_0 | \mathcal{H}' | \phi_0 \rangle$$

where ϕ_0 denotes the non-interacting ground state. Specific to the case of Coulomb interaction,

$$\langle \phi_0 | \mathcal{H}' | \phi_0 \rangle = \frac{1}{2} \sum_{k_1, k_2, q \neq 0, \sigma, \sigma'} V_q \langle \phi_0 | c^{\dagger}_{k,\sigma} c^{\dagger}_{k,\sigma'} c_{k_2+q,\sigma'} c_{k_1-q,\sigma} | \phi_0 \rangle \tag{5.94}$$

where $V_q = 4\pi e^2/q^2$. Possible pairings include the first and the fourth terms, and second and the third terms in equation (5.94). While the former corresponds to $q = 0$, since $q = 0$ is excluded from the sum, this pairing yields no contribution.

Let us look at the Fock term later, which yields a non-zero contribution that renormalizes the band energies. Instead, let us look at one of the second order contribution of the second order ($n = 2$) diagrams (figure 5.4).

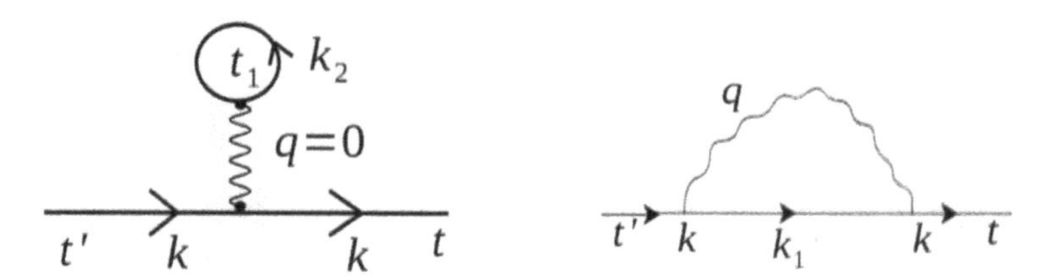

Figure 5.3. Hartree term (left), and Fock term (right) are shown corresponding to a Coulomb potential.

[9] It is quite relevant to mention that the disconnected diagrams, that is, the last two diagrams in figure 5.2 do not contribute, as they cancel with the normalization term, that is the denominator, $\langle \phi_0 | S(\infty, -\infty) | \phi_0 \rangle$. See reference [2] for details.

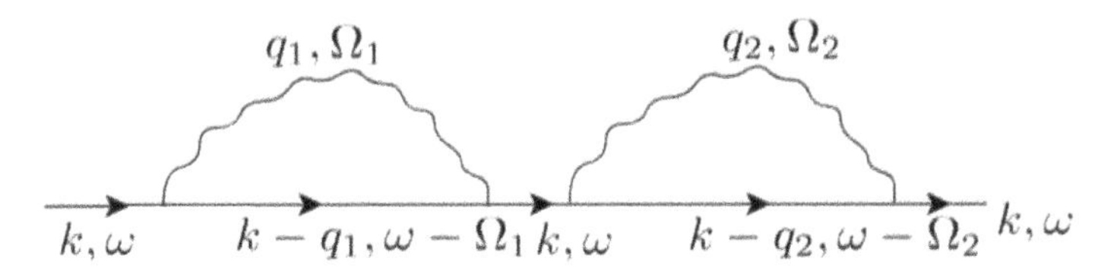

Figure 5.4. A typical second order diagram.

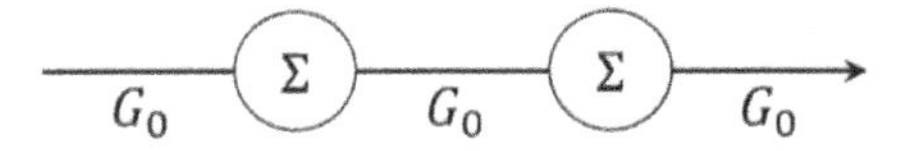

Figure 5.5. Diagrammatic representation of the self-energy.

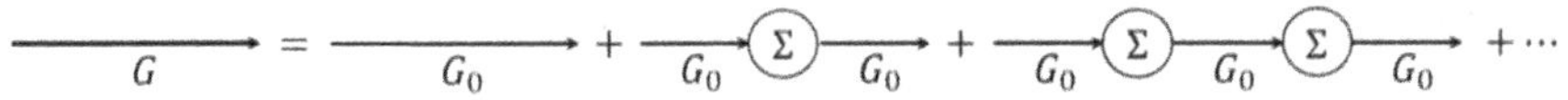

Figure 5.6. Diagrammatic representation for the full Green's function.

It yields a contribution,

$$G_0(k, \omega)\Sigma(k, \omega)G_0(k, \omega)\Sigma(k, \omega)G_0(k, \omega)$$

where $\Sigma(k, \omega)$ is denoted as the self-energy and is defined by,

$$\Sigma(k, \omega) = \int \frac{d\Omega}{2\pi}\frac{1}{2}V_q G_0(k - q, \omega - \Omega). \tag{5.95}$$

Diagrammatically, the self-energy may be shown as in figure 5.5.

Thus for an infinite series one can sum up all the terms (with G denoting the full Green's function) (figure 5.6).

One can write the above diagrammatic representation as,

$$\begin{aligned} G(k, \omega) &= G_0(k, \omega) + G_0(k, \omega)\Sigma(k, \omega)G_0(k, \omega) \\ &\quad + G_0(k, \omega)\Sigma(k, \omega)G_0(k, \omega)\Sigma(k, \omega)G_0(k, \omega) + \cdots \\ &= G_0(k, \omega)\sum_{n=0}^{\infty}\left[G_0(k, \omega)\Sigma(k, \omega)\right]^n \\ &= \frac{G_0(k, \omega)}{1 - G_0(k, \omega)\Sigma(k, \omega)} = \frac{1}{G_0^{-1}(k, \omega) - \Sigma(k, \omega)}. \end{aligned} \tag{5.96}$$

If we recall that,

$$G_0(k, \omega) = \frac{1}{\omega - \xi(k) + i\eta_k} \quad \text{with} \quad \eta_k = \eta\ \text{sgn}(\xi_k) \tag{5.97}$$

which yields,

$$G(k, \omega) = \frac{1}{\omega - \xi(k) - \Sigma(k, \omega) + i\eta_k} \quad \text{with} \quad \eta_k = \eta\ \text{sgn}(\xi_k). \tag{5.98}$$

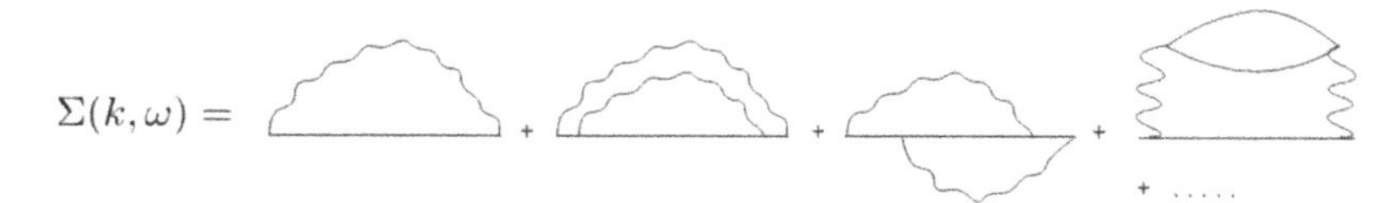

Figure 5.7. Schematic representations for a few second order diagrams are shown.

This can be shown diagrammatically in figure 5.7 where diagrams at all orders should be included. Here, of course, we show diagrams up to second order, that is, $n = 2$.

There are a few important consequences of the Dyson series.

(i) There are fewer important diagrams to calculate.

(ii) The interacting Green's function (G) has the same form as the non-interacting Green's function (G_0) with ξ_k replaced by $\xi_k - \Sigma(k, \omega)$. If in a special case, $\Sigma(k, \omega)$ depends only on k, then it is just renormalization of the energy spectrum. We shall see that in a while.

Let us discuss the self-energy in a little more detail. $\Sigma(k, \omega)$ is equivalent to an effective potential which incorporates all the interaction energies, such as, the electronic correlations, electron–phonon, electron-impurity scattering effects, etc. It leads to the concept of quasiparticles as follows. When a particle propagates through the system, it interacts with the environment, and thus gets '*dressed*'. This dressed particle is called a quasiparticle which acts as the elementary excitation of the system. The *dressing* serves to screen the interaction between the particles, thereby making it weak. The self-energy accounts for the difference in energy of a quasiparticle from that of a free particle. The quasiparticles are characterized by finite lifetime, as opposed to free particles, which have infinite lifetime. The quasiparticles also have a mass different from the bare mass of the particles.

5.5.1 Self-energy for a two-site chain: an example

Consider a two-site system whose Hamiltonian is given by,

$$\mathcal{H} = -t(c_1^\dagger c_2 + \text{h. c. }) + V_0(c_1^\dagger c_1 + c_2^\dagger c_2) \tag{5.99}$$

where t denotes the hopping strength, V_0 is the onsite energy, and (1, 2) refer to the two sites. The matrix representation of the above Hamiltonian in the site basis $(c_1^\dagger \quad c_2^\dagger)$ is,

$$\mathcal{H} = \begin{pmatrix} V_0 & -t \\ -t & V_0 \end{pmatrix}.$$

This yields the energies as, $\epsilon = V_0 \pm t$. Because of the 2×2 structure written above, the Green's function is also a 2×2 matrix in the site basis, and has a form (we write this explicitly for the retarded Green's function),

$$G^{\mathrm{R}}(t, t') = \begin{pmatrix} G_{11}^{\mathrm{R}}(t, t') & G_{12}^{\mathrm{R}}(t, t') \\ G_{21}^{\mathrm{R}}(t, t') & G_{22}^{\mathrm{R}}(t, t') \end{pmatrix} \tag{5.100}$$

where the diagonal elements, namely, G_{11}^{R} and G_{22}^{R} are written as,

$$G_{ii}^{R}(t, t') = -i\Theta(t - t')\left\langle\left\{c_i(t), c_i^{\dagger}(t')\right\}\right\rangle. \tag{5.101}$$

Similarly, the off diagonal elements are,

$$G_{ij}^{R}(t, t') = -i\Theta(t - t')\langle\left\{c_i(t), c_j^{\dagger}(t')\right\}\rangle. \tag{5.102}$$

The EOM for the Green's function can now be obtained as,

$$i\frac{d}{dt}G_{ij}^{R}(t, t') = i(-i)\frac{d}{dt}\Theta(t - t')\langle\{\dot{c}_i(t), c_j^{\dagger}(t')\}\rangle - i\Theta\langle\{ic_i(t), \dot{c}_j^{\dagger}(t')\}\rangle. \tag{5.103}$$

The operators follow the usual Heisenberg's EOM, that is,

$$\dot{c}_i(t) = \frac{1}{i}[c_i(t), \mathcal{H}].$$

Hence the EOM requires further Green's function to be calculated. Calculating the commutator $[c_i(t), \mathcal{H}]$ with $i = 1, 2$,

$$[c_i, \mathcal{H}] = \sum_i [c_i, (V_0 c_i^{\dagger} c_i - t c_i^{\dagger} c_j - t c_j^{\dagger} c_i)].$$

The first term yields,

$$[c_i, \sum_i V_0 c_i^{\dagger} c_i] = \sum_i V_0 c_i.$$

Similarly,

$$[c_i, -t\sum_j c_j^{\dagger} c_{j+1}] = -t\sum_j(\delta_{i,j}c_{j+1} + \delta_{i,j+1}c_j) = -t\sum_{j=\pm 1} c_{i+j}.$$

Now for the second term in equation (5.103) $\langle\{\dot{c}_i(t), c_j^{\dagger}(t')\}\rangle$ we get,

$$\sum_i V_0\{c_i(t), c_j^{\dagger}(t')\} - t\sum_{j=\pm 1}\{c_{i+j}(t), c_j^{\dagger}(t')\}.$$

Thus the EOMs of the elements of the retarded Green's function can be written as,

$$\begin{aligned}\left(i\frac{d}{dt} - V_0\right)G_{11}^{R}(t, t') &= \delta(t - t') - tG_{21}^{R}(t, t')\\ \left(i\frac{d}{dt} - V_0\right)G_{21}^{R}(t, t') &= -\,tG_{11}^{R}(t, t').\end{aligned} \tag{5.104}$$

Thus we get a set of coupled equations that need to be solved. Doing a Fourier transform to the frequency domain, one gets,

$$\begin{aligned}(\omega - V_0 + i\eta)G_{11}^{R}(\omega) &= 1 - tG_{21}^{R}(\omega)\\ (\omega - V_0 + i\eta)G_{21}^{R}(\omega) &= -\,tG_{11}^{R}(\omega).\end{aligned} \tag{5.105}$$

In a similar manner, the EOM for the advanced Green's function can be obtained by changing the sign of the imaginary term $i\eta \rightarrow -i\eta$.

Let us see in the following how we get a Dyson equation for this rather simple case. Define the undressed Green's function at each site,

$$G_1^{0\mathrm{R}}(\omega) = G_2^{0\mathrm{R}}(\omega) = \frac{1}{\omega - V_0 + i\eta} = G^{0\mathrm{R}}(\omega) \quad \text{(say)}.$$

This yields,

$$G_{11}^{\mathrm{R}}(\omega) = G^{0\mathrm{R}}(\omega) - G^{0\mathrm{R}}(\omega)tG_{21}^{\mathrm{R}}(\omega)$$

and

$$G_{21}^{\mathrm{R}}(\omega) = -G^{0\mathrm{R}}(\omega)tG_{11}^{\mathrm{R}}(\omega).$$

In a matrix notation,

$$G^{\mathrm{R}} = G^{0\mathrm{R}} + G^{0\mathrm{R}}VG^{\mathrm{R}} \tag{5.106}$$

where,

$$V = \begin{pmatrix} 0 & -t \\ -t & 0 \end{pmatrix}$$

is a matrix that connects the individual sites. Please note that without this term, there would be no difference between G^{R} and $G^{0\mathrm{R}}$, where the latter denotes the Green's function at individual sites.

In a general sense, V is called the self-energy that distinguishes between the dressed and the undressed Green's functions. It is usually written with $\Sigma(\omega)$ and accounts for the renormalization of the single-particle states. Without the interaction effects, such as the case here, the self-energy has a simple form. While for an interacting problem, it is a challenging task.

5.5.2 Hartree–Fock approximation

Let us now look at the Fock term, it requires $k_1 = k_2 + q$ and $\sigma = \sigma'$. Further, the fermionic loop brings in a negative sign. Consider evaluation of equation (5.95). Since we are discussing equal time Green's function, the associated frequency, ω has to be vanishingly small. The integral in equation (5.95) is,

$$\int \frac{d\Omega}{2\pi} G_0(k - q, \omega - \Omega) = \int \frac{d\Omega}{2\pi} \frac{1}{\omega - \Omega - \xi_{k-q} + i\eta_k}$$

The integral can be computed using the residue theorem. Note that there is a simple pole at $\Omega = -\xi_{k-q} - i\eta_k$ which yields,

$$i\int \frac{d\Omega}{2\pi} G_0(k - q, \omega - \Omega) = -1 = \Theta(-\xi_{k-q}) = -n(k - q). \tag{5.107}$$

Thus the self-energy at the Hartree–Fock level is,

$$\begin{aligned}\Sigma(k, \omega) &= -\int \frac{d^3q}{(2\pi)^3} V_q \; n(k-q)\\ &= -\int \frac{d^3q}{(2\pi)^3} V_{k-q} \; n(q) \quad \text{with} \quad k-q \to q\\ &= -\int_{|q|<k_F} \frac{d^3q}{(2\pi)^3} \frac{4\pi e^2}{|k-q|^2}\\ &= -\frac{e^2 k_F}{\pi}\left[1 + \frac{1-(k/k_F)^2}{2(k/k_F)} \ln\left|\frac{k_F+k}{k_F-k}\right|\right].\end{aligned} \tag{5.108}$$

At the Hartree–Fock level, $\Sigma(k, \omega)$ simply renormalizes the non-interacting band dispersion.

5.6 Finite temperature Green's function

The principle difference between the zero temperature and the non-zero temperature formalisms stems from the requirement of the averaging to be done over a plethora of excited states, unlike the zero temperature case, where the sum only involves a unique ground state. Further, it is clear from classical statistical mechanics that a Boltzmann weight, namely a factor $e^{-\beta\mathcal{H}}$ denotes the probability that a system described by a Hamiltonian $\mathcal{H}$ in thermal equilibrium occupies a state with an energy E, E being the eigenvalue of $\mathcal{H}$. T is the absolute temperature, and k_B is the Boltzmann constant. On the other hand, the time evolution of a state (irrespective of a single-particle or a many-body state) is given by $e^{-i\mathcal{H}t/\hbar}$ which is an oscillatory function[10]. This mismatch in the argument of the exponential induces difficulties in the systematic perturbation expansion of the Green's function in terms of the interaction term $\mathcal{H}'$. However, significant simplifications arise when imaginary time is introduced in dealing with system properties, instead of the real ones.

The definition of the Green's function at finite temperature is given by,

$$G(k, t, t') = \frac{\mathrm{Tr}[e^{-\beta\mathcal{H}} c_{k\sigma}(t) c^{\dagger}_{k\sigma}(t')]}{\mathrm{Tr}[e^{-\beta\mathcal{H}}]} \tag{5.109}$$

Tr denotes the trace which is the sum over the diagonal elements, that is, $\sum_n \langle n|\cdots\cdots|n\rangle$

$$c_{k\sigma}(t) = e^{i\mathcal{H}t} c_{k\sigma}(0) e^{-i\mathcal{H}t} \tag{5.110}$$

where $\mathcal{H} = \mathcal{H}_0 + \mathcal{H}'$. It may be noted that $\mathcal{H}'$ appears in both $e^{\pm i\mathcal{H}t}$ and $e^{-\beta\mathcal{H}}$. Thus the expansion of $\mathcal{H}$ in terms of $\mathcal{H}'$ is included in $e^{\pm i\mathcal{H}t}$ in the form of a S-matrix. A natural question is: should we consider a similar expansion for $e^{-\beta\mathcal{H}}$?

[10] The underlying assumption is that the Hamiltonian is not an explicit function of time.

Matsubara [7] realized that β $(=\frac{1}{k_B T})$ may be considered as the complex time and hence the arguments of the two exponentials can be combined together. However, finally what is formulated is just the converse, that is, time is treated as the complex temperature. Basically, the whole idea is to treat t and β (inverse temperature) as the real and imaginary parts of a complex variable. Thus, it will require a unique expansion of the S-matrix.

Another motivation for the Matsubara formalism is provided by examining the thermal occupation numbers for bosons $n_B = (\frac{1}{e^{\beta\hbar\omega_q}-1})$ and for fermions $n_F = \left(\frac{1}{e^{\beta\xi_k}+1}\right)$. These occupation numbers can be expanded in the form of a series given by[11],

$$n_F(\xi_k) = \frac{1}{e^{\beta\xi_k+1}} = \frac{1}{2} + \frac{1}{\beta}\sum_{n=-\infty}^{\infty}\frac{1}{(2n+1)\frac{i\pi}{\beta} - \xi_k} \tag{5.111}$$

and,

$$n_B(\omega_q) = \frac{1}{e^{\beta\hbar\omega_q-1}} = -\frac{1}{2} + \frac{1}{\beta}\sum_{n=-\infty}^{\infty}\frac{1}{\frac{2ni\pi}{\beta} - \hbar\omega_q} \tag{5.112}$$

where $n = 0, \pm1, \pm2, \ldots$. It is convenient to define the frequencies at the poles,

(i) $\omega_n = (2n+1)\pi/\beta$ for fermions.

(ii) $\omega_n = 2n\pi/\beta$ for bosons.

That is, the fermionic frequencies have poles at odd multiples of π/β and for bosons, the corresponding poles are at even multiples of π/β (including zero).

The sums appearing in the thermal occupation numbers for the fermions and the bosons hint towards a series of poles on the imaginary axis or a branch cut (see [6] for details) that needs to be dealt with. In order to proceed, let us look at the form of the S-matrix, which was central to the discussion of the Green's function at $T = 0$. We shall show how an operator can be evolved in imaginary time τ. The time evolution is written as,

$$O(\tau) = e^{\tau\mathcal{H}}O(0)e^{-\tau\mathcal{H}} \tag{5.113}$$

where O is an arbitrary operator. We have removed the cap on the top of the operators for notational simplicity. An important point requires a mention here. In general one uses a grand canonical ensemble, that is, $\mathcal{H} \to \mathcal{H} - \mu N$, where μ is the chemical potential, and N denotes the number of particles. Sometimes, $\mathcal{H} - \mu N$ is denoted by K, however, we shall continue using $\mathcal{H}$ in our discussion. The EOM can be written as,

[11] Any meromorphic function may be expanded as a summation over its poles, and residues at the poles. See reference [6] for details.

$$\frac{d}{d\tau}O(\tau) = -[O(\tau), \mathcal{H}]. \tag{5.114}$$

In analogy with the zero temperature formalism, the S-matrix can be written as,

$$S(\tau, \tau') = e^{\mathcal{H}_0\tau}e^{-\mathcal{H}(\tau-\tau')}e^{-\mathcal{H}_0\tau} \tag{5.115}$$

where, $\mathcal{H} = \mathcal{H}_0 + \mathcal{H}'$. Inverting the above equation,

$$e^{-\mathcal{H}(\tau-\tau')} = e^{-\mathcal{H}_0\tau}S(\tau, \tau')e^{\mathcal{H}_0\tau}. \tag{5.116}$$

It may be noted that $S(\tau, \tau')$ is not unitary, that is,

$$S^\dagger(\tau, \tau')S(\tau, \tau') \neq \mathbb{1}.$$

However, one still has,

$$S(\tau, \tau) = \mathbb{1}.$$

Further, the multiplication of the matrices obeys,

$$S(\tau_1, \tau_2)S(\tau_2, \tau_3) = e^{\mathcal{H}_0\tau_1}e^{-\mathcal{H}(\tau_1-\tau_2)}e^{\mathcal{H}\tau_2}e^{-\mathcal{H}(\tau_2-\tau_3)}e^{\mathcal{H}_0\tau_3} = S(\tau_1, \tau_3). \tag{5.117}$$

Also,

$$S(\tau_1, \tau_2) = S^{-1}(\tau_2, \tau_1). \tag{5.118}$$

Thus, the time-evolved operator can now be written in terms of S-matrix as,

$$\begin{aligned} O(\tau) &= e^{\mathcal{H}\tau}e^{-\mathcal{H}_0\tau}O(0)e^{\mathcal{H}_0}e^{-\mathcal{H}\tau} \\ &= S(0, \tau)\tilde{O}(\tau)S(\tau, 0) \end{aligned} \tag{5.119}$$

where,

$$\tilde{O}(\tau) = e^{\tau\mathcal{H}_0}O(0)e^{-\tau\mathcal{H}_0}$$

Thus most of the relationships that we have learnt for the $T = 0$ case, almost hold. This is true as well for the EOM of the S-matrix, namely,

$$\begin{aligned} \frac{d}{d\tau}S(\tau, \tau') &= e^{\tau\mathcal{H}_0}(\mathcal{H}_0 - \mathcal{H})e^{-\mathcal{H}(\tau-\tau')}e^{-\tau\mathcal{H}_0} \\ &= -\mathcal{H}'(\tau)S(\tau, \tau'). \end{aligned} \tag{5.120}$$

As earlier, the EOM for the S-matrix involves the interaction term, $\mathcal{H}'$. The integral solution of equation (5.120) is,

$$S(\tau, \tau') = T_\tau e^{-\int_{\tau'}^{\tau} d\tau''\mathcal{H}'(\tau'')} \tag{5.121}$$

τ is complex time, or equivalently the temperature (temperature enters through β), with T_τ being the complex time ordering operator. Thus, we have,

$$S(\beta, 0) = e^{-\int_{\tau'}^{\tau} d\tau''\mathcal{H}'(\tau'')}. \tag{5.122}$$

We are now equipped to write down the single-particle Green's function at finite temperature, namely,

$$\mathcal{G}_k(\tau, \tau') = -\frac{1}{Z}\langle c_k(\tau)c_k^\dagger(\tau')\rangle\Theta(\tau - \tau') + \frac{1}{Z}\langle c_k^\dagger(\tau')c_k(\tau)\rangle\Theta(\tau' - \tau). \tag{5.123}$$

Introducing the time ordering operator,

$$\mathcal{G}_k(\tau, 0) = \mathcal{G}_k(\tau) = -\frac{1}{Z}\langle e^{-\beta\mathcal{H}}T_\tau c_k(\tau)c_k^\dagger(0)\rangle - \frac{1}{Z}\,\mathrm{Tr}\,[e^{-\beta\mathcal{H}}T_\tau c_k(\tau)c_k^\dagger(0)] \tag{5.124}$$

where,

$$Z = \mathrm{Tr}\,(e^{-\beta\mathcal{H}}) = \mathrm{Tr}\,[e^{-\beta\mathcal{H}_0}S(\beta, 0)]$$

is the partition function. These are called Matsubara Green's functions [7]. We shall prefer writing as a 'Trace' instead of using $\langle..\rangle$ as the denominator contains Z which evidently contains a 'Trace'. Also, take note of the extra minus sign in front of the definition of $\mathcal{G}$ which was not there in the $T = 0$ formalism. Further, instead of writing $\mathcal{G}$ the momentum space, one can use real space variables $\mathbf{r}$, $\mathbf{r}'$.

A little effort will cast the Green's function in the form familiar to the $T = 0$ case,

$$\mathcal{G}_k(\tau) = -\frac{1}{Z}\,\mathrm{Tr}\,[e^{-\beta\mathcal{H}_0}S(\beta, 0)S(0, \tau)c_k(\tau)S(\tau, 0)c_k^\dagger(\tau')]. \tag{5.125}$$

Using the time ordering operator,

$$\begin{aligned}\mathcal{G}_k(\tau) &= -\frac{1}{Z}\,\mathrm{Tr}\,[e^{-\beta\mathcal{H}_0}T_\tau S(\beta, \tau)S(0, \tau)c_k(\tau)S(\tau, 0)c_k^\dagger(0)]\\ &= -\frac{1}{Z}\,\mathrm{Tr}\,[e^{-\beta\mathcal{H}_0}T_\tau c_k(\tau)c_k^\dagger(0)S(\beta, 0)].\end{aligned} \tag{5.126}$$

Using the definition of Z,

$$\mathcal{G}_k(\tau) = -\frac{\langle T_\tau c_k(\tau)c_k^\dagger(0)S(\beta, 0)\rangle}{\langle S(\beta, 0)\rangle}. \tag{5.127}$$

For the non-interacting problem, the Green's function takes the form,

$$\begin{aligned}\mathcal{G}_k^{(0)}(\tau) &= \langle T_\tau c_k(\tau)c_k^\dagger(0)\rangle\\ &= \frac{1}{i\omega_n - \xi_k}\end{aligned} \tag{5.128}$$

where $\xi_k = \epsilon_k - \mu$, ϵ_k being the band energy.

There is a further point that deserves special mention (which will also be taken up later). The Matsubara Green's function written above can be analytically continued to yield the retarded Green's function at $T = 0$, via,

$$i\omega_n = \omega + i\eta$$

where $\eta = 0^+$. Thus the knowledge of the Matsubara function yields the retarded Green's function at $T = 0$ via analytic continuation. However, the converse does not happen, that is, starting from the zero temperature formalism, it is not possible to access the finite temperature properties. We shall return to this topic a little later.

5.6.1 Properties of the Matsubara Green's function

Remember the real time that appears in the argument of the exponential in the zero temperature formalism makes the phase periodic. The Matsubara Green's function, $\mathcal{G}(\tau)$ has a similar periodicity. In fact, for the fermionic case, it is anti-periodic, while for bosons, it is a periodic function.

Consider $\tau < 0$, in which case,

$$\mathcal{G}_k(\tau) = \frac{1}{Z}\mathrm{Tr}\,[e^{-\beta\mathcal{H}}c_k^\dagger(0)c_k(\tau)]. \tag{5.129}$$

Using the cyclic invariance of 'Trace' twice[12], one obtains,

$$\mathcal{G}_k(\tau) = -\mathcal{G}_k(\tau + \beta). \tag{5.130}$$

The proof of this is simple. Recognizing $e^{-\beta\Omega_G} = \mathrm{Tr}[e^{-\beta\mathcal{H}}]$, where Ω_G is called as the grand potential ($= -\ k_\mathrm{B}T\ \ lnZ$). The Matsubara Green's function can be written as,

$$\begin{aligned}\mathcal{G}_k(\tau) &= \frac{e^{\beta\Omega_G}}{Z}\mathrm{Tr}\,[(e^{\tau\mathcal{H}}c_k e^{-\tau\mathcal{H}})e^{-\beta\mathcal{H}}]\\ &= \frac{e^{\beta\Omega_G}}{Z}\mathrm{Tr}\,[(e^{-\beta\mathcal{H}}e^{\beta\mathcal{H}})(e^{\tau\mathcal{H}}c_k e^{-\tau\mathcal{H}})e^{-\beta\mathcal{H}}c_k^\dagger]\\ &= \frac{1}{Z}\mathrm{Tr}\,[c_k(\tau+\beta)c_k^\dagger(0)]\\ &= -\ \mathcal{G}_k(\tau+\beta).\end{aligned} \tag{5.131}$$

The last line follows since $-\beta < \tau$, hence we should have $\tau + \beta > 0$. For $\tau > 0$, the same steps can be repeated to yield,

$$\mathcal{G}_k(\tau - \beta) = -\mathcal{G}_k(\tau). \tag{5.132}$$

It may also be noted that the value of τ is enclosed between $[-\beta:+\beta]$. For bosons, the relationship is periodic, that is,

$$\mathcal{G}_k(\tau) = \mathcal{G}_k(\tau + \beta). \tag{5.133}$$

The main advantages for the bounded value of τ and the anti-periodicity of the Matsubara Green's function are that it can be expanded in the Fourier series, such as,

$$\mathcal{G}_k(\tau) = \frac{1}{\beta}\sum_{n=-\infty}^{+\infty} e^{-i\omega_n\tau}\mathcal{G}_k(i\omega_n) \tag{5.134}$$

[12] $\mathrm{Tr}[ABC] = \mathrm{Tr}[BCA] = \mathrm{Tr}[CAB]$.

where $\omega_n = (2n+1)\frac{\pi}{\beta}$ for fermions ($\omega_n = \frac{2\pi n}{\beta}$ for bosons). The anti-periodicity for the fermions is automatically taken care of as,

$$e^{-i\omega_n\beta} = e^{-i(2n+1)\pi} = -1.$$

Further,

$$\mathcal{G}_k(i\omega_n) = \int_0^\beta d\tau e^{i\omega_n\tau}\mathcal{G}_k(\tau). \tag{5.135}$$

One important conclusion arises: the domain of definition of the Matsubara Green's function is a restriction of the complex time τ to have values, such that, $-\beta < \tau + \beta$. In perturbation expansion, we shall never need $\mathcal{G}_k(\tau)$ beyond this interval. We shall see later that there are frequency sums that have to be computed for the branch cuts present therein.

5.6.2 Matsubara Green's function and the retarded propagator at $T = 0$

We have mentioned in passing about a connection between the Matsubara Green's function and the retarded propagator via analytic continuation. We discuss this more elaborately in the following and establish the spectral representation of $\mathcal{G}$. Consider,

$$\mathcal{G}_k(\tau) = -\frac{1}{Z}\operatorname{Tr}[e^{-\beta\mathcal{H}}c_k(\tau)c_k^\dagger(0)]\Theta(\tau) + \frac{1}{Z}\operatorname{Tr}[e^{-\beta\mathcal{H}}c_k^\dagger(0)c_k(\tau)]\Theta(-\tau). \tag{5.136}$$

Fourier transforming,

$$\mathcal{G}_k(i\omega_n) = \int_0^\beta d\tau e^{i\omega_n\tau}\mathcal{G}_k(\tau). \tag{5.137}$$

Assume $\omega_n > 0$, the contour is illustrated by blue in figure 5.8, where the contour of integration is deformed within the domain. Here $\mathrm{Re}(t) = \mathrm{Im}(\tau) > 0$.

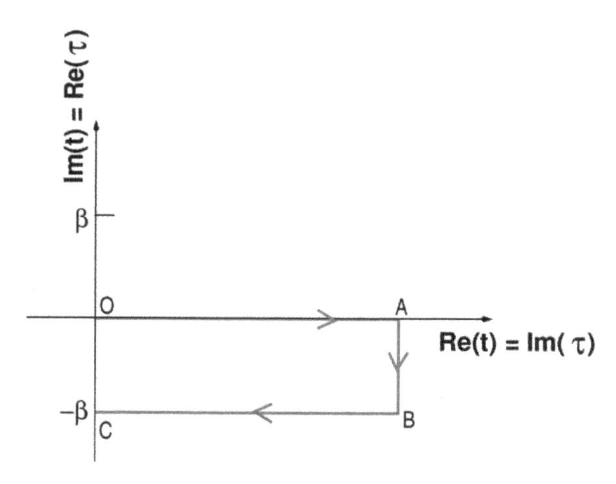

Figure 5.8. The contour of integration is shown by blue colour for evaluation of the integral in equation (5.137).

For $\mathrm{Im}(\tau) \to \infty$, the contribution from the small vertical segment will be vanishingly small owing to the decaying exponential, $e^{i\omega_n\tau}$. Now the integral becomes,

$$\begin{aligned}\mathcal{G}_k(i\omega_n) = &\frac{i}{Z}\int_0^{\infty} dt\left\{-\mathrm{Tr}\left[e^{i\mathcal{H}t}c_k^S e^{-i\mathcal{H}t}c_k^{\dagger S}\right]\right\}e^{i\omega_n(it)}\\ &+\frac{i}{Z}\int_0^{\infty} dt\left\{-\mathrm{Tr}\left[e^{i\mathcal{H}(t-i\beta)}c_k^S e^{-i\mathcal{H}(t-i\beta)}c_k^{\dagger S}\right]e^{i\omega_n i(t-i\beta)}\right\}\end{aligned} \tag{5.138}$$

where c_k^S and $c_k^{\dagger S}$ are the operators in the Schrödinger representation. Also, we have used,

$$e^{i\omega_n i(-i\beta)} = e^{i\omega_n\beta} = -1$$

owing to the odd integral values of ω_n. Also, we have used,

$$\int_0^{\infty}\cdots = -\int_{\infty}^{0}.$$

The two integrals can be combined to yield,

$$\begin{aligned}\mathcal{G}_k(i\omega_n) &= \frac{i}{Z}\int -\mathrm{Tr}\left[c_k^{\dagger}(0)c_k(t)\right]e^{i(i\omega_n)t}\\ &= -i\int dt\langle c_k^{\dagger}(0)c_k(t)\rangle e^{i(i\omega_n)t}.\end{aligned} \tag{5.139}$$

Thus, we recover the retarded Green's function at $T = 0$ by analytically continuing in the frequency domain, that is,

$$G^{\mathrm{R}}(k,\omega) = \lim_{i\omega_n\to\omega+i\eta}\mathcal{G}(k, i\omega_n). \tag{5.140}$$

In analogy of the $T = 0$ formalism, the spectral representation of the Matsubara Green's function can be written as,

$$\mathcal{G}(k, i\omega_n) = \int_{-\infty}^{+\infty}\frac{d\omega'}{2\pi}\frac{A(k,\omega')}{i\omega_n - \omega'}. \tag{5.141}$$

The Matsubara Green's function is shown schematically in figure 5.9.

5.6.3 Matsubara frequency sums

Let us calculate a generic quantity that involves sum over the Matsubara frequencies. Consider a sum,

$$S = \frac{1}{\beta}\sum_{\omega_n} f(i\omega_n) \tag{5.142}$$

where f is some arbitrary function. To compute S, we may consider the integral,

$$I = \int_{\mathrm{c}}\frac{dz}{2\pi i}f(z)n(z) \tag{5.143}$$

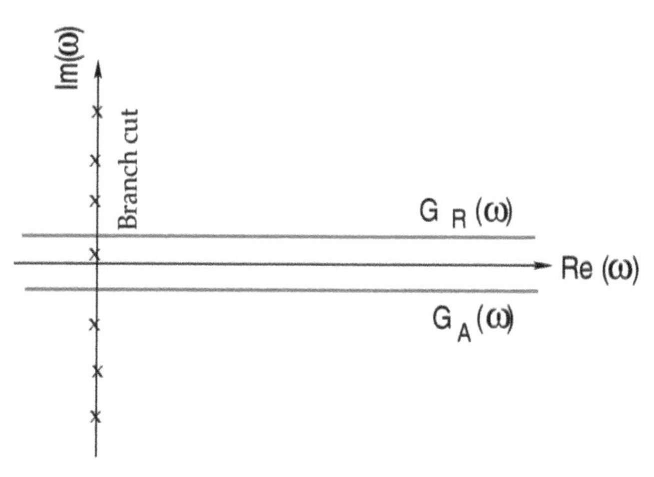

Figure 5.9. Analytic structure of $G(\omega)$ in the complex frequency plane is shown. $G(\omega)$ reduces to $G^{\mathrm{R}}(\omega)$, $G^{\mathrm{A}}(\omega)$ or $\mathcal{G}(i\omega_n)$ depending on ω.

where $n(z)$ is the thermal occupation number given by,

$$n(z) = \frac{1}{e^{\beta z} \pm 1} \tag{5.144}$$

where the +ive sign is applicable for fermions and −ive sign for bosons. $n(z)$ has simple poles at imaginary frequencies $z = i\omega_n$. The value of the integral I is related to S for a given path C.

Let us take a concrete example. Consider,

$$S = \frac{1}{\beta}\sum_{\omega_n}\frac{e^{i\omega_n\eta}}{i\omega_n - \xi}; \qquad \eta = 0^+. \tag{5.145}$$

The corresponding integral I takes the form,

$$I = \oint_{\mathrm{c}} \frac{dz}{2\pi i}\frac{e^{z\eta}}{z - \xi}n(z) \tag{5.146}$$

where C is a circle of radius R ($R \to \infty$) around $z = 0$ in the complex plane. In the limit $z \to \infty$, $e^{z\eta}$, with $\mathrm{Re}(z) < 0$ will be very small, so that the integral vanishes along R (Jordan's lemma [6]). Now evaluate I using Cauchy's residue theorem. Note that there are two kinds of poles, namely a simple pole at $z = \xi$, and the ones coming from the Matsubara frequencies, ω_n. Hence,

$$\begin{aligned} I &= \sum_n \text{Residue } (i\omega_n) + \text{Residue } (z = \xi) \\ &= \pm S + n(\xi) = 0. \end{aligned} \tag{5.147}$$

Here −ive sign is applicable to fermions, and +ive sign is for bosons. This yields,

$$S = \frac{1}{\beta}\sum_{i\omega_n}\frac{e^{i\omega_n\eta}}{i\omega_n - \xi} = n_{\rm F}(\xi) \quad \text{fermions} \\ = -\, n_{\rm B}(\xi) \quad \text{bosons} \tag{5.148}$$

As a second example, consider the bare particle–hole susceptibility (also known as the charge susceptibility), namely,

$$\chi^{(0)}(q, i\omega_n) = -\frac{2}{\beta}\sum_{k,\Omega_m}\mathcal{G}_0(k, i\omega_n)\mathcal{G}_0(k + q, i\omega_n + i\Omega_m) \tag{5.149}$$

where the non-interacting Green's functions are written as,

$$\mathcal{G}_0(k, i\omega_n) = \frac{1}{i\omega_n - \xi_k} \\ \mathcal{G}_0(k + q, i\omega_n + i\Omega_m) = \frac{1}{i\omega_n + i\Omega_m - \xi_{k+q}}. \tag{5.150}$$

Putting them in the above sum, and using the form of S from equation (5.148) for fermions, one gets,

$$\chi^{(0)}(q, i\omega_n) = -\int dk\sum_{\Omega_m}\frac{n(\xi_k) - n(\xi_{k+q})}{i\Omega_m + \xi_k - \xi_{k+q}}. \tag{5.151}$$

5.7 Summary and outlook

At the outset, we have introduced the topic of second quantization, and have discussed writing down generic Hamiltonians comprising one- and two-body terms The process of symmetrization of the many-particle wavefunction is achieved by writing down the operators in the second quantized notation. Hence we have introduced the propagator, where obtaining the Green's function for a many-particle system is of primary interest. A number of useful and relevant concepts, such as, S-matrix, time ordering, Gell-Mann and Low theorem, representations in quantum mechanics, etc are elaborately discussed. A perturbation expansion of the Green's function is shown to yield a technique for dealing with the scattering processes in interacting systems. The technique involves obtaining the Feynman diagrams via using Wick's theorem. For the simplicity in discussion, we have restricted ourselves to the six first order diagrams. The need to go beyond the first order perturbation theory is emphasized, and a particularly simple solution emerges at the infinite order, which leads us to the concept of self-energy and the Dyson equation. The self-energy is computed at the Hartree–Fock level that is shown to renormalize the band energies for the Coulomb potential. Finally, to establish relevance with experiments, we go to finite temperatures and discuss the Matsubara Green's function, where the latter is shown to yield retarded Green's function at $T = 0$ under analytic continuation to real frequencies. The trick to compute sums over the Matsubara frequencies is demonstrated for simple problems.

References

[1] Altland A and Simmons B 2010 *Condensed Matter Field Theory* (Cambridge: Cambridge University Press)
[2] Mahan G D 2013 *Many Particle Physics* 3rd edn (Berlin: Springer)
[3] Fock V A 1932 *Z. Phys.* **75** 622–47
[4] Green G 2008 *An essay on the mathematical analysis to the theories of electricity and magnetism* (arXiv:0807.0088)
[5] Stakgold I and Holst M J 2011 *Green's Functions and Boundary Value Problems* 3rd edn (Reading, MA: Addison-Wesley)
[6] Brown J W and Churchill R V 2014 *Complex Variables and Applications* 9th edn (New York: McGraw-Hill)
[7] Matsubara T 1955 *Prog. Theor. Phys.* **14** 351

IOP Publishing

Saurabh Basu

Chapter 6

Superconductivity

6.1 Introduction

In solid state physics, many of the phenomena that we are familiar with occur because of the interparticle interactions among the charge carriers, thereby resulting in an ordered state. The effects of such interactions are best perceptible at low temperature. With increase in temperature, the thermal motion of the carriers gains prominence, and outdoes the ordering process. Superconductivity denotes one such ordered phase of matter which stabilizes at low temperature, and vanishes when the temperature is increased beyond a certain critical point.

In the year 1908, H Kamerlingh-Onnes [1] at the low temperature lab in Leiden, Netherlands successfully liquified helium (He). At normal atmospheric pressure, the boiling point of He was found to be 4.2 K, thereby making exploration of material properties feasible at low temperatures. Studying electronic conductivity (or resistivity) of metals seemed like a normal choice, as metals are primarily characterized by their electrical resistance. Measurements done at low temperatures may yield the following possibilities:

(i) the resistance vanishes gradually with decreasing temperature;
(ii) it may result in a small but finite value at very low temperature;
(iii) it may have a minimum at low temperature, and finally show an upturn before diverging at very low temperatures.

In particular, the last possibility receives support on physical grounds in the sense that at sufficiently low temperature, the carriers are likely to be bound to their respective atoms. Thus their ability to move around and contribute to the conductivity vanishes.

K Onnes realized the importance of studying conductance characteristics of high purity metals at low temperatures. Initially he started with gold (Au) and platinum (Pt), mainly because they are noble metals and available in pure form. However, he shifted his attention towards mercury (Hg) which can be obtained in a highly pure form via multiple distillations. Precisely at the boiling point of liquid He, he found

doi:10.1088/978-0-7503-3031-2ch6

that the resistance of an ultra-clean Hg sample sharply vanishes, and at further lower temperatures, the resistance becomes immeasurably small (see figure 6.1). The temperature at which the resistance disappears is called the transition temperature, T_c. Below that a new state of matter emerges, which either completely expels the magnetic field or, in some cases, traps magnetic flux having values that are in integer multiples of the flux quantum, Φ_0 ($\Phi_0 = 2.07 \times 10^{-15}$ Wb). This was indeed a surprise, since laws of electromagnetic induction predict that ideal conductors retain, and do not expel the magnetic field that is trapped inside.

A complete theoretical understanding of the microscopic phenomena had to wait about half a century till the theoretical derivation by Bardeen, Cooper and Schrieffer (BCS) in 1957 came into existence, for which they were awarded the Nobel prize in 1972. They realized at the transition point, the electrons pairwise condense into a new phase which is a coherent matter wave with well-defined phase relationship among the pairs. These electrons form pairs mediated via quantized lattice vibrations, namely the phonons.

For nearly eight decades after the experimental discovery of superconductivity in Hg in 1908, it remained a low temperature phenomena till Bednorz and Müller discovered the onset of superconductivity in the copper oxide planes at larger temperatures. As the

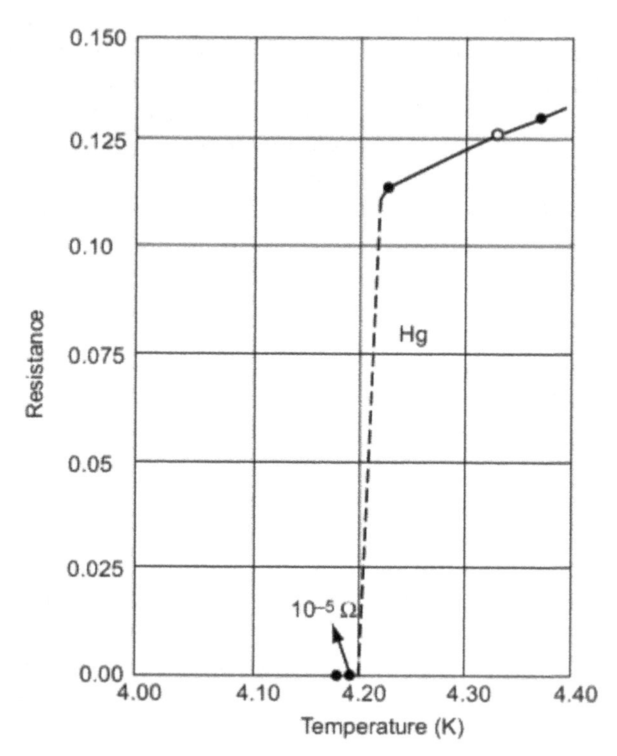

Figure 6.1. The fall of the electrical resistance of Hg (in Ω) as a function of temperature is shown. Near 4.2 K, there is a sharp fall of resistance, below which the resistance becomes almost zero (after K Onnes' original experiment [1]).

years passed, discoveries of a large number of copper oxide superconductors were made with larger and larger critical temperatures, which was thought to have far reaching consequences in terms of their industrial applications. There was moderate success on that front, however, there were many surprises, which arose as the experimental data started coming in on the physical properties of these superconductors. Quite likely, the normal state[1] of these copper oxide materials is very different than those for the conventional superconductors that were completely understood within the BCS theory. Since the beginning point of building up a theory, which in this case is the normal state, is missing, or at least, is unlike the conventional metallic state, a complete understanding of these superconductors remained elusive.

6.1.1 Historical developments

Empirical rules have been suggested from time to time to enable identification of possible new superconducting materials. Rules proposed by Matthias [2] in the 1960s are especially noteworthy. One of them is based on the number of valence electrons. It is found that the materials with on an average 5 to 7 electrons per atom show relatively high T_c's. For example, RuMo with an average of 7 electrons shows $T_c = 10.6$ K, whereas, ruthenium (Ru) with an average of 8 electrons shows $T_c = 0.5$ K, and molybdenum (Mo) with an average of 6 electrons shows $T_c = 1$ K. Later on, a structural dependence was seen on T_c. A-15 structure or the β-tungsten shows high T_c. V_3Si and Nb_3Ge are examples of this structure and they show $T_c = 17.1$ K and 23.2 K, respectively. For quite a few years, this 23.2 K barrier was never crossed. Among the organic materials, the intercalated compounds of graphite (e.g. C_8K) become superconducting below 1 K, while K_3C_{60} has $T_c \approx 18$ K. Generally alkali salts of C_{60} have been found to be superconducting with T_c ranging from 18 to 35 K.

If we refer to the periodic table and consider the top row of the middle section of the periodic table, that is, group IVB (Cr, Mn) till group VIII (Fe, Co, Ni), they have strong magnetic character. While along the same row, Ge, As, Se or some of the neighbouring elements show superconductivity. Since both the phenomena have electronic origin, and their physical properties are quite distinct, it forces us to wonder what could be the genesis of this difference. It probably lies in the extent of the wavefunction of the electrons: for magnets, the wavefunction has small spatial extent, that is spread over a smaller number of atoms, while for superconductors, the spread is large, thereby promoting pairing phenomena between the carriers.

6.1.2 Physical properties

Before we proceed to discuss the physical properties of superconductors, we wish to make the plans clear for our upcoming discussion. We shall mainly mention salient features of the superconductors that make them worthy of study. However, most of the ideas and concepts are introduced with minimal amount of derivation. For a thorough understanding of some of the properties, one has to wait for the BCS theory, a topic that we discuss somewhat elaborately afterwards.

[1] The normal state refers to the metallic state which is expected as superconductivity is lost.

Let us rewind the ongoing discussion for more details. The resistance of a metal drops to zero very sharply, within a temperature window of $\Delta T \sim 10^{-5}$ K below a certain critical temperature, T_c. The superconducting state is characterized by zero electrical resistance. Besides, there is no change in crystal structure as verified by x-ray diffraction above and below the transition temperature. Thus to list out the properties, the state is characterized by:

(i) infinite electrical conductivity, that is, $\sigma \to \infty$;
(ii) finite current density, j = finite;
(iii) the specific heat as a function of temperature shows a jump at $T = T_c$, and thus the transition to the superconducting state involves latent heat, and hence the transition from the superconducting state to the normal state is a second-order phase transition;
(iv) vanishing values of the electric field, $E \to 0$ inside the superconducting sample;
(v) constant magnetic field, $B \to$ constant inside a superconductor.

Clearly the last two statements defy the laws of classical electrodynamics. This can be understood in the following way. Using Ohm's law one can write, $\mathbf{j} = \sigma \mathbf{E}$ for j having a finite value and $\sigma \to \infty$, E has to be zero. Hence, using Maxwell's equation,

$$\nabla \times \mathbf{E} = -\frac{\partial \mathbf{B}}{\partial t} \quad \text{or,} \quad \mathbf{B} = \text{constant}.$$

6.1.3 Meissner effect

The complete and sudden vanishing of the electrical resistance below a certain temperature raised a lot of questions. Some physical phenomena that were operative to render electrical resistance above T_c, were suddenly becoming ineffective. Meissner and Ocshenfeld (1933) meanwhile discovered that a superconductor completely excludes an external magnetic field. They measured the magnetic field distributed outside the superconducting materials (such as lead (Pb) or tin (Sn)) which are cooled below their respective transition temperatures while the magnetic field is switched on. The corresponding results could not be explained by superconductors being just resistance-less metals.

Above a certain critical magnetic field (of the order of a few Oersted), there is no expulsion of the magnetic flux, where superconductivity disappears and the material reverts back to its normal resistive state and the magnetic field fully penetrates it. This also endorses a close link between magnetism and superconductivity.

The exclusion of the magnetic field from a superconductor takes place regardless whether the material becomes superconducting before or after the external field is applied. At equilibrium, the external field is cancelled in its interior by screening fields produced by the skin current.

The total exclusion of the magnetic field from inside the superconductors assigns a property known as *perfect diamagnetism*. It can be understood as follows,

$$B = \mu_0(H_{\text{ext}} + M).$$

Since $B = 0$ inside a superconductor, the induced magnetization M cancels the external field, H_{ext}. Thus, the magnetic susceptibility, χ becomes,

$$\chi = \frac{M}{H_{\text{ext}}} = -1.$$

Thus, no known material is more diamagnetic than a superconductor. Typical diamagnetic susceptibility of metals is about 10^{-5}–10^{-6}. Thus an usual diamagnetic material only expels a small fraction (about 10^{-5}) of the external field, while a superconductor does it completely. However, even if we claim that the magnetic field is totally expelled from a superconductor, the fact is that it enters only up to a certain distance, called penetration depth.

6.1.4 Perfect conductors and superconductors

In order to differentiate a perfect conductor (which may show a gradual vanishing of the resistance as $T \to 0$), and a superconductor, we resort to field-cooled (FC) and zero-field-cooled (ZFC) techniques that are schematically shown in figures 6.2 and 6.3. If we consider cooling perfect conductors and superconductors in the absence of a magnetic field (left panels of figures 6.2 and 6.3), they behave identically when an external magnetic field is applied. That is, they expel the field by developing surface current which prohibits the field lines from penetrating beyond a certain distance (discussed below). However, in the presence of an external field, the behaviour of a conductor and that of a superconductor are quite distinct. When a conductor is cooled to a perfect conducting state in the presence of an external field, the flux lines get trapped inside, while for a superconductor the flux lines are expelled. Upon withdrawal of the field (panel (g) in figures 6.2 and 6.3), a superconductor is left with no memory, where for a perfect conductor, the flux lines form a loop, and thus it retains the field in its memory.

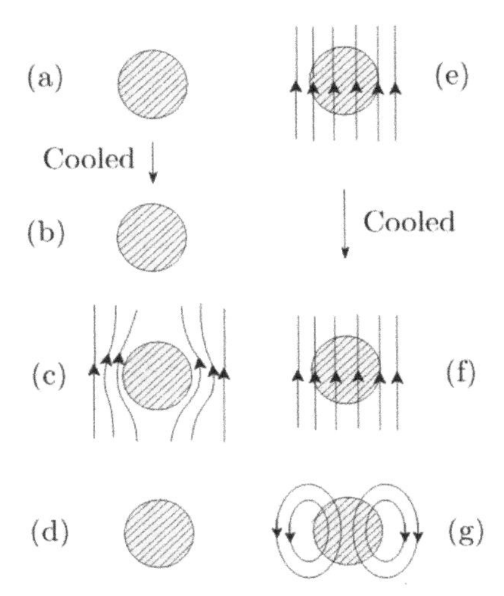

Figure 6.2. The ZFC (left panel) and FC (right panel) cases for a perfect conductor are shown schematically.

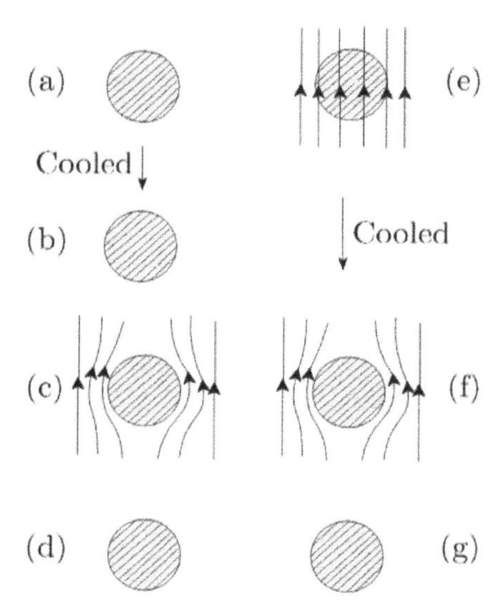

Figure 6.3. The ZFC (left panel) and FC (right panel) cases for a superconductor are shown schematically.

6.1.5 Electrodynamics of superconductors: London theory

Before the discovery of Meissner effect in superconductors, Becker *et al* [3] analysed the electrodynamic behaviour of perfect conductors using a simple free-electron model. According to it, the electrons accelerate under the application of an electric field, **E**. The argument for resistance-less motion is that whenever the electrons faced a resistance, the applied electric field would accelerate them steadily according to Newton's laws,

$$m^* \frac{d\mathbf{v}}{dt} = e^* \mathbf{E} \tag{6.1}$$

where m^*, e^* are the effective mass and the net charge, respectively. It was found later that $e^* = 2e$ and $m^* = 2m$, which were subsequently called superelectrons or Cooper pairs. If there are n_s numbers of superelectrons moving with a mean local velocity v_s, then the supercurrent density is given by,

$$\mathbf{J}_s = n_s e^* \mathbf{v}_s = -n_s |e^*| \mathbf{v}_s \tag{6.2}$$

The minus sign signifies that the supercurrent moves in an opposite direction to that of the superelectrons. Substituting equation (6.2) in equation (6.1) one gets,

$$\mu_0 \lambda_L^2 \frac{\partial \mathbf{J}_s}{\partial t} = \mathbf{E} \tag{6.3}$$

where, $\lambda_L = \sqrt{\frac{m^*}{\mu_0 n_s (e^*)^2}}$. λ_L has the dimension of length. Equation (6.3) is called the first London equation. Taking the curl of both sides of equation (6.3),

$$\mu_0 \lambda_L^2 \frac{\partial}{\partial t}(\nabla \times \mathbf{J}_s) = \nabla \times \mathbf{E} = -\frac{\partial \mathbf{B}}{\partial t}. \tag{6.4}$$

Using the following Maxwell's equation,

$$\nabla \times \mathbf{B} = \mu_0 \epsilon_0 \frac{\partial \mathbf{E}}{\partial t} + \mu_0 \mathbf{J}$$

we have,

$$\begin{aligned} \lambda_L^2 \frac{\partial}{\partial t}(\nabla \times (\nabla \times \mathbf{B})) &= -\frac{\partial \mathbf{B}}{\partial t} \\ \lambda_L^2 \nabla^2 \left(\frac{\partial \mathbf{B}}{\partial t}\right) &= \frac{\partial \mathbf{B}}{\partial t}. \end{aligned} \tag{6.5}$$

This implies that the rate of the **B** field (or the **B** field itself) will fall off exponentially inside a superconductor to a trapped field B_0. This clearly contradicts the Meissner effect, that is, the magnetic field inside the specimen is zero, irrespective of the initial condition.

Thus the brothers F London and H London [4] suggested that since the macroscopic theory of a perfect conductor makes a correct prediction about superconductor for the special case $B_0 = 0$, thus it might be reasonable to assume that the magnetic behaviour of a superconductor may be correctly described according to Meissner effect not only to $\frac{d\mathbf{B}}{dt}$ but also to **B** itself, that is,

$$\begin{aligned} \lambda_L^2 \nabla^2 \mathbf{B} = \mathbf{B} \qquad & [\nabla \times \mathbf{B} = \mu_0 \mathbf{J}_s] \\ \text{which implies,} \quad \mathbf{B} = -\mu_0 \lambda_L^2 \nabla \times \mathbf{J}_s. & \end{aligned} \tag{6.6}$$

By introducing a magnetic vector potential defined by $\nabla \times \mathbf{A} = \mathbf{B}$ and choosing a proper gauge, such as a transverse gauge, namely, $\nabla \cdot \mathbf{A} = 0$, the current density can be written as,

$$\mathbf{J}_s = -\frac{1}{\mu_0 \lambda_L^2} \mathbf{A} \qquad [\mathbf{B} = \nabla \times \mathbf{A}]. \tag{6.7}$$

This is called the second London equation, and is reminiscent of Ohm's law. This is of course true for simply connected superconductors. The gauge condition demands

$$\nabla \cdot \mathbf{A} = 0 \Rightarrow \mathbf{A} \cdot \hat{n} = 0$$

on the boundary surface of the superconductor, $\hat{n}$ being the unit drawn normal.

Now, using Maxwell's equation,

$$\begin{aligned} \nabla \times \mathbf{E} &= -\frac{\partial \mathbf{B}}{\partial t} \\ \nabla \times \left(\mathbf{E} - \mu_0 \lambda_L^2 \frac{\partial \mathbf{J}_s}{\partial t}\right) &= 0 \\ \text{or,} \quad \mathbf{E} - \mu_0 \lambda_L^2 \frac{\partial \mathbf{J}_s}{\partial t} &= \nabla \phi \end{aligned} \tag{6.8}$$

ϕ is a scalar. One important comment is due here. There is no component of $\nabla\phi$ in the direction of $\mathbf{J}_s$, that is,

$$\nabla\phi \cdot \mathbf{J}_s = 0 \tag{6.9}$$

as $\mathbf{J}_s \cdot \mathbf{E} = 0$ from first London equation (6.3). This is correct as otherwise it would mean energy dissipation in a constant magnetic field which we know to be incorrect as it violates conservation of energy.

6.1.6 Penetration depth

Let us prepare a setup to solve equation (6.5) for a superconductor. Refer to figure 6.4 where a semi-infinite superconducting sample is placed along the y-axis. We are interested in studying the variation of the magnetic field only in the x-direction along which it has a finite width (axis shown in figure 6.4). In figure 6.4 the variation is expected to be along the x-direction, we can convert it into a one-dimensional equation, that is, by assuming,

$$\frac{\partial B}{\partial y} = \frac{\partial B}{\partial z} = 0$$

which yields,

$$\frac{dB_x}{dt} = \alpha\frac{\partial^2}{\partial x^2}\left(\frac{dB_x}{dt}\right).$$

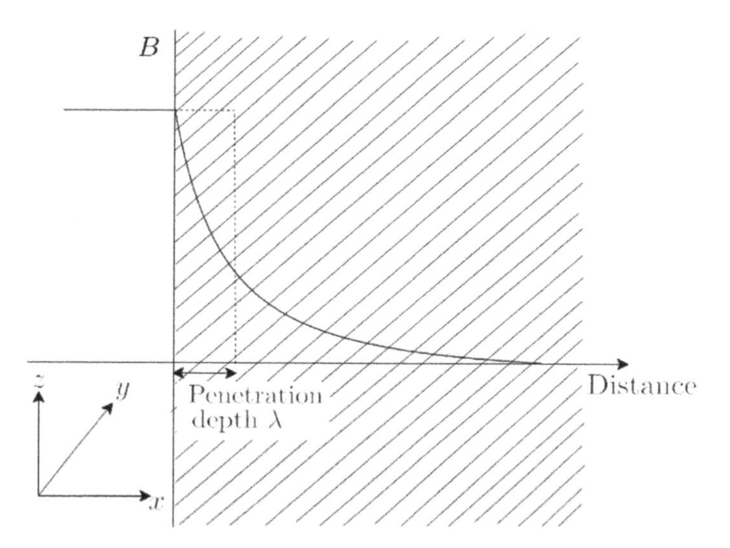

Figure 6.4. Schematic demonstration of the behaviour of an external magnetic field is presented. The field strength falls by a factor $1/e$ inside the superconductor, which is defined as the penetration depth.

Integrating over time yields,

$$B_x = \alpha \frac{\partial^2}{\partial x^2} B_x$$

$$B(x) = B_{\text{ext}} e^{-x/\lambda_L}$$

where an exponentially growing solution, even though mathematically admissible, is dropped, and

$$\alpha = \lambda_L^2$$

λ_L is of the order of 500Å for commonly known superconductors. This is how the magnetic field varies inside a superconducting sample. At $x = 0$, the value of the magnetic field just outside the surface is B_{ext}, which decays exponentially inside the sample.

6.1.7 Flux quantization

In the mid-1930s [4] it was realized that superconductivity is inherently a quantum phenomenon that manifests at macroscopic scales. The description thus requires a wavefunction which can be written as,

$$\psi(\mathbf{r}, t) = |\psi| e^{i\theta} \quad (6.10)$$

where θ is a real scalar function representing the phase of the wavefunction and $|\psi|^2 = n_s$, where n_s is the density of superelectrons. The canonical momentum for a particle of charge e^* and mass m^* in a magnetic field is given by,

$$\mathbf{p} = m^* \mathbf{v}_s + e^* \mathbf{A}. \quad (6.11)$$

Putting $\mathbf{p} = -i\hbar\nabla$ and writing down the Schrödinger equation,

$$\hbar \nabla \theta \psi = (m^* \mathbf{v}_s + e^* \mathbf{A}) \psi \quad (6.12)$$

yields,

$$\frac{\hbar}{e^*} \nabla \theta = \mu_0 \lambda_L^2 \mathbf{j}_s + \mathbf{A} \quad (6.13)$$

where $\lambda_L = \sqrt{\frac{m^*}{\mu_0 n_s e^{*2}}}$ is the London penetration depth (described later), and $\mathbf{j}_s$ is the supercurrent density defined via, $\mathbf{j}_s = n_s e^* \mathbf{v}_s$. Since $\mathbf{A}$ enjoys gauge freedom, the (local) phase of the wavefunction can be changed in the following manner, so as to keep the velocity unchanged.

$$\theta \to \theta' = \theta + \frac{e^*}{\hbar} \mathbf{A}.$$

It may be noted that equation (6.13) is consistent with Maxwell's equation, which can be checked by taking a curl of the equation, and noting that $\mathbf{B} = \nabla \times \mathbf{A}$. Now taking a derivative of equation (6.13) with respect to time, one gets,

$$\frac{\hbar}{e^*}\nabla\left(\frac{\partial\theta}{\partial t}\right) = \mu_0\lambda_L^2\frac{\partial \mathbf{j}_s}{\partial t} + \frac{\partial \mathbf{A}}{\partial t}. \tag{6.14}$$

Using the first of the London equations, namely,

$$\frac{\partial \mathbf{j}_s}{\partial t} = \frac{\mathbf{E}}{\mu_0\lambda_L^2}$$

one arrives at

$$\mu_0\lambda_L^2\frac{\partial \mathbf{j}_s}{\partial t} = \mathbf{E}. \tag{6.15}$$

Hence, with

$$\mathbf{E} = -\nabla\phi - \frac{\partial A}{\partial t}$$

one gets,

$$\frac{\hbar}{e^*}\frac{\partial\theta}{\partial t} = -\phi. \tag{6.16}$$

Thus the time variation of the local phase denotes a scalar potential ϕ.

There is a feature of the superconductors that deserves special attention, which is quantization of the magnetic flux. Consider a multiply connected superconductor, and consider a closed loop C lying entirely in the superconducting regime, and encircling the hole (marked in figure 6.5). Thus any open surface bounded by C contains regions that are partially superconducting, and partially in a non-superconducting (normal) region. Applying Stokes theorem on equation (6.13) one obtains,

$$\frac{\hbar}{e^*}\oint_c \nabla\theta \cdot d\mathbf{l} = \Phi_s + \oint_c \mu_0\lambda_L^2\mathbf{j}_s \cdot d\mathbf{l} \tag{6.17}$$

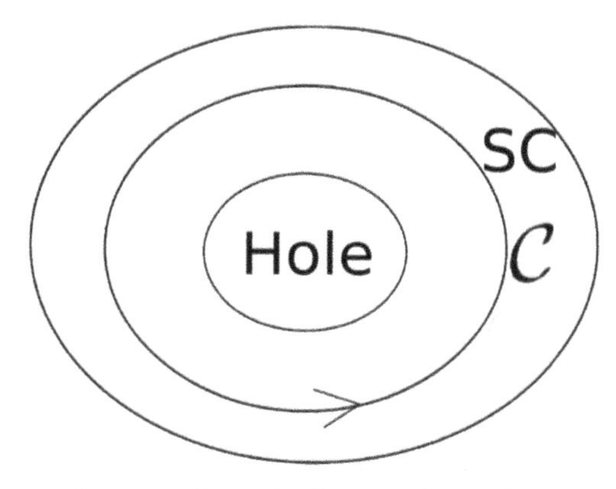

Figure 6.5. A simply connected superconductor is shown with a '*hole*' within. C denotes a contour of integration in equation (6.13).

Φ_s is the magnetic flux defined by,

$$\Phi_s = \oint \mathbf{B} \cdot d\mathbf{s}. \tag{6.18}$$

The requirement that ψ is a single valued function at each point, that is,

$$\phi e^{i\theta(\mathbf{r},\, t)} = \phi e^{i\{\theta(\mathbf{r},\, t)+2n\pi\}}$$

indicates that the LHS of equation (6.17) has the form,

$$\frac{\hbar}{e^*}\int_c \nabla\theta \cdot d\mathbf{l} = n\Phi_0 \qquad n = 0, \pm 1, \pm 2, \cdots \tag{6.19}$$

where Φ_0 is called the flux quantum

$$\Phi_0 = \frac{2\pi\hbar}{|e^*|} = 2.07 \times 10^{-15} \quad \text{Wb} \tag{6.20}$$

$n = 0$ corresponds to simply connected superconductor.

Hence integrating over a closed loop C should be the same after one complete rotation from where one has started, except for a phase of 2π, or a multiple of 2π. A flux quantum has this value for $e^*=2e$. So the supercurrent is carried by a pair of electrons. In fact,

$$\oint \nabla\theta \cdot d\mathbf{l} = 2\pi n = -\frac{2e}{\hbar}\Phi.$$

Thus the magnetic flux penetrating through a superconductor can only appear in multiples of the superconducting flux quantum Φ_0, which is often referred to as fluxoid (or fluxon). It may be noted that the value of the fluxoid is extremely small. Experimentally, the flux quantization was measured by Deaver and Fairbank in 1961 [5]. Importantly, the involvement of Planck's constant ($\hbar$) is a signature of quantum effects existing at macroscopic scales.

In the right-hand side of equation (6.17) we can introduce a fluxoid Φ_L,

$$\Phi_L = \int_s \mathbf{B} \cdot d\mathbf{s} + \oint_c \mu_0\lambda_L^2\mathbf{j}_s \cdot d\mathbf{l} \tag{6.21}$$

Thus within approximation of the London theory,

$$\begin{aligned}\frac{\partial\Phi_L}{\partial t} &= \frac{d}{dt}\int_s \mathbf{B} \cdot d\mathbf{s} + \oint_c \mu_0\mathbf{E} \cdot d\mathbf{l} \\ &= \int_s \left(\frac{\partial\mathbf{B}}{\partial t} + \nabla \times \mathbf{E}\right) \cdot d\mathbf{s} \\ &= 0\end{aligned} \tag{6.22}$$

where the last step is obtained using Maxwell's equation. Thus the fluxoid Φ_L is constant with time. It also does not depend upon the exact shape of the contour C, as long as it encompasses the hole, but only once. If there is no hole, then the fluxoid is zero.

6.1.8 Non-local electrodynamics

The current density equation in electrodynamics, namely,

$$\mathbf{J}_s = n_s e \mathbf{v}_s$$

is a local equation in the sense that it relates the current density at a point $\mathbf{r}$ to the velocity of the charge carriers at that point. Thus it clearly ignores the spatial structure of the electric field. An improvement to this can be done by considering the current density at $\mathbf{r}$ due to the variation of the electric field within a shell around the point $\mathbf{r}$, with a finite radius. On similar grounds, Pippard argued that the current density at a point $\mathbf{r}$ should depend on $\mathbf{E}(\mathbf{r}')$. Further he conjectured that the wavefunction of the electrons must have a characteristic dimension, namely, ξ which is called the coherence length. A simple uncertainty calculation may be sufficient to yield an estimate of the value of ξ in typical superconductors. We know that electrons within an energy range $k_B T_c$ play a dominant role in the pairing phenomena. The momenta of these electrons have an uncertainty given by,

$$\Delta p \simeq \frac{\Delta E}{v_F} \simeq \frac{k_B T_c}{v_F}.$$

Thus, their position uncertainty, using Heisenberg's uncertainty principle is,

$$\Delta x \simeq \frac{\hbar}{\Delta p} \simeq \frac{\hbar v_F}{k_B T_c}.$$

This yields,

$$\Delta x \simeq \xi = a \frac{\hbar v_F}{k_B T_c}$$

a is usually of the order of unity. In particular, $a = 0.8$ in BCS theory [6].

Hence, Pippard suggested that the simple equation for the current density written above should be modified as,

$$J_s(\mathbf{r}) = -(\text{const}) \int \frac{\mathbf{R}[\mathbf{R} \cdot \mathbf{A}(\mathbf{r}')]}{R^4} e^{-R/\xi} d\mathbf{r}' \tag{6.23}$$

where, $\mathbf{R} = \mathbf{r} - \mathbf{r}'$, with $\mathbf{r}$ being the source point and $\mathbf{r}'$ being the field point, or where the observation is made.

6.2 Magnetic phase diagram of superconductors

When a superconductor is used to form a circuit with a battery, and a steady state is established, all the currents passing through the superconductor are supercurrents. Normal currents due to motion of charged particles contribute zero, since no voltage difference can be sustained in a homogeneous superconductor (because otherwise a current will induce a B which is zero). Experiments show that all supercurrents flow near the type-I superconductor's surface within a thin layer characterized by the penetration depth. These surface supercurrents run so that the B field is cancelled in the interior of a conductor.

Thus, let us review the magnetic phase diagram of superconductors. It is clear by now that the superconducting phase is a stable phase in a certain range of magnetic fields and temperatures. For higher fields and temperatures, the normal metallic state becomes more stable. The critical magnetic field (H_c) beyond which the material cannot be superconducting even if it is cooled below T_c (see figure 6.6). The value of H_c depends on temperature, since it is easier to quench a superconductor at higher temperatures, than at lower temperatures. An empirical relation for the same can be written as,

$$H_c = H_0\left[1 - \left(\frac{T}{T_c}\right)^2\right] \qquad \text{where,} \qquad H_0 = H_{ext}(T = 0). \tag{6.24}$$

The corresponding M versus H phase diagrams are shown in figure 6.7. There is a sharp fall of the (negative) magnetization ($-M$) as a function of the external field, and a sharp rise of the magnetic induction denotes a phase transition from a superconducting to a normal phase.

In type-II superconductors, the phase diagram is more complex. When small magnetic fields are applied, surface currents develop that screen the magnetic field from penetrating homogeneously into the sample. This is termed as the Meissner

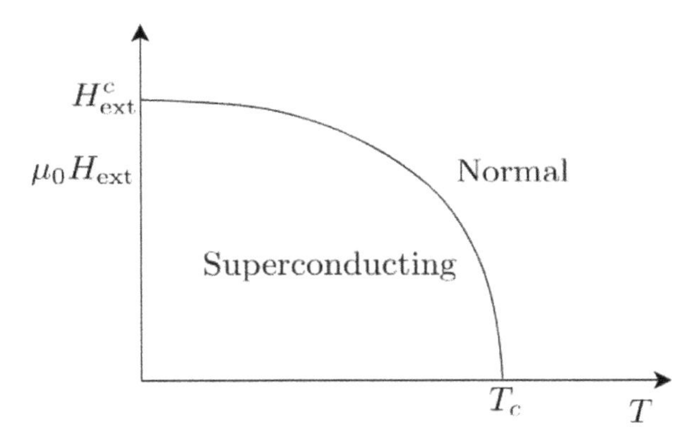

Figure 6.6. The dependence of the critical field, H_C as a function of temperature T is shown.

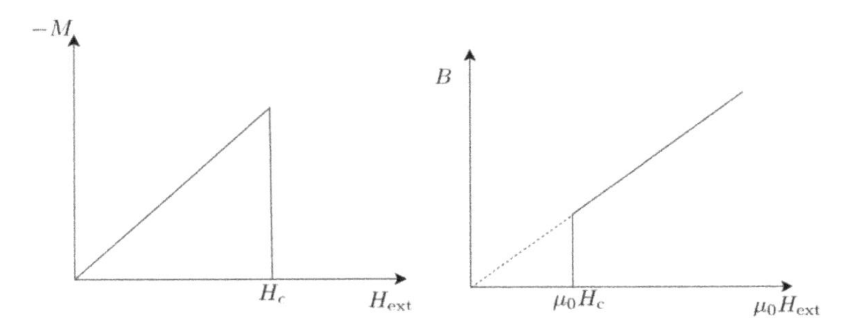

Figure 6.7. Magnetic phase diagrams for type-I superconductors are shown. There is a direct transition from the superconductor to the normal phase in type-I superconductors.

state, which persists till a certain critical value, called the first critical magnetic field, H_{c1}. Beyond this value, the system enters into a mixed state, where the magnetic flux lines, in the form of vortices, penetrate the superconducting sample. Thus there is a partial screening of the magnetic field.

Beyond H_{c1}, more and more vortices appear insider the superconductor, and eventually when the external field reaches a second critical value, namely, H_{c2} the material becomes a normal metal (figure 6.8). The mixed phase is called the Abrikosov phase, where it is energetically favourable for the vortices to self-organize into a hexagonal array. The scanning tunnelling microscopy (STM) measurements[2] confirm the existence of such a vortex array, which minimize energy by arranging into hexagonal structure in which the vortices line up. These structures are called Abrikosov lattices. The magnetic susceptibility of a superconductor is -1 below H_c and has a value 10^{-5} above H_c. This is shown in the following figures.

From here we get a characteristic length called coherence length ξ. ξ is defined as the length scale over which the wavefunction or the order parameter does not vary much. The ratio of these two characteristic length scales is given by,

$$\kappa = \frac{\lambda}{\xi}. \tag{6.25}$$

For typical superconductors, $\lambda = 500\text{Å}$, $\xi \approx 5000\text{Å}$. Thus $\kappa < 1$.

In 1957, Abrikosov found that for some class of superconductors $\kappa > 1$. This is used as a distinguishing criterion for the type I and type II superconductors, in the following sense.

$\kappa < 1$ type I superconductors
$\kappa > 1$ type II superconductors.

For type II superconductors, the flux lines penetrate the sample till a certain threshold magnetic field. A threshold value of κ for which such flux penetration

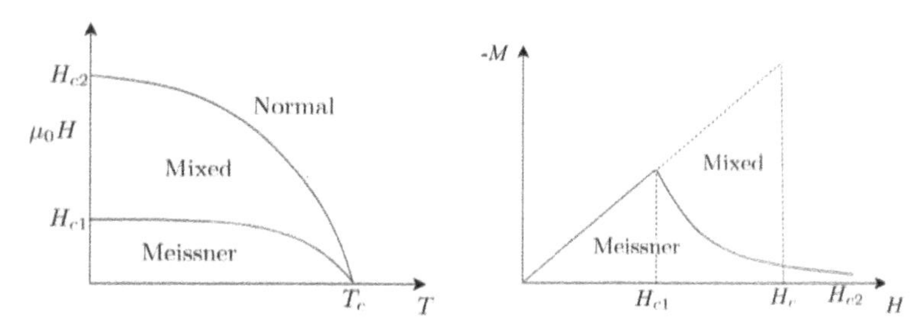

Figure 6.8. Magnetic phase diagrams for type-II superconductors are shown. In type-II materials, there is an intermediate phase that interrupts a direct transition, where the flux lines penetrate.

[2] The working principle of STM is based on the measurement of electronic tunnelling current that depends exponentially on the distance between the tip and the sample. The position of a tip on top of a flat surface of the sample can be controlled with subatomic resolution.

starts to occur is $\kappa = \kappa_c \approx \frac{1}{\sqrt{2}}$. At this value of κ, the flux penetration starts at lower critical field H_{c1} and reaches $B_{\text{ext}} = H_{\text{ext}}$ at $H_{\text{ext}} = H_{c2}$.

Even experiments on the high-T_c superconductors confirmed many of the features in the magnetic phase diagram as discussed above. For example, the quantization of vortices ($\Phi_0 = \frac{h}{2e}$), Abrikosov phase, etc are observed in experiments. In the type-II superconductors, the identification of the upper critical field has particularly been controversial, quite unlike the conventional superconductors, where the onset of strong diamagnetism and superconducting critical currents occur at the same H_{c2}, whereas in the high-T_c superconductors, they are found to occur at quite different values. There are several other features which are distinct, possibly because of defect induced pinning mechanism, etc. We shall not continue this discussion here any further, and will briefly come back to it at the end of the chapter. However, certain contrasting properties of these superconductors compared to the conventional ones will be mentioned during the course of discussion.

6.2.1 Thermodynamics of superconductors

The thermodynamics route is usually the easiest route to understand phase transition. Consider Gibbs free energy of a superconductor. If its magnetization is M and magnetic induction H, then the work done in bringing the superconductor into a region with magnetic induction H_{ext} from far away (where $H_{\text{ext}} = 0$) is given by,

$$W = -\mu_0 \int_0^{H_{\text{ext}}} M dH = \mu_0 \frac{H_{\text{ext}}^2}{2} \tag{6.26}$$

as $M = -H$ for a superconductor.

This is the extra Gibbs free energy of a superconductor. Let us call this g_s (per unit volume)

$$g_s(T, H_{\text{ext}}) = g_s(T, 0) + \frac{\mu_0 H_{\text{ext}}^2}{2} \tag{6.27}$$

or,

$$g_s(T, B_{\text{ext}}) = g_s(T, 0) + \frac{B_{\text{ext}}^2}{2\mu_0} \tag{6.28}$$

In the normal state, the magnetization is a very small value, thus the magnetic work term is vanishingly small. Hence

$$g_n(T, H_{\text{ext}}) = g_n(T, 0) \tag{6.29}$$

$g_s(T, H_{\text{ext}})$ increases quadratically with H_{ext}. For some H_{ext}, the normal and superconducting states have the same free energy. Beyond this critical value, the normal state is more stable. Equating equations (6.27) and 6.29 at $H_{\text{ext}} = H_c$ (see figure 6.9)

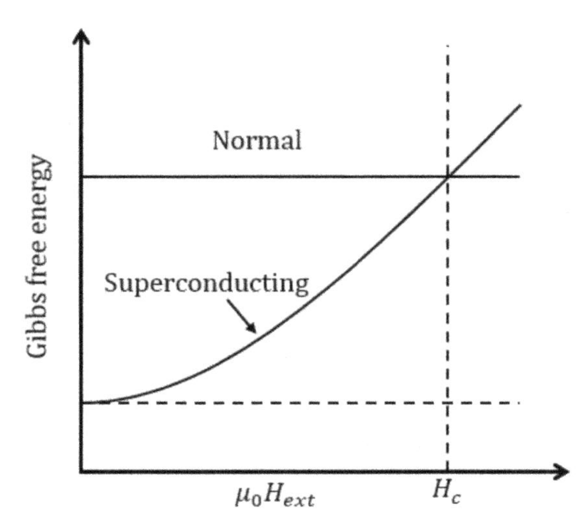

Figure 6.9. The Gibbs free energy is indicated schematically as a function of the external magnetic field. The energies for the superconducting and the normal states intersect at some value of the magnetic field, beyond which the normal state stabilizes.

$$g_n(T, 0) = g_s(T, 0) + \frac{\mu_0 H_c^2}{2}. \tag{6.30}$$

The RHS is positive, and hence a superconducting state is more stable below H_c. For example, $H_c = 0.08$ T for Pb at $T = 0$ [7]. Thus at $T = 0$, the superconducting state is stabilized by 4.25×10^{-25} Joule/mole. This is a really small amount, however, interestingly this small number stabilizes the superconducting state.

6.2.2 Specific heat

The study of specific heat is in general interesting because it provides a good measure of the range of applicability of phonon mediated superconductivity. Above the transition temperature, the specific heat, C_n follows Debye's theory. In general for a normal metal, at temperatures $T < \theta_D$, the specific heat consists of:

(i) electronic contribution, γT;
(ii) phonon contribution, AT^3;
(iii) Schottky contribution, a/T^2.

Schottky contribution can mostly be neglected and hence,

$$C_n = \gamma T + AT^3. \tag{6.31}$$

Thus, $C_n/T = \gamma + AT^2$. C_n/T versus T^2 thus denotes a linear plot.

At $T = T_c$, the contribution of electronic and phonon specific heat is,

$$\frac{C_{ph}}{C_e} = \frac{1}{\beta}\left(\frac{AT_c^2}{\gamma}\right)$$

where β is the ratio of the number of conduction electrons to that of atoms. From free-electron theory,

$$A = 234\frac{R}{\theta_D^3}.$$

[8] Remember,

$$C_{ph} = \frac{12\pi^4}{5}R\left(\frac{T}{\theta_D}\right)^3 = 234R\left(\frac{T}{\theta_D}\right)^3$$

Whereas,

$$\gamma = 4.93\frac{R}{T_F}$$

and,

$$C_e = \frac{1}{2}\pi^2 R\left(\frac{T}{T_F}\right) = 4.93R\left(\frac{T}{T_F}\right).$$

Thus,

$$\frac{C_{ph}}{C_e} = \left(\frac{47.5}{\beta} \cdot \frac{T_F}{\theta_D^3}\right)T_c^2$$

β can be set to be of the order of 1. For most of the conventional superconductors, $T_c \ll \theta_D$ So $C_e \gg C_{ph}$.

If the conduction electrons have effective mass, m^*, which is defined by the inverse of the curvature of the band structure, and that differs from bare mass m, then the conduction electron specific heat is given by,

$$\gamma = \left(\frac{m^*}{m}\right)\gamma_0$$

where γ_0 is the bare electron counterpart of γ.

$$\gamma_0 = \frac{\pi^2 R}{2T_F}.$$

Thus

$$\frac{m^*}{m} = \frac{\gamma}{\gamma_0} = \frac{2\gamma T_F}{\pi^2 R}.$$

It is worth mentioning here that for a typical high temperature superconductor, T_F is pretty low, $T_F \sim 4000$ K, at least an order lower than conventional elemental superconductors, such as for Cu, $T_F \sim 80\,000$ K. The discrepancy is probably due to the large values of the effective masses.

With C_{es} numerically obtained, we can integrate to find the change in internal energy $U(T)$ as we decrease temperature from T_c. At $T = T_c$ it must be the same as the normal value $U_{en}(0) + \frac{1}{2}\gamma T_c^2$. So

$$U_{es}(T) = U_{en}(0) + \frac{1}{2}\gamma T_c^2 - \int_T^{T_c} C_{es} dT.$$

From the internal energy and entropy we may compute the free energy using,

$$F_{es}(T) = U_{es}(T) - TS_{es}(T).$$

The critical field is determined by

$$\mu_0 \frac{H_c^2(T)}{2} = F_{en}(T) - F_{es}(T)$$

where,

$$F_{en}(T) = U_{en}(0) - \frac{1}{2}\gamma T^2.$$

To compute the thermodynamic quantities, we explicitly consider the temperature dependence of the specific heat of a superconductor, and explore its behaviour at the normal-superconductor phase transition. There is a jump in the specific heat at $T = T_c$ as we have seen earlier.

The Gibb's free energy per unit volume for a superconductor in a magnetic field is written as,

$$G = U - TS - HM \tag{6.32}$$

neglecting the usual PV term. Since,

$$dU = TdS + HdM. \tag{6.33}$$

This is the change of internal energy in the presence of a magnetic field, from (6.32) and (6.33),

$$dG = -SdT - MdH. \tag{6.34}$$

Thus integrating (6.34),

$$G_s(H) = G_s(0) + \frac{H^2}{2}. \tag{6.35}$$

Along the critical curve where the SC and the normal metal are in equilibrium,

$$G_n = G_s(0) + \frac{H^2}{2}. \tag{6.36}$$

Where G_s and G_n are the Gibb's free energy for the superconducting and the normal states. From equation (6.34)

$$\left(\frac{\partial G}{\partial T}\right)_H = -S. \tag{6.37}$$

Thus, at equilibrium,

$$S_n - S_s = -H_c\frac{dH_c}{dT}. \tag{6.38}$$

S_s denotes the entropy of an SC in zero field. Since $\frac{dH_c}{dT}$ is always negative, the entropy of the normal state is always greater than that of the superconducting state. Thus,

$$\begin{aligned}\Delta C = C_s - C_n &= -T\frac{d}{dT}(S_s - S_n)\\ &= TH_c\frac{d^2H_c}{dT^2} + T\left(\frac{dH_c}{dT}\right)^2.\end{aligned} \tag{6.39}$$

At $T \to T_c$, $H_c \to 0$,

$$\Delta C = T_c\left(\frac{dH_c}{dT}\right)^2. \tag{6.40}$$

From (6.38) at $T = T_c$, $H_c = 0$, there is no latent heat of transition $\Delta S = 0$, but according to (6.40) $\Delta C =$ finite. So the transition is second order (at $T = T_c$). However, away from T_c, the phase transition has a latent heat and is a first-order phase transition. The behaviour of entropy (S), specific heat (C) (scaled by γT_c), internal energy (U) and Helmholtz free energy (F) (scaled by γT_c^2) are shown as a function of reduced temperature (T/T_c are shown in figures 6.10 and 6.11). The discontinuity of C at $T = T_c$ is a noteworthy feature, and denotes a signature of a superconducting phase transition.

However, this is not true for high temperature superconductors. In fact for ($La_{0.9}Sr_{0.1}CuO_{4-\delta}$) and ($YBa_2Cu_3O_{7-\delta}$) $AT_c^3 \gg \gamma$, so the phonon term dominates at $T = T_c$. The distinct behaviour of the specific heat for conventional and high-T_c superconductors as a function of temperature is shown in figure 6.12.

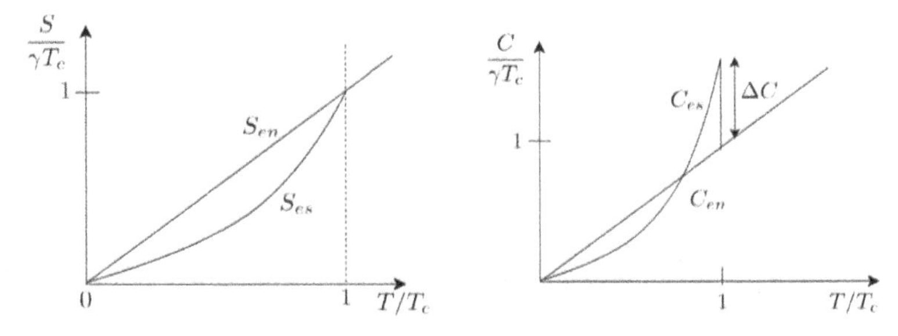

Figure 6.10. Schematic plots of entropy (S) and specific heat (C) scaled by γT_c (γ defined in text) are presented as a function of reduced temperature, $\frac{T}{T_c}$. $T/T_c < 1$ denotes the superconducting state, while $T/T_c > 1$ depicts the normal state.

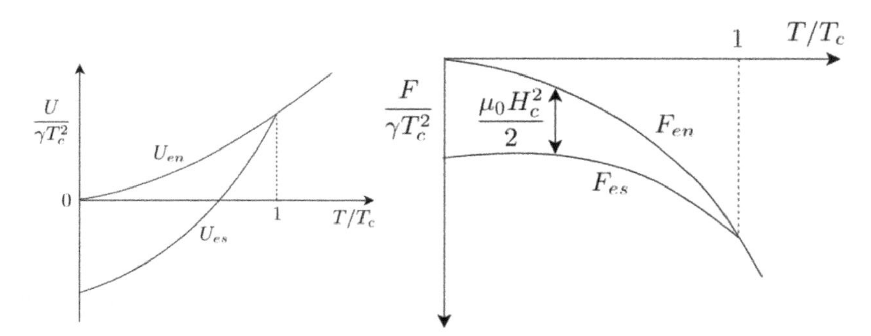

Figure 6.11. Schematic plots of entropy the internal energy and the Helmholtz free energy scaled by γT_c^2 are presented as a function of reduced temperature.

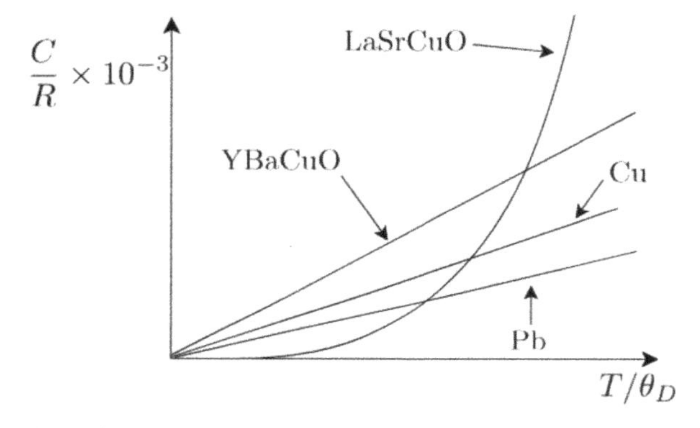

Figure 6.12. The behaviour of the specific heat is schematically shown as a function of the reduced temperature T/θ_D for both conventional (Cu and Pb) and unconventional (cuprate) superconductors.

6.2.3 Density of states

The Bogoliubov quasiparticle spectrum, E_k is easily seen to have a minimum of Δ_k for $\xi_k = 0$ (ξ_k is positive definite including zero). Thus, in addition to playing the role of the order parameter Δ_k is also the energy gap in the single-particle spectrum. To see this explicitly, we can do a change of variables from ξ_k (for the normal state energies) to E_k (denoted as the quasiparticle energies), namely,

$$N(E)dE \rightarrow N_n(\xi)d\xi$$

Since, the density of states (DOS) should remain conserved, as the number of carriers remains conserved,

$$N(E)dE = N_n(\xi)d\xi. \tag{6.41}$$

Since, the gap Δ is much smaller than the energy range ($\Delta \ll E$) over which the normal state DOS varies, we can replace,

$$N_N(\xi) \approx N_n(0) = N_0 \tag{6.42}$$

Now the quasiparticle energies are given by $E_k^2 = \Delta^2 + \xi_k^2$.

Dropping the subscripts,

$$E^2 = \Delta^2 + \xi^2$$

$$2EdE = 2\xi d\xi$$

$$\frac{d\xi}{dE} = \frac{E}{\xi} = \frac{E}{\sqrt{E^2 - \Delta^2}}.$$

Here,

$$N(E)dE = N_0 d\xi$$

$$\frac{N(E)}{N_0} = \frac{d\xi}{dE}.$$

Thus,

$$\frac{N(E)}{N(0)} = \begin{matrix} \frac{E}{\sqrt{E^2 - \Delta^2}} & \text{for } E > \Delta \\ 0 & \text{for } E < \Delta. \end{matrix}$$

Excitations with all momenta k, even those whose ξ_k fall in the gap have their energies raised above Δ. Moreover, we get a divergent DOS for $E \to \Delta$ from above.

Considering that across the phase transition, the DOS remains constant, i.e.

$$N_s(E)dE = N_n(\xi)d\xi.$$

Since we are interested in energy range of a few meV from the Fermi energy, we take $N_n(\xi) = N(\varepsilon_F) =$ a constant.

$$\frac{N_s(E)}{N(\varepsilon_F)} = \frac{d\xi}{dE} = \frac{E}{\sqrt{E^2 - \Delta^2}} \quad \text{for} \quad E > \Delta$$

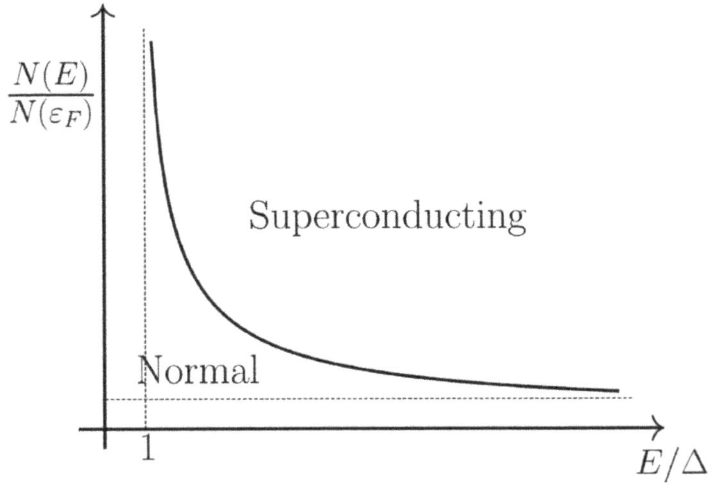

Figure 6.13. The DOS as a function of the scaled energy, E/Δ is shown. The right of the bold line denotes the superconducting phase, while on the left, it is the normal state.

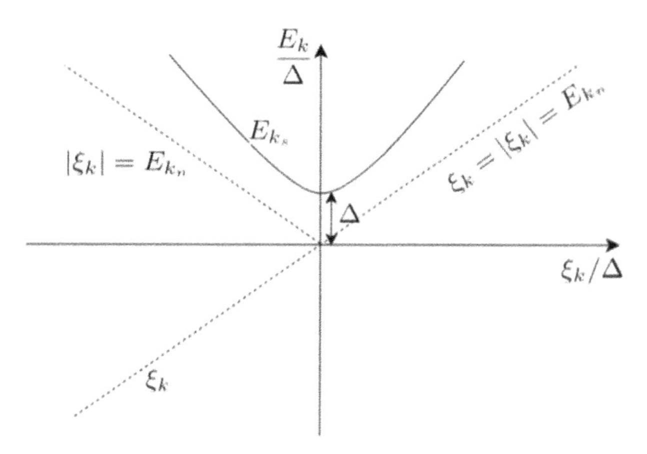

Figure 6.14. The scaled quasiparticle energy, E/Δ is plotted as function of scaled single-particle energy, ξ_k/Δ. The region above the parabola is superconducting (SC) and the region below the dotted lines denote metallic phases.

The behaviour of the DOS is plotted in figure 6.13. It is zero for $E < \Delta$, and falls off as E becomes larger than the gap, Δ. The quasiparticle energies (E_k) as a function of the single-particle energies (ξ_k) are plotted in figure 6.14 where the superconducting phase exists above the parabola, and the dotted lines denote the boundaries separating the superconducting and the normal phases.

6.3 BCS theory

6.3.1 Introduction

Even though a lot of experimental progress occurred in the field of superconductivity, a complete understanding of the pairing phenomena and consequently the formation of the superconducting state were lacking. In the following, we describe a microscopic theory that addresses all of these concerns. The BCS theory of superconductivity was formulated by Bardeen, Cooper and Schrieffer in 1957. It accurately describes many of the properties of the weak coupling superconductors. The term '*weakly coupled*' refers to the strength of the electron–phonon coupling that is considered as the origin of pairing in this class of superconductors, as we shall see below. The basic idea of the BCS theory is that the electrons in metal form bound pairs below a certain critical temperature, called the transition temperature. Not all the electrons take part in this pairing process, only the ones that reside close to the Fermi surface, and within a distance of the Debye energy, ω_D. The paired state of all of these electrons needs a many-body description.

6.3.2 Isotope effect

Fröhlich in 1950 [9] realized that the electrons could interact by exchanging phonons, and in such a process, the interaction could be attractive. That was the first proposal that superconductivity can originate from electron–phonon interaction. But how do the phonons come into the picture? The dependence on phonon parameters was experimentally demonstrated by the fact that the transition temperature, T_c is a function of the ionic mass for different isotopes of the same metal (that becomes a superconductor at lower temperatures). The effect of T_c is shown to have the form,

$$\Delta T_c / T_c = -\frac{1}{2}\left(\frac{\Delta M}{M}\right) \tag{6.43}$$

where M denotes the ionic mass, and ΔM is the difference in masses between the isotopes. This is called the isotope effect. It played a significant role in unravelling the issue concerning the attractive interaction among the charge carriers that eventually leads to superconductivity. Thus, we shall see that the contribution of the lattice excitations or the phonons is central to the formation of bound pairs. It will be shown later that physical quantities, such as the energy of the bound pairs, T_c, etc involve phonon frequencies, and thus should display isotope effect.

The isotope effect is more conveniently represented in terms of the isotope coefficient, α via the relation,

$$T_c \sim M^{-\alpha} \tag{6.44}$$

where α is given by,

$$\alpha = -\frac{M}{\Delta M}\frac{\Delta T_c}{T_c}. \tag{6.45}$$

In general, α has a value 0.5, however, it may vary from one material to another.

6.3.3 Origin of attractive interaction

We consider interaction between two electrons, one with ($\mathbf{k}$, $\uparrow$) and the other ($-\mathbf{k}$, $\downarrow$). One may consider a spin independent interaction, but since it is not relevant to the context, we can neglect the spin indices for future discussions. There are two mechanisms by which two electrons can interact: they can interact directly or via a third party, which in this case is phonons. We shall explore these two cases in the following.

Case I

One obvious process is a direct electron–electron interaction mediated via the Coulomb forces. Let us consider the initial state as $|i\rangle$ (before scattering) and the final state $|f\rangle$ after scattering. Let us denote this interaction term as $H^{\text{dir}}_{\text{e–e}}$. The matrix elements of such a term between the states can be written as,

$$\langle i|H^{\text{dir}}_{\text{e–e}}|f\rangle = \int e^{i\mathbf{k}\cdot\mathbf{r}} U_c(\mathbf{r}) e^{-i\mathbf{k}'\cdot\mathbf{r}} d^3\mathbf{r}. \tag{6.46}$$

Assuming both $|i\rangle$ and $|f\rangle$ are plane wave states. This term cannot yield a negative contribution and hence does not contribute to the formation of a bound state. The energies involved in this case can be enumerated as:

(i) initial: $\epsilon_i = 2\xi_\mathbf{k}$ $\xi_\mathbf{k} = \frac{\hbar^2 k^2}{2m} - \mu$;

(ii) final: $\epsilon_f = 2\xi_{\mathbf{k}'}$ $\mathbf{k}' = \mathbf{k} + \mathbf{q}$.

Case II

Now let us consider an indirect process. Consider scattering of two electrons via exchange of a phonon. Thus there is an intermediate process involved, and hence it is a second-order process. There are two ways that such a second-order process can occur:

(i) electron 1 with momentum $\mathbf{k}$ emits a phonon, of frequency ω_q which is later re-absorbed by electron 2 with momentum $-\mathbf{k}$ (see left panel of figure 6.15);

(ii) electron 2 with momentum $-\mathbf{k}$ emits a phonon with frequency ω_q which is later re-absorbed by electron 1 with momentum $\mathbf{k}$ (see right panel of figure 6.15).

Since both these processes are equally probable, they need to be incorporated with equal weight. The corresponding energies involved in the second case are:

(i) initial state (before scattering): $\epsilon_i = 2\xi_{\mathbf{k}}$;

(ii) final state (after scattering): $\epsilon_f = 2\xi_{\mathbf{k}'}$.

Even if the initial and the final states are the same as in case I, the intermediate state is different. However, the corresponding energies are still the same, namely,

$$\epsilon_{\text{int}}^{(i)} = \xi_{\mathbf{k}'} + \xi_{\mathbf{k}} + \hbar\omega_q$$
$$\epsilon_{\text{int}}^{(ii)} = \xi_{\mathbf{k}} + \xi_{\mathbf{k}'} + \hbar\omega_q.$$

Now the matrix element of $H_{\text{e-e}}^{\text{indirect}}$ can be calculated as,

$$\begin{aligned} \langle i|H_{\text{e-e}}^{\text{indirect}}|f\rangle &= \sum_{\text{int}} \frac{\langle i|H_{\text{e-e}}^{\text{indirect}}|int\rangle\langle int|H_{\text{e-e}}^{\text{indirect}}|f\rangle}{(E_{i,f} - E_{\text{int}})^2} \\ &= \sum_{\text{int}} \langle i|H_{\text{e-e}}^{\text{indirect}}|int\rangle \frac{1}{2}\left[\frac{1}{E_f - E_{\text{int}}} - \frac{1}{E_i - E_{\text{int}}}\right] \\ &\quad \times \langle int|H_{\text{e-e}}^{\text{indirect}}|f\rangle \end{aligned} \tag{6.47}$$

where the summation is over the intermediate states. Note that the energy denominators can be written as,

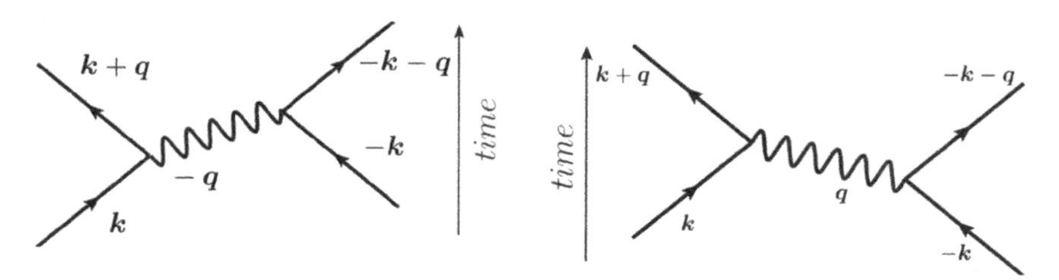

Figure 6.15. Feynman diagrams are shown for the electron–electron interaction mediated via phonons. The left panel shows electron 1 emitting a phonon which is captured by electron 2 at a later time. The right panel shows electron 2 emitting a phonon which is captured by electron 1 at a later time.

$$\frac{1}{E_f - E_{\text{int}}} = \frac{1}{2\xi_{\mathbf{k}'} - \xi_{\mathbf{k}} - \xi_{\mathbf{k}'} - \hbar\omega_q}$$
$$\frac{1}{E_i - E_{\text{int}}} = \frac{1}{2\xi_{\mathbf{k}} - \xi_{\mathbf{k}'} - \xi_{\mathbf{k}} - \hbar\omega_q} \tag{6.48}$$

where, $\xi_{\mathbf{k}'} - \xi_{\mathbf{k}} = \hbar\omega$. Thus,

$$\begin{aligned} \langle i|H_{\text{e-e}}^{\text{indirect}}|f\rangle &= \frac{1}{\omega - \omega_q} - \frac{1}{\omega + \omega_q}|\widetilde{V_c}(q)|^2 \\ &= \frac{2\omega_q}{\omega^2 - \omega_q^2}|\widetilde{V_c}(q)|^2. \end{aligned} \tag{6.49}$$

The matrix elements are positive definite, so we are interested to know the sign of the energy denominator. For $\omega < \omega_q$, it is negative, and hence attractive. This is known as Cooper's instability. In real physical situations, the Coulomb term gets weakened by screening, which eventually takes a form,

$$V(\mathbf{q}, \omega) = \frac{4\pi e^2}{q^2 + k_0^2} + \frac{4\pi e^2}{q^2 + k_0^2}\frac{\omega_q}{\omega^2 - \omega_q^2}. \tag{6.50}$$

A schematic plot of the interaction term as a function of the electron energy, ω is plotted in figure 6.16. In a small region, between the points A and B marked in figure 6.16, $V(\mathbf{q}, \omega)$ becomes negative and hence attractive.

6.3.4 The BCS ground state

The ground states of a free Fermi gas correspond to the filled states with energy ϵ below ϵ_F and all states empty beyond $\epsilon > \epsilon_F$. This Fermi gas of electrons becomes unstable against the formation of at least one bound pair, regardless of how weak the interaction is, so long as it is attractive [10].

It is well known that the binding does not ordinarily occur in a two-body problem in two or three dimensions, until the strength of the potential exceeds a finite

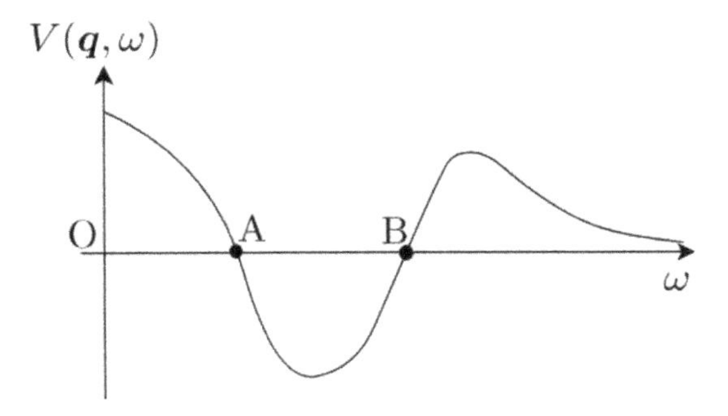

Figure 6.16. A schematic plot of the interaction energy $V(\mathbf{q}, \omega)$ plotted as a function of the electron energy ω. The region between A and B is where the interaction becomes attractive.

threshold value [11]. So, in order to see how this binding comes about, we consider a simple model of two electrons added to the Fermi sea at $T = 0$. The restriction imposed here is that these two electrons interact with each other, but not with those inside the Fermi sea, except, of course due to the exclusion principle.

Thus, we seek a two-particle wavefunction for our purpose. Bloch argued that the lowest energy state for a two-particle system has zero total momentum. Thus, the two electrons have equal and opposite momenta. This means the orbital wavefunction, $\psi_0(\mathbf{r}_1, \mathbf{r}_2)$ should be of the form,

$$\psi_0(\mathbf{r}_1, \mathbf{r}_2) = \sum_k g_k e^{i\mathbf{k}\cdot\mathbf{r}_1} e^{-i\mathbf{k}\cdot\mathbf{r}_2}. \tag{6.51}$$

Taking into account the asymmetry of the total wavefunction with respect to the exchange of two electrons, ψ_0 contains even functions, such as, $\cos[\mathbf{k}\cdot(\mathbf{r}_1 - \mathbf{r}_2)]$ with an antisymmetric spin part of the form, $[|\uparrow\downarrow\rangle - |\downarrow\uparrow\rangle]$, or a sum of products of $\sin[\mathbf{k}\cdot(\mathbf{r}_1 - \mathbf{r}_2)]$ with a symmetric triplet function of the form, $[|\uparrow\uparrow\rangle, |\downarrow\downarrow\rangle]$ and $[|\uparrow\downarrow\rangle + |\downarrow\uparrow\rangle]$. In a two-particle system, we expect the singlet to have lower energy because the cosinusoidal dependence of the orbital wavefunction $\mathbf{r}_1 - \mathbf{r}_2$ gives a larger probability for the electrons to be near to each other. Thus, we zero in on a two-electron singlet wavefunction of the form,

$$\psi_0(\mathbf{r}_1, \mathbf{r}_2) = \left[\sum_{k>k_\mathrm{F}} g_k \cos[\mathbf{k}\cdot(\mathbf{r}_1 - \mathbf{r}_2)]\right][|\uparrow\downarrow\rangle - |\downarrow\uparrow\rangle]. \tag{6.52}$$

This can be plugged into the Schrödinger equation to yield,

$$(E - 2\epsilon_\mathbf{k})g_\mathbf{k} = \sum_{k' > k_\mathrm{F}} V_{\mathbf{k}\mathbf{k}'} g_{\mathbf{k}'}. \tag{6.53}$$

The above equation is true for any value of k. ϵ_k are the single-particle energies (of the plane wave type), and $V_{kk'}$ are the matrix elements of the interaction potential,

$$V_{\mathbf{k}\mathbf{k}'} = \frac{1}{V}\int V(\mathbf{r}) e^{i(\mathbf{k}-\mathbf{k}')\cdot\mathbf{r}} d\mathbf{r} \tag{6.54}$$

where $\mathbf{r}$ denotes the spatial distance between the two electrons. $V_{\mathbf{k}\mathbf{k}'}$ describes a scattering of a pair of electrons with momenta $(\mathbf{k}'_\uparrow, -\mathbf{k}'_\downarrow)$ to $(\mathbf{k}_\uparrow, -\mathbf{k}_\downarrow)$. We have to solve for the amplitude, $g_\mathbf{k}$ such that the total energy $E < 2\epsilon_\mathrm{F}$, such that a bound pair exists.

It is hard to carry out calculations for a general form of $V_{kk'}$. Cooper introduced a notable simplification,

$$\begin{aligned} V_{kk'} &= -V \quad \text{for} \quad \varepsilon < \epsilon_{k,k'} < \epsilon_\mathrm{F} + \hbar\omega_D \\ &= 0 \quad \text{otherwise} \end{aligned} \tag{6.55}$$

where V denotes the strength of the electron–electron interaction. This implies,

$$g_\mathbf{k} = V\sum_{\mathbf{k}'} \frac{g_{\mathbf{k}'}}{2\epsilon_k - E} \tag{6.56}$$

summing over the momentum index k,

$$\frac{1}{V} = \sum_{k>k_F} \frac{1}{2\epsilon_k - E}. \tag{6.57}$$

Note that this summation over k is for k values that are larger than the Fermi wavevector. We replace the summation by an integration, via using the DOS at the Fermi level, namely, $N(\epsilon_F)$, which is constant for a particular material, and hence can be brought out of the integral. Hence,

$$\frac{1}{V} = N(\epsilon_F)\int_{\epsilon_F}^{\epsilon_F+\hbar\omega_D} \frac{d\epsilon}{2\epsilon - E} = \frac{1}{2}N(\epsilon_F)\ln\left(\frac{2\epsilon_F - E + 2\hbar\omega_D}{2\epsilon_F - E}\right). \tag{6.58}$$

For conventional superconductors,

$$N(\epsilon_F)V < 0.3.$$

This allows applicability of weak coupling approximation valid for $N(\epsilon_F)V \ll 1$. This yields,

$$\begin{aligned} \frac{2}{N(\epsilon_F)V} &= \ln\left(\frac{2\epsilon_F - E + 2\hbar\omega_D}{2\epsilon_F - E}\right) \\ \text{or,}\quad \frac{2\epsilon_F - E + 2\hbar\omega_D}{2\epsilon_F - E} &= e^{2/N(\epsilon_F)V} \\ 2\epsilon_F - E + 2\hbar\omega_D &= (2\epsilon_F - E)e^{2/N(\epsilon_F)V} \\ E &= 2\epsilon_F - 2\hbar\omega_D e^{-2/N(\epsilon_F)V}. \end{aligned} \tag{6.59}$$

This is the expression for the total energy of the system of two electrons in the vicinity of a Fermi sea. Thus with respect to the Fermi surface of two electrons, the energy is negative, and hence it corresponds to a bound state. A bound state of two electrons is thus formed even for an infinitesimal interaction strength. Also, the form says that the binding energy is not analytic at $V = 0$, so it cannot be expanded in powers of V. Thus, any finite order perturbation theory would not be able to yield this result, and thus the theory is non-perturbative.

As for the wavefunction, it is dependent on the relative coordinate, $\mathbf{r} = \mathbf{r}_1 - \mathbf{r}_2$, and is proportional to,

$$\sum_{k>k'} \frac{\cos(\mathbf{k}\cdot\mathbf{r})}{2\xi_{\mathbf{k}} + E'}$$

where, $\xi_{\mathbf{k}} = \epsilon_k - \epsilon_F$, and $E' = 2\epsilon_F - E > 0$. E' can now be called the binding energy relative to $2\epsilon_F$. A few comments are in order.

(i) The amplitude of the wavefunction, namely, $(2\xi_k + E')^{-1}$ has its maximum value $1/E'$ when $\xi_k = 0$, that is, for electrons at the Fermi level. It falls off at positive values of ξ_k. Thus, it elucidates the dominant role of the electrons at the Fermi energy on the pairing mechanism.

(ii) The electronic states within an energy range E' about ϵ_F are involved in the formation of the bound state.
(iii) Note that $E' < \hbar\omega_D$ for $N(\epsilon_F)V < 1$, this makes sure that the detailed behaviour of $V_{kk'}$ is not important, and the assumption of the form in equation (6.55) will suffice.
(iv) Also, the small energy range allows estimation of the range of the Cooper pairs via uncertainty principle, namely $\xi_0 = \frac{a\hbar v_F}{kT_c}$. Thus the pairs are highly overlapping.

6.3.5 Statistical description of the BCS ground state

What we have learnt so far is that the filled Fermi sea becomes unstable to the formation of a bound pair. In s-wave superconductivity, a large number of bound pairs form and a system condenses into ground state below the same transition temperature, T_c. Let us postulate a paired many-body paired ground state of the form (since it is a ground state, we assume it to be a singlet state),

$$|\psi_0\rangle = \sum_{k>k_F} g_k c^\dagger_{k\uparrow} c^\dagger_{-k\downarrow}|0\rangle \tag{6.60}$$

where $|0\rangle$ does not denote vacuum, rather it represents a filled Fermi sea. Suppose there are M electrons, and we have to involve N electrons of those to form $\frac{N}{2}$ pairs, the number of ways $\frac{N}{2}$ pairs can be formed from M electrons is,

$$\frac{M!}{\left(M - \frac{N}{2}\right)!\left(\frac{N}{2}\right)!} \approx (10^{20})^{20} \qquad \text{for} \quad M = 10^{23}. \tag{6.61}$$

Thus a solution of the problem demands that $(10^{20})^{20}$ number of g_k be solved, which is certainly an impossible task. Thus a statistical description must be employed here. Further, since the number of particles is large, no serious error will be done if one works with grand canonical ensemble, where only the average number of particles, that is, $\bar{N}$ is fixed.

To circumvent a significantly difficult problem, the BCS theory postulated a ground state of the form,

$$|\psi_G\rangle = \prod_{k=k_1,k_2,\cdots k_M} (u_k + v_k c^\dagger_{k\uparrow} c^\dagger_{-k\downarrow}|\psi_0\rangle). \tag{6.62}$$

The probability that a paired state $(k_\uparrow, -k_\downarrow)$ is occupied is denoted by $|v_k|^2$, while it is unoccupied is given by $|u_k|^2$ with the condition,

$$|u_k|^2 + |v_k|^2 = 1 \qquad \forall \ k. \tag{6.63}$$

Take an example of two states $\mathbf{k}_1$ and $\mathbf{k}_2$, then the wavefunction amplitudes represent,

$$u_{k_1}, u_{k_2} \to 0 \text{ pair}$$
$$u_{k_1}, v_{k_2} \to 1 \text{ pair in } (\mathbf{k}_2, -\mathbf{k}_2)$$
$$v_{k_1}, v_{k_2} \to 2 \text{ pairs in } (\mathbf{k}_1, -\mathbf{k}_1), (\mathbf{k}_2, -\mathbf{k}_2).$$

We can calculate the average number of particles in the following way,

$$\begin{aligned}
\bar{N} = \langle \hat{N}_{OP} \rangle &= \left\langle \sum_{\mathbf{k},\sigma} \hat{n}_{\mathbf{k},\sigma} \right\rangle \\
&= \langle \psi_G | \sum_{\mathbf{k}} c_{k\uparrow}^{\dagger} c_{k\uparrow} + c_{k\downarrow}^{\dagger} c_{k\downarrow} | \psi_G \rangle \\
&= 2 \sum_{k} \langle \psi_G | c_{k\uparrow}^{\dagger} c_{k\uparrow} | \psi_G \rangle \\
&= 2 \sum_{k} \langle \psi_0 | (u_k^* + v_k^* c_{-k\downarrow} c_{k\uparrow}) c_{k\uparrow}^{\dagger} c_{k\uparrow} (u_k + v_k c_{k\uparrow}^{\dagger} c_{-k\downarrow}^{\dagger}) \\
&\quad \times \prod\nolimits_{l \neq k} (u_l^* + v_l^* c_{-l\downarrow} c_{l\uparrow})(u_l + v_l c_{l\uparrow}^{\dagger} c_{-l\downarrow}^{\dagger}) | \psi_0 \rangle.
\end{aligned}$$

In principle u_k and v_k are complex quantities. We have also used $\langle A\phi || \psi \rangle = \langle \phi | A^{\dagger} | \psi \rangle$. Let us look at various terms.

First, let us look at the term for $l \neq k$, namely,

$$|u_l|^2 + u_l^* v_l c_{l\uparrow}^{\dagger} c_{-l\downarrow}^{\dagger} + v_l^* u_l c_{-l\downarrow} c_{l\uparrow} + |v_l|^2 c_{-l\downarrow} c_{l\uparrow} c_{l\uparrow}^{\dagger} c_{-l\downarrow}^{\dagger} \tag{6.64}$$

where we have taken the expectation with $|\psi_0\rangle$. The middle two terms (second and the third terms) yield zero as they change occupancy of the state with lth pair (first term creates and the second term destroys). The last term only yields a normalization because it creates and destroys a pair. Thus the $l \neq k$ term yields,

$$|u_l|^2 + |v_l|^2$$

which, by normalization is equal to 1.

When the same procedure is applied to $l = k$ term, we shall have two terms surviving with $u_k v_k$ that will cancel out. Further,

$$|u_k|^2 c_{k\uparrow}^{\dagger} c_{k\uparrow} | \phi_0 \rangle = 0$$

since there are no states to annihilate for $k > k_{\mathrm{F}}$. Thus we are left with only $|v_k|^2$.

Thus the average number of particles yields,

$$\bar{N} = 2 \sum_{k} |v_k|^2 .$$

The factor 2 appears for the summation over spin. Further, our assumption that this reinforces $|v_k|^2$ indeed denotes probability of occupied states.

To calculate some more statistical quantities that may be of physical relevance, we estimate the RMS fluctuations of $\bar{N}$, we calculate,

$$\begin{aligned}\langle (N-\bar{N})^2\rangle &= \langle N^2 - 2N\bar{N} + \bar{N}^2\rangle \\ &= \langle N^2\rangle - \bar{N}^2 \\ &= 4\sum_k u_k^2 v_k^2.\end{aligned} \tag{6.65}$$

This is a positive definite quantity. In fact, v_k as a function of k goes from 1 to zero and u_k goes from $0 \to 1$ in an energy range given by $k_B T_c$. Thus the sum above is proportional to $\frac{T_c}{T_F}\bar{N}$

6.4 The variational calculation

We write down a generic Hamiltonian that encodes the kinetic energy and a two-particle interaction term as in the following,

$$\mathcal{H} = \sum_{k,\sigma} \xi_{k\sigma} c_{k\sigma}^\dagger c_{k\sigma} + \sum_{k,l} V_{kl} c_{k\uparrow}^\dagger c_{-k\downarrow}^\dagger c_{-l\downarrow} c_{l\uparrow} = \hat{K} + \hat{V} \tag{6.66}$$

where $\hat{K}$ and $\hat{V}$ denote the kinetic and the interaction energies, and $\xi_k = \epsilon_k - \mu$. k, l are the momenta indices. As exact solutions are hard to arrive at, and perturbation theory will not be of relevance for reasons discussed earlier, we shall resort to a variational calculation, that is, minimize the energy with respect to the tunable parameters, which are the coefficients u_k and v_k. That is,

$$\delta\left[\langle \psi_G | \sum_{k,\sigma} \xi_k n_{k\sigma} + \sum_{k,l} V_{kl} c_{k\uparrow}^\dagger c_{-k\downarrow}^\dagger c_{-l\downarrow} c_{l\uparrow} | \psi_G\rangle\right]. \tag{6.67}$$

Let us look at the kinetic energy term. We have already computed it earlier, namely,

$$\langle \hat{K}\rangle = \langle \psi_G | \sum_{k,\sigma} \xi_k n_{k\sigma} | \psi_G\rangle = 2\sum_k |v_k|^2 \xi_k. \tag{6.68}$$

Similarly for the potential energy term,

$$\langle \hat{V}\rangle = \sum_{k,l} V_{kl} u_k^* v_l^* u_l v_k \tag{6.69}$$

where we have used the normalization constraint, that is, $|u_k|^2 + |v_k|^2 = 1 \ \forall \ k$.

Doing a variational calculation with two variables is a difficult job, which, however, is simplified owing to the normalization condition. This yields an identification of (u_k, v_k) with $(\sin\theta_k, \cos\theta_k)$ which facilitates the Hamiltonian to be dependent only on one parameter, namely θ_k. In particular, we take,

$$u_k = \sin\theta_k, \ v_k = \cos\theta_k. \tag{6.70}$$

Thus we can minimize energy with respect to θ_k, which implies,

$$\frac{\partial}{\partial\theta_k}\left[\sum_{k'} \xi_{k'}(1 + \cos 2\theta_{k'}) + \frac{1}{4}\sum_{k',l} V_{k'l} \sin 2\theta_{k'} \sin 2\theta_l\right] = 0 \tag{6.71}$$

$$-2\xi_k \sin 2\theta_k + \sum_l V_{kl} \cos 2\theta_k \sin 2\theta_l = 0. \tag{6.72}$$

Now define,

$$\Delta_k = -\frac{1}{2}\sum_l V_{kl} \sin 2\theta_l = -\frac{1}{2}\sum_l V_{kl} \sin 2\theta_l. \tag{6.73}$$

Using this in the above equation can be written as,

$$2\xi_k \sin 2\theta_k = \frac{1}{2}\sum_l V_{kl} \cos 2\theta_k \sin 2\theta_l = -2\Delta_k \cos 2\theta_k. \tag{6.74}$$

This yields,

$$\tan 2\theta_k = -\frac{\Delta_k}{\xi_k}, \quad \text{implying that,} \quad \frac{\sin 2\theta_k}{\cos 2\theta_k} = -\frac{\Delta_k}{\xi_k}. \tag{6.75}$$

Using the definitions,

$$\begin{aligned} 2u_k v_k = \sin 2\theta_k &= \frac{\Delta_k}{\sqrt{\xi_k^2 + \Delta_k^2}} \\ v_k^2 - u_k^2 = \cos 2\theta_k &= -\frac{\xi_k}{\sqrt{\xi_k^2 + \Delta_k^2}}. \end{aligned} \tag{6.76}$$

We have,

$$v_k^2 - u_k^2 = -\frac{\xi_k}{E_k}$$

where, $E_k = \sqrt{\xi_k^2 + \Delta^2}$, and, of course the normalization condition, namely, $u_k^2 + v_k^2 = 1$. Thus the definition for the coherence factors can be written as,

$$\begin{aligned} v_k^2 &= \frac{1}{2}\left(1 - \frac{\xi_k}{E_k}\right) = \frac{1}{2}\left[1 - \frac{\xi_k}{\sqrt{\Delta^2 + \xi_k^2}}\right] \\ u_k^2 &= \frac{1}{2}\left(1 + \frac{\xi_k}{E_k}\right) = 1 - v_k^2. \end{aligned} \tag{6.77}$$

This is a choice which seems to be fitting all definitions. There is a close resemblance between the v_k^2 for the BCS ground state at $T = 0$, and the free Fermi distribution at $T = T_F$ (see figure 6.17). Also, v_k^2 falls off as $1/\xi_k^2$ for $\xi_k \gg \Delta$. If one remembers, g_k^2 falls off in a similar fashion in the simple treatment of a single Cooper pair presented earlier. Also we have chosen $\cos 2\theta_k$ with a negative sign, because if ξ_k is large (that is, $\epsilon_k \gg \mu$) v_k should go to zero.

Hence, Δ_k assumes a form,

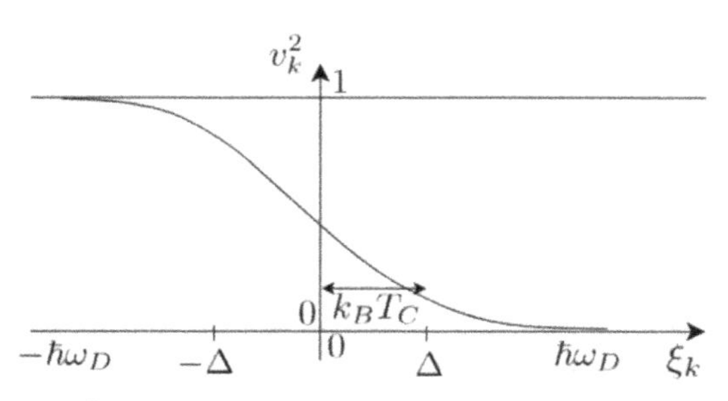

Figure 6.17. The behaviour of v_k^2 (probability of states occupied by a pair of momentum k and $-k$) is shown as a function of ξ_k $(=\epsilon_k - \mu)$.

$$\begin{aligned}\Delta_k &= -\frac{1}{2}\sum_l V_{kl}\sin 2\theta_l \\ \Delta_k &= -\frac{1}{2}\sum_l V_{kl}\frac{\Delta_l^2}{\sqrt{\Delta_l^2+\xi_l^2}}.\end{aligned} \tag{6.78}$$

A trivial solution of the above equation is $\Delta = 0$. This implies, $2u_kv_k = 0$, and $v_k^2 - u_k^2 = -1$. Since, $u_k^2 + v_k^2 = 1$, it implies that $v_k^2 = 0$, which in turn means a normal state and hence there are no pairs. The above discussion clearly indicates that Δ_k can be used as order parameters for the superconducting transition, which when finite implies a superconducting state, and $\Delta_k = 0$ denotes a normal state.

Let us recall Cooper's original assumption on the nature of the interaction, that is,

$$V_{kl} = -V \quad \text{if} \quad |\xi_k - \xi_l| < \hbar\omega_D.$$

Hence putting it in equation (6.73),

$$\begin{aligned}\Delta_k &= \frac{1}{2}V\sum_l \sin 2\theta_l \quad \text{for} \quad |\xi_k - \xi_l| < \hbar\omega_D \\ &= 0 \qquad \text{otherwise.}\end{aligned} \tag{6.79}$$

From the above equation, it is clear that the order parameter, Δ is independent of the momentum index, k. Substituting $\sin 2\theta_l$ from equation (6.76), one may cancel Δ from both sides to get,

$$\begin{aligned}1 &= \frac{V}{2}\sum_l \frac{1}{\sqrt{\Delta^2+\xi_l^2}} \\ \frac{1}{V} &= \int_0^{\hbar\omega_D} \frac{N(\varepsilon_F)d\xi}{2\sqrt{\Delta^2+\xi^2}} \\ \frac{2}{N(\varepsilon_F)V} &= \int_0^{\hbar\omega_D} \frac{d\xi}{\sqrt{\Delta^2+\xi^2}}.\end{aligned} \tag{6.80}$$

Integrating and solving for Δ,

$$2\sinh^{-1}\left(\frac{\xi}{\Delta}\right)_0^{\hbar\omega_D} = \frac{2}{N(\varepsilon_F)V} \tag{6.81}$$

$$\text{Or, } 2\sinh^{-1}\left(\frac{\hbar\omega_D}{\Delta}\right) = \frac{2}{N(\varepsilon_F)V}. \tag{6.82}$$

Reorganizing,

$$\Delta = \frac{\hbar\omega}{\sinh(1/N(\varepsilon_F)V)}. \tag{6.83}$$

For a weak coupling superconductor, $N(\varepsilon_F)V \ll 1$, one arrives at,

$$\Delta = 2\hbar\omega_D e^{-1/N(\varepsilon_F)V}. \tag{6.84}$$

This is the expression for the energy difference between the superconducting state and the normal state. Thus this much of energy has to be supplied in order to break a superconducting state, and achieve a normal (metallic) state. Suppose a thermal excitation facilitates such a transition, which yields, $\Delta \sim k_B T_c$; *a priori*, from BCS theory one gets (will be shown immediately afterwards),

$$k_B T_c \approx 1.13\ \hbar\omega_D \exp[-1/N(\varepsilon_F)V] \tag{6.85}$$

which only changes the factor '2' in equation (6.84) to 1.13 in equation (6.85). Having found Δ, we may simply compute the coefficients u_k and v_k which should specify the optimal BCS wavefunction.

6.4.1 Temperature dependence of the gap

We have introduced that the energy gap (which also acts as the order parameter) in a superconductor is defined by,

$$\Delta_k = -\sum_{k'} V_{kk'}\langle c_{-k'\downarrow}c_{k'\uparrow}\rangle \tag{6.86}$$

Δ_k allows us to write the mean field BCS Hamiltonian,

$$\mathcal{H} = \sum_{k,\sigma}\xi_k c_{k\sigma}^\dagger c_{k\sigma} - \sum_{k,\sigma}(\Delta_k c_{k\uparrow}^\dagger c_{-k\downarrow}^\dagger + \Delta_k^* c_{-k\downarrow}c_{k\uparrow}). \tag{6.87}$$

We have neglected a constant term in the above equation. This Hamiltonian can be diagonalized by Bogoliubov–Valatin transformation, where the creation and the annihilation operators are written in terms of the quasiparticle operators as,

$$c_{k\uparrow} = u_k^*\gamma_{k_0} + v_k\gamma_{k_1}^* \tag{6.88}$$

$$c_{-k\downarrow}^\dagger = -v_k^*\gamma_{k_0} + u_k\gamma_{k_1}^*. \tag{6.89}$$

Being a linear combination of c_k and $c_k^\dagger$, γ_k obeys fermionic anticommutator relations and thus the relationships are canonical. This along with the normalization condition yields,

$$\begin{aligned}\mathcal{H} = &\sum_k \xi_k[(|u_k|^2 - |v_k|^2)(\gamma_{k0}^\dagger \gamma_{k0} + \gamma_{k1}^\dagger \gamma_{k1}) + 2|v_k|^2 + 2u_k^* v_k^* \gamma_{k1}\gamma_{k0} + 2u_k v_k \gamma_{k0}^\dagger \gamma_{k0}^\dagger] \\ &+ \sum_k (\Delta_k u_k v_k^* + \Delta_k^* u_k^* v_k)(\gamma_{k0}^\dagger \gamma_{k0} + \gamma_{k1}^\dagger \gamma_{k1} - 1) + (\Delta_k v_k^{*2} - \Delta_k^* u_k^{*2})\gamma_{k1}\gamma_{k0} \\ &+ (\Delta_k^* v_k^2 - \Delta_k u_k^2)\gamma_{k0}^\dagger \gamma_{k1}^\dagger.\end{aligned} \tag{6.90}$$

Finally, one gets,

$$\mathcal{H} = \sum_k E_k(\gamma_{k0}^\dagger \gamma_{k0} + \gamma_{k1}^\dagger \gamma_{k1}) + \sum_k (\xi_k - E_k + |\Delta_k|^2) \tag{6.91}$$

where $E_k = \sqrt{\xi_k^2 + |\Delta|^2}$. The second term is just a constant. So that the Hamiltonian is diagonal in the basis of the quasiparticle operators[3].

For the gap function, Δ, in terms of the quasiparticle operators after a little bit of algebra yields,

$$\Delta_k = -\sum_{k'} V_{kk'} u_{k'}^* v_{k'} \langle 1 - \gamma_{k_0'}^\dagger \gamma_{k_0'} - \gamma_{k_1'}^\dagger \gamma_{k_1'} \rangle. \tag{6.92}$$

At zero temperature, $\langle \gamma_{k0}^\dagger \gamma_{k0} \rangle = 1$, however, at non-zero temperatures,

$$\langle \gamma_{k0}^\dagger \gamma_{k0} \rangle = f(E_k) = \frac{1}{e^{\beta E_k} + 1}$$

where $f(E_k)$ denotes the Fermi distribution function for the quasiparticles. Since expectation value of each of the number operators yields a Fermi function, one gets,

$$\langle 1 - \gamma_{k0}^\dagger \gamma_{k0} - \gamma_{k1}^\dagger \gamma_{k1} \rangle = 1 - 2f(E_k).$$

Thus, for the gap function,

$$\begin{aligned}\Delta_k &= -\sum_{k'} V_{kk'} u_{k'}^* v_{k'} (1 - 2f(E_k)) \\ &= -\sum_{k'} V_{kk'} \frac{\Delta_{k'}}{2E_{k'}} \tanh\left(\frac{\beta E_{k'}}{2}\right).\end{aligned}$$

With $V_{kk'} = -V$,

$$\frac{1}{V} = \frac{1}{2}\sum_k \frac{\tanh(\beta E_k/2)}{E_k}. \tag{6.93}$$

[3] In order for the quasiparticle operators (γ) to diagonalize, the idea is to make coefficients of $\gamma\gamma$ or $\gamma^\dagger\gamma^\dagger$ equal to zero and retain $\gamma^\dagger\gamma$. This happens when $2\xi_k u_k v_k + \Delta_k^* v_k^2 - \Delta_k u_k^2 = 0$.

The above equation holds a clue to the temperature dependence of the energy gap, $\Delta(T)$. Converting the sum into an integral via introducing the DOS,

$$\frac{1}{N(\epsilon_F)V} = \int_0^{\beta_c\hbar\omega_D/2} \frac{\tanh x}{x} dx, \qquad \text{where,} \quad x = \frac{\beta E_k}{2}$$

$$\int_0^{\beta_c\hbar\omega_D/2} \frac{\tanh x}{x} dx = \ln\left(\frac{2e^{\gamma}}{\pi}\beta_c\hbar\omega_D\right) \tag{6.94}$$

where γ denotes the Euler's constant, which has a value, $\gamma = 0.577\cdots$ and thus $\frac{2e^{\gamma}}{\pi} \simeq 1.14$. This yields an expression for the transition temperature, T_c above which superconductivity vanishes,

$$k_B T_c = 1.13\hbar\omega_c e^{-1/N(\epsilon_F)V}. \tag{6.95}$$

One may recall that earlier we have obtained an expression for the energy gap, namely,

$$\Delta = 2\hbar\omega_D e^{-1/N(\epsilon_F)V}$$

which was subsequently equated to $k_B T_c$. Also, a relationship between the energy gap at zero temperature to that of T_c is obtained as,

$$\frac{\Delta(0)}{k_B T_c} = \frac{2}{1.13} = 1.764. \tag{6.96}$$

Thus the gap at $T = 0$ is indeed comparable to the magnitude of $k_B T_c$. The numerical factor has been tested in many experiments and found to be reasonable. The value of $2\Delta(0)$ varies between $3k_B T_c$ to $4.5k_B T_c$ for the conventional (that is, BCS) superconductors, and is mostly clustered around the BCS value $3.5k_B T_c$.

Finally, to arrive at an explicit temperature dependence of the gap function, we consider the integral form of the gap equation, namely,

$$\frac{1}{N(\varepsilon_F)V} = \int_0^{\hbar\omega_D} \frac{\tanh\left[\frac{1}{2}\beta(\xi^2 + \Delta^2)^{1/2}\right]}{(\xi^2 + \Delta^2)^{1/2}} d\xi. \tag{6.97}$$

Now $\Delta(T)$ needs to be computed numerically by self-consistently solving the above equation. For weak coupling superconductors, for which $\hbar\omega_D \gg k_B T_c$, $\Delta(T)/\Delta(0)$ is a universal function of (T/T_c) which decreases monotonically from 1 at $T = 0$ to zero at $T = T_c$. With $\Delta(T)$ determined, the quasiparticle energies can be written as,

$$E_k = \sqrt{\xi_k^2 + \Delta^2(T)}.$$

In the vicinity of $T = 0$, the temperature variation is exponentially slow because of $e^{-\Delta/k_B T} \approx 0$ (see figure 6.18). Hence the tan hyperbolic term is insensitive to T and stays very close to 1. Physically speaking, Δ remains nearly a constant until a significant number of quasiparticles are thermally excited.

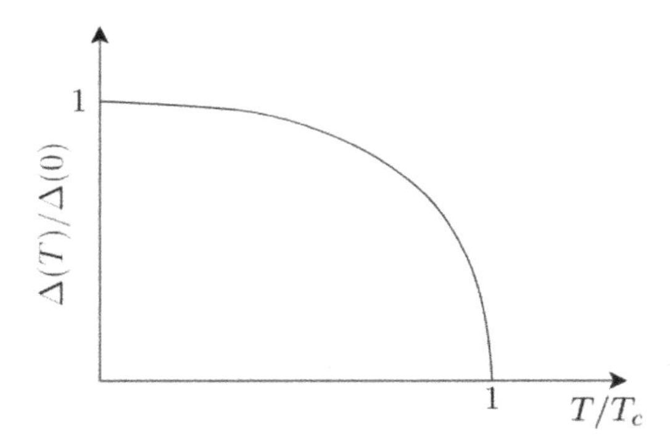

Figure 6.18. The scaled gap function $\Delta(T)/\Delta(0)$ is numerically obtained and plotted as a function of the reduced temperature, T/T_c. The area enclosed by the curve denotes the superconducting phase.

On the other hand, for $T \sim T_c$, $\Delta(T)$ vanishes with a vertical slope approximately as,

$$\frac{\Delta(T)}{\Delta(0)} \approx 1.74\left[1 - \left(\frac{T}{T_c}\right)\right]^{1/2} \text{for } T \to T_c \quad \text{from below.}$$

This square root dependence of the Δ on T is a hallmark feature of all mean field theories. For example, the magnetization vanishes in a similar fashion with temperature in molecular field theory of ferromagnetism.

6.4.2 Thermodynamics from BCS theory

The energy of a superconductor is obtained by explicitly solving the BCS Hamiltonian, which yields,

$$E = 2\sum_{k>k_F} \xi_k v_k^2 + 2\sum_{k<k_F} |\xi_k| u_k^2 + \sum_{kk'} V_{kk'} u_k v_k u_{k'} v_{k'} \tag{6.98}$$

where,

$$v_k^2 = \frac{1}{2}\left(1 - \frac{\xi_k}{\sqrt{\xi_k^2 + \Delta_k^2}}\right)$$

with $\xi_k = \varepsilon_k - \mu$, and $\Delta_k = -\sum_{k'} V_{kk'} u_{k'} v_{k'}$. Further, the form of the attractive interaction is,

$$V_{kk'} = \begin{cases} -V & \text{for both } \quad |\xi_k| \text{ and } |\xi_{k'}| < \hbar\omega_D \\ 0 & \text{otherwise.} \end{cases}$$

Also, we have obtained that,

$$\Delta \simeq 2\hbar\omega_D \exp[-1/N(\varepsilon_F)V].$$

Thus the energy difference between the superconducting and the normal states is given by,

$$E_s - E_n = -\frac{1}{2}N(\varepsilon_F)\Delta^2\left[1 - \exp\left(-\frac{2}{N(\varepsilon_F)V}\right)\right] \tag{6.99}$$

which in the weak coupling limit assumes a form,

$$E_s - E_n = -\frac{1}{2}N(\varepsilon_F)\Delta^2$$

To enumerate the jump in the specific heat from BCS theory, we consider the entropy for the SC state given by [12],

$$S_s = -2k_B\sum_k\left[(1 - f_k)\ln(1 - f_k) + f_k \ln f_k\right]. \tag{6.100}$$

Further, in terms of the probability of a microstate being occupied, the entropy has a form,

$$S = -k_B\sum p_i \ln p_i. \tag{6.101}$$

Following this, the specific heat can be determined by,

$$C_{es} = T\frac{dS_s}{dT} = T\frac{dS_s}{d\beta}\cdot\frac{d\beta}{dT} = T\frac{dS_s}{d\beta}\cdot\left(-\frac{1}{k_BT^2}\right) = -\beta\frac{dS_s}{d\beta}$$

now,

$$\begin{aligned} C_s &= 2\beta k_B\sum_k\frac{\partial f_k}{\partial\beta}\ln\frac{f_k}{1 - f_k} \\ &= -2\beta^2k_B\sum_k E_k\frac{\partial f_k}{\partial\beta} = -2\beta^2k_B\sum_k E_k\frac{df_k}{d(\beta E_k)}\left(E_k + \beta\frac{dE_k}{d\beta}\right) \\ &= -2\beta k_B\sum_k\left(\frac{df_k}{dE_k}\right)\left(E_k^2 + \frac{1}{2}\beta\frac{d\Delta^2}{d\beta}\right). \end{aligned} \tag{6.102}$$

The first term is the contribution of the quasiparticles to the specific heat, while the second term is due to the temperature dependent gap function. The features of this specific heat are as follows:

(i) C_{es} is exponentially small at $T \ll T_c$, where the excitation energy Δ is much greater than k_BT. This explains the exponential dependence at low temperatures, namely $C_s \sim T_c e^{-T/T_c}$.

(ii) Near T_c, the gap, $\Delta(T) \to 0$, when one can replace E_k by $|\xi_k|$, the first term reduces to

$$C_s(T) = \gamma T = \frac{2\pi^2}{3}N(\varepsilon_F)k_B^2T$$

which is continuous at $T = T_c$.

(iii) The second term is finite below T_c, where $\frac{d\Delta^2}{dT}$ is large but is zero above T_c, thereby giving rise to a discontinuity, ΔC in the electronic specific heat at T_c. This discontinuity is characteristic of a second-order phase transition.

The jump in the specific heat is enumerated as,

$$\begin{aligned}\Delta C &= (C_s - C_n)|_{T=T_c} \\ &= N(\varepsilon_F)k_B\beta^2\frac{d(\Delta^2)}{d\beta}\int_{-\infty}^{\infty}\left(\frac{df}{d|\xi|}\right)d\xi \\ &= N(\varepsilon_F)\left(-\frac{d\Delta^2}{dT}\right)_{T=T_c}.\end{aligned} \tag{6.103}$$

Here, $\frac{df}{d|\xi|} = \frac{df}{d\xi}$ as f is an even function of ξ. Further, using[4]

$$\Delta(T) = 1.74\Delta(0)(1 - \frac{T}{T_c})^{1/2}$$

the magnitude ΔC can be computed to yield,

$$\Delta C = 9.4N(\varepsilon_F)k_B^2T_c$$

where $\Delta(0) = 1.76k_BT_c$. The normalized discontinuity, usually measured in experiments is given by,

$$\frac{\Delta C}{C_{en}} = \frac{9.4}{2\pi^2/3} \approx 1.43$$

$C_{es}(T)$ can be found numerically from the parent expression (equation (6.102)). $C_{es}(T)$ can be integrated to obtain $U_{es}(T)$. The clue is that at $T = T_c$, $U_{es}(T)$ must be same as the normal value $U_{en}(0) + \frac{1}{2}\gamma T_c^2$.

6.5 Electromagnetic considerations

Let us consider a system of N electrons subjected to a magnetic field, $\mathbf{B}$ which is described by a vector potential. The momentum for each of these particles described by, $i = 1, \cdots, N$ under an external vector potential, $\mathbf{A}(\mathbf{r}, t)$ can be replaced by,

$$\mathbf{p}_i \rightarrow \mathbf{p}_i + e\mathbf{A}(\mathbf{r}).$$

This yields a Hamiltonian,

$$\mathcal{H} = \frac{1}{2m}\sum_{i=1}^{N}(\mathbf{p}_i - e\mathbf{A}(\mathbf{r}_i))^2 + \hat{V} = \mathcal{H}_0 + \mathcal{H}' \tag{6.104}$$

[4] This is the numerical solution of the gap function, $\Delta(T)$ shown in figure 6.18.

where, $\mathcal{H}_0$ is the many-body Hamiltonian with $\mathcal{H}_0 = \sum_i \frac{p_i^2}{2m}$. To the linear order in $\mathbf{A}$, the coupling to the external probe can be written as,

$$\mathcal{H}' = -\frac{e}{2m}\sum_{i=1}^{N}\left[\mathbf{p}_i \cdot \mathbf{A}(\mathbf{r}_i) + \mathbf{A}(\mathbf{r}_i) \cdot \mathbf{p}_i\right]. \tag{6.105}$$

With the choice of a transverse gauge, that is, $\nabla \cdot \mathbf{A} = 0$, both the terms yield the same contribution. Hence we can write,

$$\mathcal{H}' = -\frac{e}{m}\sum_{i=1}^{N}\mathbf{p}_i \cdot \mathbf{A}(\mathbf{r}_i). \tag{6.106}$$

This greatly simplifies the analysis, as will be clear in the following discussion.

The total current operator can be written as,

$$\mathbf{j}(\mathbf{r}) = \frac{e}{2}\sum_{i=1}^{N}(\mathbf{v}_i\delta(\mathbf{r} - \mathbf{r}_i) + \delta(\mathbf{r} - \mathbf{r}_i)\mathbf{v}_i) \tag{6.107}$$

with the velocity operator given by,

$$\mathbf{v}_i = \frac{1}{m}(\mathbf{p}_i - \frac{e}{c}\mathbf{A}(\mathbf{r}_i)) \tag{6.108}$$

and the δ-functions denote the positions of the particles. The above form for current can be split into,

$$\mathbf{j}(\mathbf{r}) = \mathbf{j}_p(\mathbf{r}) + \mathbf{j}_d(\mathbf{r}) \tag{6.109}$$

where $\mathbf{j}_p$ and $\mathbf{j}_d$ denote the paramagnetic and diamagnetic current densities, and are given by the following.

$$\mathbf{j}_p(\mathbf{r}) = \frac{e}{2m}\sum_{i=1}^{N}\left(\mathbf{p}_i\delta(\mathbf{r} - \mathbf{r}_i) + \delta(\mathbf{r} - \mathbf{r}_i)\mathbf{p}_i\right). \tag{6.110}$$

In terms of the fermionic operators,

$$\mathbf{j}_p(\mathbf{q}) = \frac{e\hbar}{m}\sum_{\mathbf{k},\sigma}(\mathbf{k} + \mathbf{q}/2)c_{k\sigma}^{\dagger}c_{k+q,\sigma} \tag{6.111}$$

where $\mathbf{k}$ and $\mathbf{q}$ denote momentum indices. Next, the diamagnetic current density is written as,

$$\mathbf{j}_d(\mathbf{r}) = -\frac{e^2}{m}\mathbf{A}(\mathbf{r})\sum_{i=1}^{N}\delta(\mathbf{r} - \mathbf{r}_i) \tag{6.112}$$

which again in terms of the fermionic operators can be expressed via,

$$\mathbf{j}_d(\mathbf{q}) = -\frac{e^2}{m}\mathbf{A}_\mathbf{q}\sum_{\mathbf{k},\sigma} c^\dagger_{k\sigma}c_{k\sigma} = -\frac{ne^2}{m}\mathbf{A}_\mathbf{q}. \tag{6.113}$$

Using these definitions, the coupling to the vector potential is expressed as,

$$\mathcal{H}' = -\frac{1}{\Omega}\int d\mathbf{r}\ \mathbf{j}_p(\mathbf{r})\cdot\mathbf{A}(\mathbf{r}) = -\frac{1}{\Omega}\sum_\mathbf{q}\mathbf{j}_p(-\mathbf{q})\cdot\mathbf{A}(\mathbf{q}) \tag{6.114}$$

where Ω denotes the volume and can be set to unity for convenience. It may be noted that the external probe $\mathbf{A}$ couples only to the paramagnetic current $\mathbf{j}_p$ in the Hamiltonian, $\mathcal{H}'$. $\mathbf{j}_d$ is already linear in $\mathbf{A}$. So it is only needed to evaluate $\mathbf{j}_p$ at the linear response level, and hence add to $\mathbf{j}_d$ to get the total current response, such that,

$$\langle \mathbf{j}\rangle = \langle \mathbf{j}_p\rangle + \langle \mathbf{j}_d\rangle. \tag{6.115}$$

According to linear response theory one can write,

$$\langle j_\alpha\rangle(\mathbf{q},\omega) = \chi_{\alpha\beta}(\mathbf{q},\omega)A_\beta(\mathbf{q},\omega) - \frac{e^2}{m}\left\langle\sum_{k\sigma} c^\dagger_{k\sigma}c_{k\sigma}\right\rangle A_\alpha(\mathbf{q},\omega) \tag{6.116}$$

where ω denotes the frequency and the current–current response function is given by,

$$\chi_{\alpha\beta} \equiv \chi^P_{j_\alpha j_\beta}(\mathbf{q},\omega)$$

with the density appearing in the second term,

$$n \equiv \left\langle\sum_{k\sigma} c^\dagger_{k\sigma}c_{k\sigma}\right\rangle.$$

For obtaining the Meissner effect, we are interested in the zero momentum (long wavelength) and zero frequency (static) response of $\chi_{\alpha\beta}$, that is, for $\omega = 0$, $q \to 0$. One finally gets for the current density,

$$\begin{aligned}\langle j_\alpha\rangle(\mathbf{q}\to 0,\omega = 0) &= \chi_{\alpha\beta}(\mathbf{q}\to 0,\omega = 0)A_\beta(\mathbf{q}\to 0,\omega = 0)\\ &\quad -\frac{ne^2}{m}A_\alpha(\mathbf{q}\to 0,\omega = 0).\end{aligned}$$

For computing conductivity, we have to consider the finite frequency limit ($\omega \neq 0$)[5]. Consider the electric field (the scalar potential is set to zero because of the gauge condition),

$$\mathbf{E} = -\frac{\partial\mathbf{A}}{\partial t} = i\omega\mathbf{A}. \tag{6.117}$$

[5] For the dc conductivity, we can take $\omega \to 0$.

Thus,

$$\langle j_\alpha \rangle = \sigma_{\alpha\beta} E_\beta = i\omega\sigma_{\alpha\beta} A_\beta \tag{6.118}$$

which subsequently yields,

$$\sigma_{\alpha\beta} = \frac{1}{i\omega}\left[\chi_{\alpha\beta}(q = 0, \omega) - \frac{ne^2}{m}\,\delta_{\alpha\beta}\right] \tag{6.119}$$

here,

$$\sigma_{\alpha\beta} = \Re\sigma_{\alpha\beta} + \Im\sigma_{\alpha\beta}. \tag{6.120}$$

Finally at the linear response level, the Kubo formula for the conductivity tensor is written as,

$$\sigma_{\alpha\beta}(\omega) = \frac{1}{i\omega}\left[\chi_{j_p\alpha j_p\beta}(q = 0, \omega) - \frac{ne^2}{m}\,\delta_{\alpha\beta}\right]. \tag{6.121}$$

6.5.1 Meissner effect

We demonstrate the expulsion of the magnetic field or the Meissner effect in the following.

$$\mathbf{j}_p = \frac{e\hbar}{m}\sum_k(\mathbf{k} + \mathbf{q}/2)(c^\dagger_{k\uparrow}c_{k+q\uparrow} + c^\dagger_{-k\downarrow}c_{-(k+q)\downarrow}). \tag{6.122}$$

The creation and the annihilation operators are written in terms of the quasiparticle operators via,

$$\begin{aligned} c_{\mathbf{k}\uparrow} &= u_{\mathbf{k}}\gamma_{\mathbf{k}\uparrow} + v_{\mathbf{k}}\gamma^\dagger_{-\mathbf{k}\downarrow} \\ c^\dagger_{-\mathbf{k}\downarrow} &= -\,v_{\mathbf{k}}\gamma_{\mathbf{k}\uparrow} + u_{\mathbf{k}}\gamma_{-\mathbf{k}\downarrow} \end{aligned} \tag{6.123}$$

where $u_{\mathbf{k}}$ and $v_{\mathbf{k}}$ are the coherence factors. These relationships are canonical, since they preserve the fermionic anticommutation relations. This allows the paramagnetic current to be written as,

$$\begin{aligned} j_p = \frac{e\hbar}{m}\sum_k(\mathbf{k} + \mathbf{q}/2)[(uu' + vv')(\gamma^\dagger_{k\uparrow}\gamma_{k+q\uparrow} - \gamma^\dagger_{-(k+q)\downarrow}\gamma_{-k\downarrow}) \\ + (uv' - vu')(\gamma^\dagger_{k\uparrow}\gamma^\dagger_{-(k+q)\downarrow} - \gamma_{-k\downarrow}\gamma_{k+q\uparrow})] \end{aligned} \tag{6.124}$$

where the momentum indices are dropped by redefining,

$$u = u_k \qquad v = v_k \qquad u' = u_{k+q} \qquad v' = v_{k+q}.$$

Now in the linear regime, the current–current response function is written as,

$$\chi_{\alpha\beta}(q, \omega = 0) = \sum_{m,n} \frac{e^{-\beta\omega_m}}{Z} \left\{ \frac{(j_p^\alpha(q))_{m,n}(j_p^\beta(-q))_{n,m}}{\omega_{nm}} + \text{c. c.} \right\}. \tag{6.125}$$

One can import the contribution of j_p from equation (6.124). In particular we look at different terms below.

(i) *$\gamma^\dagger\gamma$ term with $\sigma = \uparrow$*

$$\sum_{k, k'} f(u, v)(k' + q/2)^\alpha(k - q/2)^\beta \langle m|\gamma_{k'}^\dagger\gamma_{k'+q}|n\rangle\langle n|\gamma_k^\dagger\gamma_{k-q}|m\rangle \frac{1}{\omega_{nm}}$$

where $f(u, v)$ is the coherence factor determined later. The only $|n\rangle$ that contributes is given by $|n\rangle = \gamma_k^\dagger\gamma_{k-q}|m\rangle$. Thus,

$$\mathbf{k}' + \mathbf{q} = \mathbf{k}, \quad \text{and} , \quad \omega_{nm} \equiv \omega_n - \omega_m = E_k - E_{k-q}.$$

This yields the above contribution to have the form.

$$\begin{aligned} \sum_k \underbrace{f(u, v)(k - q/2)^\alpha(k - q/2)^\beta}_{E_k - E_{k-q}} &= \sum_m \frac{e^{-\beta\omega_m}}{Z} \langle m|\gamma_{k-q}^\dagger\gamma_k\gamma_k^\dagger\gamma_{k-q}|m\rangle \\ &= \langle \gamma_{k-q}^\dagger\gamma_{k-q}(1 - \gamma_k^\dagger\gamma_k)\rangle \\ &= f_{k-q}(1 - f_k). \end{aligned} \tag{6.126}$$

The above being the thermal average, we get the Fermi distribution function, f_k in the above expression.

Finally, the coherence factor, $f(u, v)$ is,

$$\begin{aligned} &= (u_{k'}u_{k'+q} + v_{k'}v_{k'+q})(u_k u_{k-q} + v_k v_{k-q}) \\ &= (u_k u_{k-q} + v_k v_{k-q})^2, \quad \text{since} \quad \mathbf{k}' = \mathbf{k} - \mathbf{q}. \end{aligned}$$

One may shift the dummy index, $\mathbf{k} \to \mathbf{k} + \mathbf{q}$ which makes the complex conjugate equal to the term calculated above. This yields the term with $\gamma^\dagger\gamma$ for $\uparrow$-spin as,

$$= \frac{2e^2\hbar^2}{m^2} \sum_k \frac{(uu' + vv')^2 f(1 - f')}{E' - E} (k + q/2)^\alpha(k + q/2)^\beta.$$

(ii) *Combination of $\gamma^\dagger\gamma$ with $\sigma = \downarrow$*

Same as (i) except for a negative sign in front of $\gamma_\downarrow^\dagger\gamma_\downarrow$ comes in $(-1)^2$ in χ and $\mathbf{k} \to \mathbf{k} + \mathbf{q}$ in u' s, v' s, f's and E' s. This yields for the $\gamma^\dagger\gamma$ term corresponding to $\downarrow$-spin,

$$= \frac{2e^2\hbar^2}{m^2} \sum_k \frac{(uu' + vv')^2 f'(1 - f)}{E - E'} (k + q/2)^\alpha(k + q/2)^\beta.$$

Subtracting (ii) from (i), that is, contributions from spin-↑ and spin-↓,

$$= \frac{2e^2\hbar^2}{m^2}\sum_k (k + q/2)^\alpha (k + q/2)^\beta (uu' + vv')^2 \frac{(f - f')}{E - E'}.$$

(iii) and (iv) *The combination of the* $\gamma^\dagger\gamma^\dagger$ and $\gamma\gamma$ terms

This comes with a coherence factor $(u_k v_{k+q} - v_k u_{k+q}) \to 0$ as $q \to 0$ so they will not contribute, however, we shall retain them for considering the finite q response. They can be written as,

$$\sum_{m,n} \frac{e^{-\beta\omega_m}}{Z} \langle m|\gamma_k^\dagger \gamma_{-(k+q)}^\dagger|n\rangle\langle n|\gamma_{-(k+q)}\gamma_k|m\rangle \frac{(-1)}{\omega_{nm}}$$

$$= \frac{(-1)\left\langle \gamma_k^\dagger \gamma_k \gamma_{-(k+q)}^\dagger \gamma_{-(k+q)}\right\rangle}{-E_k - E_{k+q}} = \frac{ff'}{E + E'}.$$

The negative sign, that is, -1 is the only difference between the $\gamma^\dagger\gamma$ and the $\gamma\gamma$ terms The other one is,

$$= \frac{(-1)\left\langle \gamma_{-(k+q)}\gamma_k \gamma_k^\dagger \gamma_{-(k+q)}^\dagger\right\rangle}{E_k + E_{k+q}} = \frac{(-1)(1 - f)(1 - f')}{E + E'}.$$

Here the coherence factor is written as,

$$(uv' - vu')(u'v - v'u) = -(uv' - vu')^2.$$

Adding contributions of (iii) and (iv)

$$= \frac{(1 - f + f')}{E + E'}(uv' - vu')^2 \times 2.$$

The current–current response function is written as,

$$\chi_{\alpha\beta}(\mathbf{q}, \omega = 0) = \frac{2e^2\hbar^2}{m^2}\sum_{\mathbf{k}} (k + q/2)^\alpha (k + q/2)^\beta \times \left[(uu' + vv')^2 \frac{(f - f')}{E - E'} + (uv' - vu')^2 \frac{(1 - f - f')}{E + E'}\right] \quad (6.127)$$

Now consider special cases of this general formula, that is, $\mathbf{q} \to 0$ response which yields,

$$(uv' - vu') \to 0 \qquad (uu' + vv') \to 1.$$

In which case,

$$\chi_{\alpha\beta}(\mathbf{q} \to 0, \omega = 0) = \frac{2e^2\hbar^2}{m^2}\sum_{\mathbf{k}} k_\alpha k_\beta \left(-\frac{\partial f_k}{\partial E_k}\right). \tag{6.128}$$

The same result can be obtained in a more direct manner [6].

In the large wavelength limit, namely, $\mathbf{q} = 0$

$$\mathbf{j}_p = \frac{e\hbar}{m}\sum_{\mathbf{k}} \mathbf{k}\left(\gamma_{k\uparrow}^\dagger \gamma_{k\uparrow} - \gamma_{-k\downarrow}^\dagger \gamma_{-k\downarrow}\right)$$

$$\langle \mathbf{j}_p \rangle = \frac{e\hbar}{m}\sum_{\mathbf{k}} \mathbf{k}\left(f_{k\uparrow} - f_{-k\downarrow}\right). \tag{6.129}$$

We must keep track of how the quasiparticle spectrum, $E_{k\uparrow}$, changes under the influence of the external Hamiltonian,

$$\begin{aligned} \mathcal{H}' &= -\frac{1}{c}\sum_{\mathbf{q}} \mathbf{j}_1(-\mathbf{q}) \cdot \mathbf{A}(\mathbf{q}) \\ &= -\frac{e\hbar}{mc}\sum_{\mathbf{k}} (\mathbf{k} \cdot \mathbf{A}_0)(\gamma_{k\uparrow}^\dagger \gamma_{k\uparrow}^\dagger - \gamma_{-k\downarrow}\gamma_{-k\downarrow}) \end{aligned} \tag{6.130}$$

where the $q = 0$ term is retained in the last step of the equation. This yields the energy spectrum that is reminiscent of spin splitting in a magnetic field. The energies are modified as,

$$E_{k\uparrow} \to E_{k\uparrow} - \frac{e\hbar}{mc}\, \mathbf{k} \cdot \mathbf{A}_0$$

Similarly,

$$E_{-k\downarrow} \to E_{-k\downarrow} + \frac{e\hbar}{mc}\, \mathbf{k} \cdot \mathbf{A}_0$$

where $\mathbf{A}_0$ is the vector potential for $q = 0$. Now,

$$\begin{aligned} f_{k\uparrow} - f_{-k\downarrow} &\approx \left(-\frac{\partial f_k}{\partial E_k}\right)\frac{2e\hbar}{mc}\, \mathbf{k} \cdot \mathbf{A}_0 \\ \langle \mathbf{j}_p(\mathbf{q} = 0)\rangle &= \frac{2e^2\hbar^2}{m^2 c}\sum_{\mathbf{k}} (\mathbf{k} \cdot \mathbf{A}_0)\, \mathbf{k}\left(-\frac{\partial f_k}{\partial E_k}\right). \end{aligned} \tag{6.131}$$

Since, by symmetry $\langle \mathbf{j}_p(\mathbf{q} = 0)$ is parallel to $\mathbf{A}_0$. Also,

$$\int \frac{d\Omega}{4\pi}(\mathbf{k} \cdot \hat{A}_0)^2 = \frac{k^2}{3}, \qquad \text{as } \langle \cos^2\theta \rangle = 1/3.$$

Therefore, $\hat{A}_0$ being the unit vector along A_0,

$$\langle \mathbf{j}_p(0) \rangle = \frac{2}{3}\frac{e^2\hbar^2}{m^2c}\sum_{\mathbf{k}} k^2\left(-\frac{\partial f}{\partial E_k}\right)\mathbf{A}_0$$

$$\text{Since,} \quad \sum_{\mathbf{k}} k^2\left(-\frac{\partial f}{\partial E_k}\right) \approx k_{\mathrm{F}}^2\ N(0)\int d\xi_k\left(-\frac{\partial f}{\partial E_k}\right) \tag{6.132}$$

where $N(0)$ is the density of states at the Fermi level. Moreover, the pre-factor of $\int d\xi_k$ is given by,

$$=\frac{2}{3}\ \frac{e^2\hbar^2}{m^2}\ k_{\mathrm{F}}^2\ \frac{3n}{4\varepsilon_{\mathrm{F}}} = \frac{ne^2}{m}.$$

Hence, the average paramagnetic current is given by,

$$\langle \mathbf{j}_p(q=0) \rangle = \frac{ne^2}{m}\int d\xi_k\left(-\frac{\partial f}{\partial E_k}\right)\mathbf{A}_0. \tag{6.133}$$

The normal state can obtained by putting $\Delta = 0$ in the expression for quasiparticle energies, that are given by, $E = \sqrt{\xi^2 + \Delta^2}$. This yields,

$$E = |\xi| \quad \text{Also,} \qquad \int d\xi\left(-\frac{\partial f}{\partial \xi}\right) = 1.$$

Finally, one gets,

$$\langle \mathbf{j}_p(q=0) \rangle = \frac{ne^2}{m}\ \mathbf{A}_0. \tag{6.134}$$

Hence, the paramagnetic response, $\langle \mathbf{j}_p \rangle$ and the diamagnetic response $\langle \mathbf{j}_d \rangle$ exactly cancel each other, thereby yielding the total induced current, $\langle \mathbf{j} \rangle = 0$. Consequently, no circulating currents are induced in a normal metal in equilibrium condition and thus the external magnetic field **H** is not shielded.

One can ignore weak Landau diamagnetic effects, which arise from the fact that,

$$\langle \mathbf{j}_p(\mathbf{q} \to 0) \rangle = \left(\frac{ne^2}{m} - \alpha q^2\right)\mathbf{A}(q) \tag{6.135}$$

where one can write,

$$\mathbf{A}_{\text{total}} = \left(\frac{ne^2}{m} - \alpha q^2\right)\mathbf{A}(q).$$

The total current density now becomes,

$$\langle \mathbf{j}(\mathbf{q} \to 0) \rangle = -\,\alpha q^2\ \mathbf{A}_{\text{total}} \qquad (\alpha > 0) \tag{6.136}$$

where α is a positive quantity, and $\mathbf{A}_{\text{ind}}$ obeys the wave equation,

$$\left(\frac{1}{c^2}\frac{\partial^2}{\partial t^2} - \nabla^2\right)\mathbf{A}_{\text{ind}} = \mu_0\langle\mathbf{j}\rangle. \tag{6.137}$$

In a static situation, the contribution of the term $\frac{1}{c^2}\frac{\partial^2}{\partial t^2} = 0$. Then,

$$q^2\mathbf{A}_{\text{ind}} = -\mu_0 \times q^2\ \mathbf{A}_{\text{total}}. \tag{6.138}$$

Using self-consistent fields,

$$\begin{aligned} \mathbf{A}_{\text{total}} &= \mathbf{A}_{\text{ind}} + \mathbf{A}_{\text{ext}} \\ \text{or, } \mathbf{A}_{\text{total}} &= \frac{\mathbf{A}_{\text{ext}}(\mathbf{q})}{1 + \mu_0\alpha} \qquad (\alpha > 0). \end{aligned} \tag{6.139}$$

Hence the total vector potential derives contribution from the external field and $\mathbf{A}_{\text{ind}}$. Taking curl to obtain the magnetic field,

$$|\mathbf{H}_{\text{total}}| = \frac{|\mathbf{H}_{\text{ext}}|}{1 + \eta} \tag{6.140}$$

where η is a positive quantity, such that $\mathbf{H}_{\text{total}} < \mathbf{H}_{\text{ext}}$ which means that the magnetic field is somewhat screened, and the system shows weak diamagnetism.

Let us now discuss the superconducting state which is characterized by $\Delta \neq 0$ and the cancellation of the paramagnetic response $\langle\mathbf{j}_p\rangle$ and the diamagnetic response $\langle\mathbf{j}_d\rangle$ is not valid any longer. In fact, as we shall see, at $T = 0$, $\langle\mathbf{j}_p\rangle = 0$ and therefore only the diamagnetic response $\langle\mathbf{j}_d\rangle$ contributes.

To this effect, let us define the ratio of the density of normal electrons to that of the total density, namely, $\frac{n_n}{n}$. This quantity will vanish in the superconducting state, or as, $T \to T_c^-$, that is, when T approaches T_c from below.

$$\begin{aligned} \frac{n_n}{n} &= \int_{-\infty}^{\infty} d\xi_k\left(-\frac{\partial f_k}{\partial \xi}\right) = 2\int_{\Delta}^{\infty} dE\ \frac{E}{\sqrt{E^2 - \Delta^2}}\left(-\frac{\partial f}{\partial E}\right) \\ &= \frac{1}{2T}\int_0^{\infty} d\xi\ \text{sech}^2\left(\frac{\sqrt{\xi^2 + \Delta^2(T)}}{2T}\right) = Y(T) \end{aligned} \tag{6.141}$$

where $Y(T)$ is known as the Yosida function. The Yosida function characterizes the response of quantities to which other pairs do not contribute, and thus represents the response of the normal fluid. Let us explore the following limits, namely, $T \ll T_c$ and $T \to T_c^-$.

(i) For $T \ll T_c$, $Y(T)$ behaves as, $Y(T) \sim e^{-\Delta(0)/T}$.

(ii) For $T \to T_c^-$, $Y(T) \approx 1 - 2\frac{(T_c - T)}{T_c}$.

The $q = 0$ components of the paramagnetic and the diamagnetic contributions are respectively written as,

$$\langle \mathbf{j}_p \rangle_{q=0} = \frac{ne^2}{m}\left(\frac{n_n}{n}\right)\mathbf{A}_0$$
$$\langle \mathbf{j}_d \rangle_{q=0} = -\frac{ne^2}{m}\mathbf{A}_0. \tag{6.142}$$

Thus,

$$\langle \mathbf{j} \rangle = \langle \mathbf{j}_p + \mathbf{j}_d \rangle = -\frac{ne^2}{m}\left(1 - \frac{n_n}{n}\right)\mathbf{A}_0 \equiv -\frac{n_s e^2}{m}\mathbf{A}_0 \tag{6.143}$$

where one can define the superelectron density, that is, $\frac{n_s}{n}$

$$\frac{\rho_s}{\rho} = 1 - \frac{\rho_n}{\rho}.$$

So,

$$\langle \mathbf{j} \rangle = -\frac{n_s e^2}{m}\mathbf{A}. \tag{6.144}$$

Thus the superelectron density can be described in terms of the Yoshida function as,

$$\frac{n_s(T)}{n} = 1 - Y(T).$$

Hence the London penetration depth (or the inverse of it) is expressed as,

$$\frac{1}{\lambda_L^2(T)} = \frac{4\pi n_s(T)e^2}{m} = \frac{1}{\lambda_L^2(0)}(1 - Y(T)). \tag{6.145}$$

Thus the temperature dependence of the penetration depth is expressed by the temperature variation of the Yoshida function, $Y(T)$ (see figure 6.19).

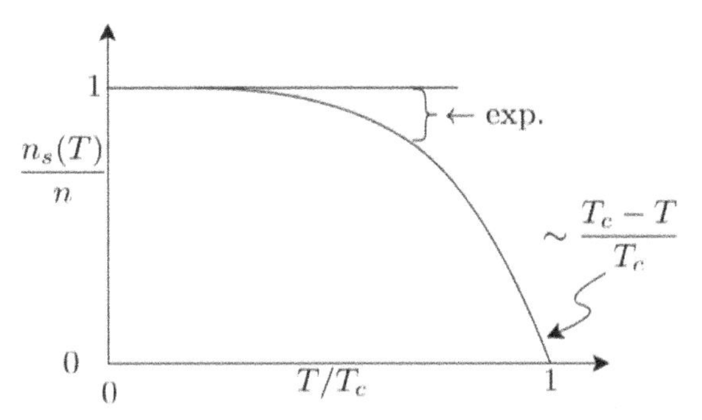

Figure 6.19. The behaviour of the superfluid density scaled by the total density, $\frac{n_s(T)}{n}$ is shown as a function of reduced temperature, T/T_c. The behaviour of the superfluid density close to $T = T_c$ is given by $(1 - \frac{T}{T_c})$.

6.5.2 Electromagnetic response in the transverse gauge

An externally applied vector potential, $\mathbf{A}_{\text{ext}}(\mathbf{r}, t)$, induces current in the system $\langle \mathbf{j}(\mathbf{r}, t)\rangle$, which in turn generates an electromagnetic field. Using Maxwell's equation we have,

$$\left(\frac{1}{c^2}\frac{\partial^2}{\partial t^2} - \nabla^2\right)\mathbf{A}_{\text{ind}}(\mathbf{r}, t) = \mu_0\langle \mathbf{j}(\mathbf{r}, t)\rangle \tag{6.146}$$

which follows from,

$$\mathbf{E} = -\frac{\partial \mathbf{A}}{\partial t} \quad \text{and} \quad \nabla \times (\nabla \times \mathbf{A}) = -\nabla^2\mathbf{A}$$

in a gauge where $\nabla \cdot \mathbf{A} = 0$. Therefore, Fourier transforming one obtains,

$$\left(q^2 - \frac{\omega^2}{c^2}\right)\mathbf{A}_{\text{ind}}(\mathbf{q}, \omega) = \mu_0\langle \mathbf{j}(\mathbf{q}, \omega)\rangle. \tag{6.147}$$

The total vector potential is written as,

$$\mathbf{A}_{\text{tot}} = \mathbf{A}_{\text{ext}} + \mathbf{A}_{\text{ind}}$$

and for the current density,

$$\begin{aligned}\langle \mathbf{j}(\mathbf{q}, \omega)\rangle = \left(\chi^T - \frac{ne^2}{m}\right)\mathbf{A}_{\text{tot}} &= -\frac{n_s e^2}{m}\mathbf{A}_{\text{ext}} \\ &= -\frac{1}{\lambda_L^2}\mathbf{A}_{\text{tot}}.\end{aligned} \tag{6.148}$$

Now,

$$\begin{aligned}&\mathbf{B} = \mu_0(\mathbf{H} + 4\pi\mathbf{M}) \\ \text{also} \quad &\mathbf{B} = \nabla \times \mathbf{A}_{\text{tot}} \\ \text{and} \quad &\mathbf{H} = \nabla \times \mathbf{A}_{\text{ext}} \quad \text{that is,} \quad \mathbf{M} = \frac{1}{\mu_0}\nabla \times \mathbf{A}_{\text{ind}}.\end{aligned} \tag{6.149}$$

In the static limit, that is, $\omega = 0$,

$$\begin{aligned}\mathbf{A}_{\text{ind}} &= \frac{4\pi}{c}\frac{1}{q^2}\langle \mathbf{j}(\mathbf{q}, \omega)\rangle \\ \text{or, } \mathbf{A}_{\text{ind}} &= \frac{-1/\lambda_L^2(T)}{q^2 + 1/\lambda_L^2(T)}\mathbf{A}_{\text{ext}}.\end{aligned} \tag{6.150}$$

Thus the susceptibility can be written as,

$$\chi = \frac{\partial M}{\partial H} = \frac{\frac{1}{\lambda_L^2(T)}}{q^2 + \frac{1}{\lambda_L^2(T)}} \quad \text{that is, } \lim_{q\to 0}\chi = -1. \tag{6.151}$$

The above equation implies the condition of perfect diamagnetism. The superconductors are known to be perfect diamagnets. It may be noted that the diamagnetic susceptibility of metals is usually of the order of 10^{-5}–10^{-6}.

6.6 Ginzburg–Landau (GL) theory

In 1937 Ginzburg and Landau [13] developed a model to describe second-order phase transition[6] that is, those transitions which involve latent heat. It is physically meaningful to state that the quantum many-body state, ψ (something similar to what we have seen in BCS theory) should be a function that minimizes the free energy, F_s of the superconducting state. Ginzburg and Landau considered F_s to be a functional of the wavefunction ψ, and hence use a variational principle to minimize F_s with respect to ψ, and obtain a set of differential equations for ψ that govern the physical properties of the superconductor in equilibrium conditions. To this end, a concept of the order parameter has been developed which smoothly vanishes at a temperature $T > T_c$, while it is non-zero for $T < T_c$. The identification of the order parameter is often obvious from the properties of second-order phase transitions. For example, for a ferromagnetic to a paramagnetic transition, the order parameter is the magnetization, which is finite below the magnetic transition temperature, T_c and goes to zero above T_c, whereas, a superconducting state has an energy gap that separates the ground state and the first excited state, as the order parameter assumes non-zero values below the superconducting transition temperature T_c which, however, vanishes above T_c.

Since the order parameter evolves continuously from zero below T_c, it is natural to expand the free energy as a power series in the order parameter. The free energy is a scalar, but the order parameter, may in general be a vector, or even a complex quantity. We shall study GL theory by constructing the free energy for the superconductor in terms of the wavefunction for the '*superelectrons*' as the order parameter (instead of the energy gap between the ground and the excited states), $\psi(\mathbf{r})$, the density of the superelectrons is $n_s = |\psi(\mathbf{r})|^2$.

Consider a homogeneous superconductor at zero external magnetic field, that is, $\mathbf{H} = 0$. Also, $\psi(\mathbf{r})$ is independent of $\mathbf{r}$. Let us expand the free energy near the transition from a normal metal to a superconductor, that is, near T_c,

$$F_s = F_n + \alpha|\psi|^2 + \frac{\beta}{2}|\psi|^4 \tag{6.152}$$

where F_s denotes the free energy density of the superconductor, and F_n being the free energy density of the normal state. Further, α, β are the phenomenological expansion coefficients characterizing the material, and they depend on temperature. The minimum of F_s is obtained via minimizing F with respect to the order parameter,

[6] It may be mentioned here that we have introduced the Ginzburg–Landau theory very briefly. In no way should it be considered as complete. For a detailed discussion, see *Theory of type-II superconductors* in reference [14].

$$\frac{dF_s}{d|\psi|^2} = 0. \tag{6.153}$$

The minimization implies that (writing the equilibrium value of ψ and ψ_0),

$$|\psi_0|^2 = -\alpha/2\beta.$$

Recall that $|\psi_0|^2$ is the density of the superelectrons that minimizes F_s. Now,

$$\begin{aligned}\frac{dF_s}{d\psi} &= 2\alpha\psi + 2\beta\psi^3 \\ &= 2\psi[\alpha + \beta\psi^2].\end{aligned} \tag{6.154}$$

Thus, the difference in energies between the normal and the superconducting states is obtained from,

$$\begin{aligned}F_n - F_s &= \alpha\left(\frac{\alpha}{\beta}\right) - \frac{\beta}{2}\left(\frac{\alpha}{\beta}\right)^2 \\ &= \frac{\alpha^2}{\beta} - \frac{\alpha^2}{2\beta} = \frac{\alpha^2}{2\beta}.\end{aligned} \tag{6.155}$$

The temperature dependence of α and β can be deciphered as follows. At the first order in temperature, that is, when T is close to T_c, let us postulate a linear behaviour of the form,

$$\alpha \simeq a(T - T_c) \tag{6.156}$$

where a is independent of temperature. This implies,

$$\begin{aligned}\alpha = 0 \text{ at } \quad (T = T_c) \\ \alpha < 0 \text{ at } \quad (T < T_c).\end{aligned} \tag{6.157}$$

Thus the order parameter assumes a form,

$$\begin{aligned}|\psi| &= 0 \quad \text{for} \quad T > T_c \\ |\psi| &= \left[\frac{a(T - T_c)}{\beta}\right]^{1/2} \quad \text{for} \quad T < T_c\end{aligned} \tag{6.158}$$

α changes sign across the phase transition, and hence β should not change sign and remain a constant at least for small deviations in temperature from the transition point.

Next consider a superconductor in an external magnetic field, $\mathbf{H}_{ext}$. The electrons interact with the external field via their momenta transforming as,

$$\begin{aligned}&\mathbf{p} \rightarrow \mathbf{p} - e\mathbf{A} \quad \text{where} \quad \mathbf{B}_{ext} = \nabla \times \mathbf{A} \\ \text{which implies,} \quad &\nabla \rightarrow \nabla - ie\mathbf{A}.\end{aligned} \tag{6.159}$$

Thus the free energies have to be written as,

$$F_s = F_n + \alpha|\psi|^2 + \frac{1}{2}\beta|\psi|^4 + \frac{\hbar^2}{2m}\left|[\nabla - \frac{ie}{\hbar c}A(\mathbf{r})]\psi(\mathbf{r})\right|^2 + \mu_0 H_{\text{ext}}^2(\mathbf{r}) \tag{6.160}$$

where $\mathbf{H}_{\text{ext}}(\mathbf{r}) = \nabla \times \mathbf{A}(\mathbf{r})$. The free energy is a functional of

$$\{\psi(\mathbf{r}), \psi^*(\mathbf{r}), \nabla\psi(\mathbf{r}), \nabla\psi^*(\mathbf{r}), H(\mathbf{r})\}$$

and each one of these is a function of the spatial variable $\mathbf{r}$. Minimizing with respect to $\psi^*(\mathbf{r})$ yields,

$$\begin{aligned}\delta F_s = &\left\{-\frac{\hbar^2}{2m}(\nabla - \frac{ie}{\hbar}A(\mathbf{r}))^2\psi(\mathbf{r}) + \alpha\psi(\mathbf{r}) + \beta|\psi(\mathbf{r})|^2\psi(\mathbf{r})\right\}\delta\psi^*(\mathbf{r}) \\ &+ \frac{\hbar^2}{2m}(\nabla - \frac{ie}{\hbar}A(\mathbf{r}))\psi(\mathbf{r})\delta\psi^*(\mathbf{r}).\end{aligned} \tag{6.161}$$

The last term can be written as, $\mu_0 H^2$, or $\mu_0(\nabla \times \mathbf{A})^2$. Further, variation with respect to ψ yields the complex conjugate of this equation. Putting $\delta F_s = 0$ for an arbitrary variation of the order parameter, namely, $\delta\psi^*(\mathbf{r})$, we get the first GL equation,

$$-\frac{\hbar^2}{2m}\left(\nabla - \frac{ie}{\hbar}A(\mathbf{r})\right)^2\psi(\mathbf{r}) + \alpha\psi(\mathbf{r}) + \beta|\psi(\mathbf{r})|^2\psi(\mathbf{r}) = 0 \tag{6.162}$$

where we have neglected the magnetic energy density due to the external field, that is, the last term. Now minimizing it with respect to $\mathbf{A}(\mathbf{r})$ yields Ampere's law,

$$\nabla \times \mathbf{H}(\mathbf{r}) = \mu_0 \mathbf{j}(\mathbf{r}) \tag{6.163}$$

where,

$$\mathbf{j}(\mathbf{r}) = -\frac{ie\hbar}{2m}[\psi^*(\mathbf{r})\nabla\psi(\mathbf{r}) - \psi(\mathbf{r})\nabla\psi^*(\mathbf{r})] - \frac{e^2}{m}|\psi|^2 A(\mathbf{r}). \tag{6.164}$$

This is the second GL equation. These two GL equations in equation (6.162) and equation (6.164) are the main triumphs of GL theory. They enunciate the variation of the order parameter, $\psi(\mathbf{r})$, and the current density, $\mathbf{j}(\mathbf{r})$.

6.6.1 Coherence length and the penetration depth

Using the GL equations obtained above, we wish to derive expressions for some physical observables that are measured in experiments, such as, the coherence length and the penetration depth. Let us first talk about the coherence length.

Consider an inhomogeneous order parameter for a system generated by the presence of the boundary in the absence of an external magnetic field[7]. The superconducting (SC) region is denoted by $x > 0$. Thus, the order parameter is zero for the interface. For deriving expression for the coherence length, putting $\mathbf{A} = 0$ in the first GL equation (6.162),

[7] To derive the coherence length, the presence of a magnetic field is not essential.

$$-\frac{\hbar^2}{2m}\frac{d^2\psi}{dx^2} + \alpha|\psi| + \beta\psi^3 = 0. \tag{6.165}$$

It may be noted that α is negative in the superconducting state, which allows us to set, $\alpha = -|\alpha|$. Now defining the coherence length as,

$$\xi^2 = \frac{\hbar^2}{2m|\alpha|}. \tag{6.166}$$

Also writing $(\beta/|\alpha|)\psi^2 = f^2$, we can write equation (6.165) as,

$$-\xi^2 f'' - f + f^3 = 0. \tag{6.167}$$

Multiply both sides by f' to get,

$$\begin{aligned} &\frac{d}{dx}\left[-\frac{\xi^2 f'^2}{2} - \frac{1}{2}f^2 + \frac{1}{4}f^4\right] = 0 \\ \text{or,}\quad &-\xi^2\frac{f'^2}{2} - \frac{1}{2}f^2 + \frac{1}{4}f^4 = \text{constant.} \end{aligned} \tag{6.168}$$

Far from the boundary, f' (or ψ') should be zero, and $f^2 = 1$. Hence $|\psi|^2 = \frac{|\alpha|}{\beta}$. Equation (6.168) now becomes,

$$\xi^2(f')^2 = \frac{1}{2}(1 - f^2)^2. \tag{6.169}$$

The solution of the above equation is given by,

$$f = \tanh\left(\frac{x}{\sqrt{2}}\right).$$

Hence,

$$\psi = \left(\frac{|\alpha|}{\beta}\right)^{1/2} \tanh\left(\frac{x}{\sqrt{2}}\right)$$

This yields that ξ is the measure of the distance over which the order parameter responds to a perturbation. Also, since $\alpha = a(T - T_c)$, the temperature dependence of the coherence length can obtained as,

$$\xi(T) = \frac{\hbar^2}{2maT_c}\left(1 - \frac{T}{T_c}\right)^{-1/2} = \xi_0\left(1 - \frac{T}{T_c}\right)^{-1/2}. \tag{6.170}$$

Thus $\xi(T)$ diverges as $(T - T_c)^{-1/2}$ as $T \to T_c$ from below. Also the temperature independent length, ξ_0 is given by,

$$\xi_0 = \frac{\hbar^2}{2maT_c}.$$

A little manipulation of the above formula yields,

$$\xi_0^2 = \frac{\hbar^2}{2m\epsilon_F}\left(\frac{\epsilon_F}{k_B T_c}\right)^2.$$

Thus ξ_0 is larger than $\left(\frac{\hbar^2}{2m\epsilon_F}\right)^{1/2}$, which is of the order of interparticle spacing, by a factor of $(\epsilon_F/k_B T_c)^2$. Remember, ϵ_F is usually of the order of a few eV (typically, 5–6 eV), and $k_B T_c$ for superconductors is of the order of a few meV. Thus ξ_0 is usually of the order of a few thousand lattice spacings.

Next, in order to arrive at an expression for the penetration depth, we examine the current expression from the second GL equation, that is equation (6.164). If we neglect the first term in comparison to the second term, then one obtains the London equation (discussed earlier), namely,

$$\mathbf{j}(\mathbf{r}) = \mu_0 \frac{1}{\lambda_L^2}\mathbf{A}(\mathbf{r}) \tag{6.171}$$

where,

$$\lambda_L^2 = \frac{mc^2}{4\pi e^2 |\psi|^2}$$

Again the temperature dependence of the penetration depth is obtained as,

$$\lambda_L = \left(\frac{mc^2\beta}{4\pi e^2 a T_c}\right)\left(1 - \frac{T}{T_c}\right)^{-1/2} \tag{6.172}$$

λ_L has a similar divergence to that of $\xi(T)$. Also, a ratio of the two can be found as,

$$\kappa = \lambda_L/\xi = \frac{mc}{e\hbar}\left(\frac{\beta}{2\pi}\right)^{1/2} \tag{6.173}$$

As discussed earlier, the value of κ demarcates between various types of superconductors, namely, the type-I and type-II superconductors, which have very different magnetic properties.

6.7 Experimental determination of energy gap

The existence of a forbidden energy regime in superconductors has given us a simple way to understand the fact that below a certain excitation energy the bound state of the electrons cannot be broken. Thus the current in the superconducting state is mediated via the cooper pairs (also referred to as the *superelectrons*), and the normal current carried by the electrons is zero. More precisely, there are no unpaired electrons in the superconducting state. Here we describe a few of the early experimental methods below to determine the energy gap.

6.7.1 Absorption of electromagnetic radiation

This technique of exploring the energy gap was reported in 1957 by Glover and Tinkham [15]. They observed infrared transmission in thin superconducting films. It was proposed earlier that the electromagnetic (EM) waves of appropriate frequencies can break the ordered state of a superconductor. Assuming a cooper pair has a binding energy given by, $E_B \sim 10^{-3}$ eV, one needs a radiation frequency $\nu \sim E_B/h = 2.4 \times 10^{11}$ Hz or 240 GHz. The wavelength of the waves is about 1 mm, for which neither generation of such frequencies, or detection methods were available in 1930s. Only two decades later, such EM waves could be produced.

In the experimental method, EM radiation is guided through a small cavity made in a superconducting material. Within the cavity, the radiation is reflected several times before it emerges through the cavity, and hence is detected. The stronger the absorption of the radiation from the walls of the cavity, the lower will be the output power. At a fixed temperature, $T < T_c$, that is, when the material is in the superconducting state, let the power in the superconducting and the normal states be I_s and I_n, respectively. $\frac{I_s - I_n}{I_n}$ is plotted as a function of frequency of the EM waves in figure 6.20. At small frequencies we see a distinct difference between detected power in the superconducting and the normal states, with the absorption being larger for the superconducting state. At certain threshold frequencies, the difference drops abruptly to zero. At even larger frequencies, the relative difference in the detected power stays at values that are vanishingly small. The abrupt drop sets in as soon as the quantum energy of the radiation is large enough to break the cooper pairs. Thus, there is an absorption edge at point C in figure 6.20. Similar absorption edges are also seen for semiconductors, however, the scale of the energy gap is much higher there, e.g. ~0.8 eV in Ge. As an example, for $h\nu > 2\Delta_0$, the energy gap practically vanishes and so does the absorption spectrum. This aids us in obtaining $2\Delta_0$, which is proportional to T_c. Some typical values for the superconductors, and their transition temperatures are given in table 6.1.

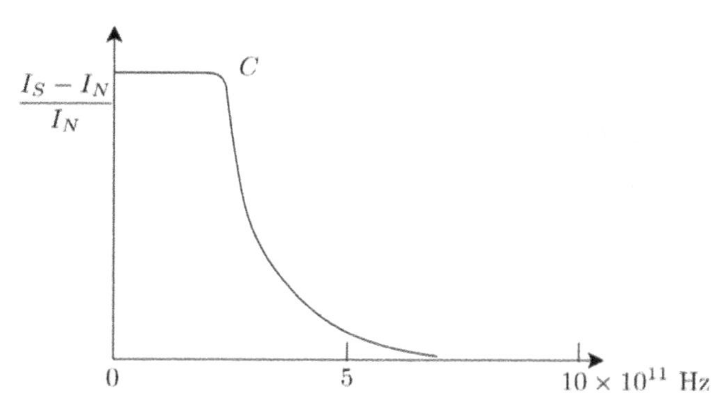

Figure 6.20. The absorption of electromagnetic radiation is shown as a function of frequency. The frequency at which the intensity almost vanishes yields the superconducting gap 2Δ.

Table 6.1. $2\Delta$and T_c are reported for a few conventional superconductors. The gap is of the order of a few meV.

	T_c	2Δ(meV)
Nb_3Sn	18	6.55
MgB_2	40	10
Rb_3C_{60}	30	12
$ErRh_4B_4$	9	3

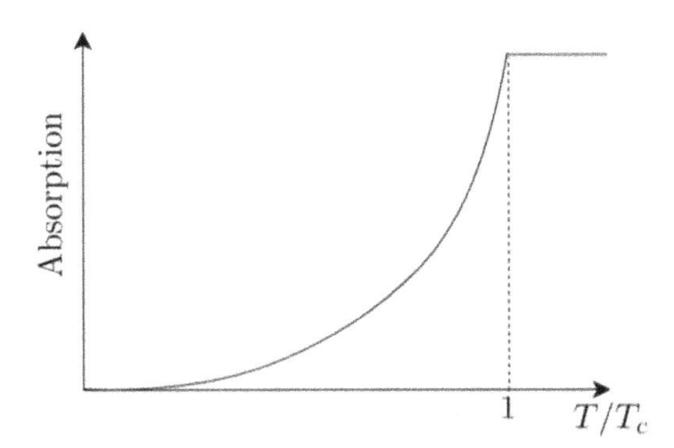

Figure 6.21. Schematic depiction of the ultrasound absorption intensity is presented as a function of T/T_c. The absorption is low in the superconducting phase owing to the absence of unpaired electrons.

6.7.2 Ultrasound absorption

Sound waves denote propagation of phonons in a material. Most measurements with sound waves are carried out with typical frequencies of the order of 10 MHz. These frequencies correspond to energies that are much smaller than the superconducting gap. Only very close to T_c, where the energy gap Δ approaches zero, do sound energies at the MHz frequency range become comparable to the gap. Unpaired electrons below T_c interact with the lattice excitations leading to damping. Below T_c, as the temperature is decreased, the number of unpaired electrons decreases rapidly, thereby causing reduced damping of the sound intensity. A schematic plot of the ultrasound absorption spectra (in arbitrary units) is presented as a function of the reduced temperature, T/T_c. As T tends to T_c from below, the Cooper pairs start dissociating yielding an increasingly large number of unpaired electrons in the system. These electrons absorb the sound waves leading to an enhancement in the absorption intensity (see figure 6.21). Since at a given temperature, the number of unpaired electrons depends on the energy gap, the latter can be determined from such absorption experiments.

6.7.3 Tunneling experiment

In 1961, I Giaeaver [16] pointed out the possibility of determining the energy gap by means of tunnelling experiments. The experiments involve observation of tunnelling current across a metal-insulator-semiconductor (MIS) junction. The metal and the insulating materials are usually taken as Al and Al_2O_3, respectively. Al_2O_3 is known to be a good insulator which can be fabricated nearly perfectly with a thickness of only a few nanometers.

One can understand the tunnelling effect without a detailed calculation. Consider a particle that incidents on a barrier with an energy lower than the barrier height. While a classical particle is unable to travel to the other side of the barrier, it is possible for a quantum mechanical particle (because of its wave nature) to do so. Considering the wave character of the particle, when a wave is incident on the surface separating two media into which it cannot enter, the wave must be totally reflected. We understand intuitively that being a wave, it can penetrate up to a small distance into the forbidden region with its amplitude decreasing exponentially. The decrease is faster and complete within the barrier if the width and height of the barrier are large. However, for a sufficiently thin barrier there is a finite possibility that the wave can penetrate into the barrier (of the order of de Broglie wavelength), and propagate into the medium (lead) on the other side. If a superconductor is introduced in one or both sides of the barrier, the current as a function of the biasing voltage ($I-V$) characteristic is bound to be different. This happens because an energy gap appears in the superconducting state. So no tunnelling current can flow below a certain critical voltage, given by, $V_c = \Delta/e$. From figure 6.22 it is clear that from V_c, the value of Δ_0 can be determined.

However, at the normal-superconductor junction there is an additional phenomenon which contributes to the tunnelling current. Here especially we are interested in a situation in which the biasing voltage is much smaller compared to the energy gap, Δ, in a superconductor. Thus, in usual circumstances, no tunnelling can occur. Thus, for an electron arriving at the interface of a normal-superconductor junction, two possible processes can take place (see figure 6.23), namely:

(i) a normal reflection in which an electron is simply reflected back giving rise to no net current;

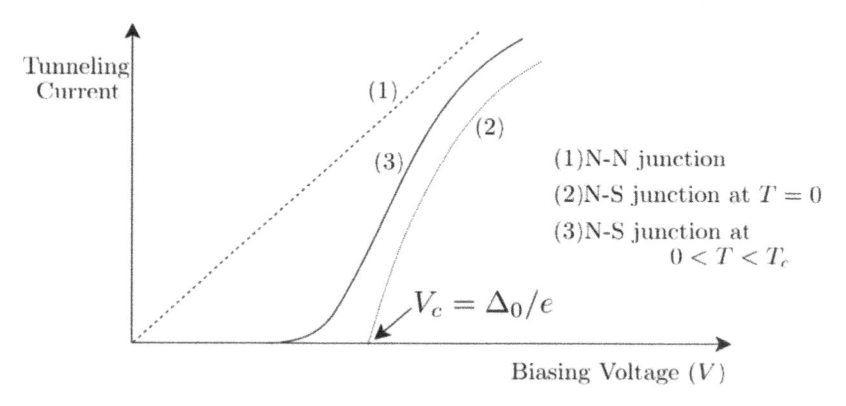

Figure 6.22. The current–voltage characteristics are shown schematically for N–N and N–S junctions at $T = 0$ and $0 < T < T_c$.

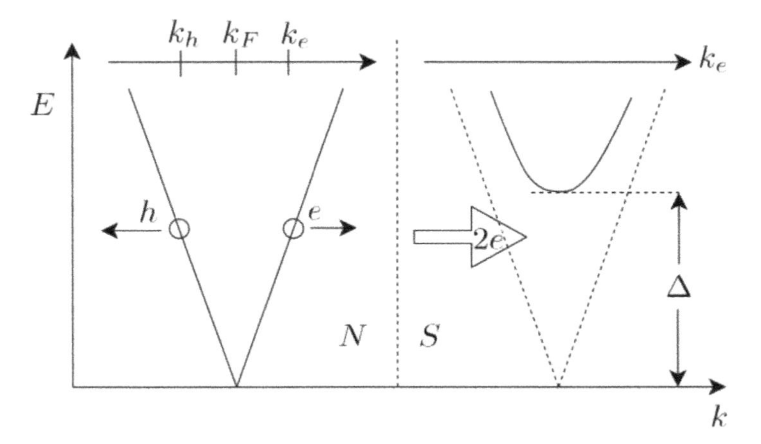

Figure 6.23. A schematic diagram of the dispersion E versus k corresponding to the Andreev reflection at the normal-superconductor (N–S) junction is shown. A Cooper pair is created at the boundary which can propagate at low bias inside a superconductor. The electron is specularly reflected as a hole on the metallic side.

(ii) Andreev reflection which is an anomalous reflection of an electron. What actually happens here is the following. An electron comes at the interface, instead of undergoing a reflection
  (a) a Cooper pair is added in the superconducting region, it has an energy Δ;
  (b) the incident electron is annihilated;
  (c) a hole is reflected back along the original path of the electron.

This is known as Andreev reflection. The incident, and the reflected quasiparticles have approximately equal wave vectors, which are given by,

$$\begin{aligned} k^{\mathrm{el}} &= k_{\mathrm{F}} + \frac{\epsilon}{\hbar v_{\mathrm{F}}} \\ k^{n} &= k_{\mathrm{F}} - \frac{\epsilon}{\hbar v_{\mathrm{F}}} \end{aligned} \tag{6.174}$$

but have opposite directions of motion as appears from the sign of the group velocity $\frac{1}{\hbar}\frac{\partial \epsilon}{\partial k}$. The momentum is conserved up to the order,

$$\hbar|k^{\mathrm{el}} - k^{n}| \leqslant \hbar/\xi$$

where $\xi = \frac{\hbar v_{\mathrm{F}}}{\pi \Delta_0}$, which is the coherence length.

The energy is also conserved in the process. The minimum energy of a Cooper pair has a value $2\epsilon_{\mathrm{F}}$, the incident electron has energy $\epsilon_{\mathrm{F}} + \epsilon$, and that of the reflected hole is $\epsilon_{\mathrm{F}} - \epsilon$. In the process, a charge $2e$ is transferred from the normal to the superconducting side of the barrier, which is equivalent to a Cooper pair being injected into the superconductor.

We can understand the tunnelling problem in the following manner. The number of electrons tunnelling per unit time with an energy E from the left (say, a metal) to the right (a superconductor) is proportional to the number of occupied states on the left, namely, $N_1(E)f(E)$, and that to the unoccupied states on the right, $N_2(E')(1 - f(E'))$. $N_1(E)$ and $N_2(E')$ denote DOSs for the metallic and the superconducting regions, respectively. In the presence of the biasing voltage, V, the energies are given by,

$$E' = E + eV$$

Thus the unoccupied states on the right are denoted by,

$$N_2(E + eV)(1 - f(E + eV)) \tag{6.175}$$

The tunnelling current is proportional to both these independent process, which can be written as,

$$I_{1\to 2} \propto \int_{-\infty}^{\infty} N_1(\varepsilon)f(\varepsilon)N_2(\varepsilon + eV)(1 - f(\varepsilon + eV))\, d\varepsilon \tag{6.176}$$

This tunnelling current is a measurable quantity in experiments.

For a normal metal–superconductor junction, the current, $I_{N\to S}$ can be obtained by replacing $N_1 \to N_N$ and $N_2 \to N_S$. A schematic plot of the tunnelling current for the junctions, namely, metal–metal (N–N) and metal–superconductor (N–S) both at $T = 0$ and for $T < T_c$ as a function of the bias voltage are presented in figure 6.22. The threshold potential yields the energy gap. A differential conductance, $\frac{dI}{dV}$, which is proportional to the density of states at the Fermi level is often obtained in experiments.

6.7.4 Unconventional superconductivity

There are several superconductors, that have been discovered over the last few decades, which challenge the well-established paradigm of conventional superconductivity. Quite a few of them, including the high-T_c cuprates, pose anomalies owing to features that are unfamiliar in the context of Fermi liquid theory. As a first signature, Uemura presented the plot of T_c versus T_F (both in log scale) that yields a straight line, while the conventional superconductors are far away from this line. This is called the Uemura plot and its freehand sketch is presented in figure 6.24.

6.7.5 High-T_c cuprates

In 1986, 75 years after the discovery of superconductivity, G Bednorz and K Muller at IBM, Zurich demonstrated superconductivity in a perovskite structured lanthanum (La) based copper oxide material which registered a T_c of about 39 K for which the Nobel prize was awarded in 1987. It was a remarkable discovery, as it later allowed chemical substitution in perovskite cuprates to push the T_c well beyond the liquid nitrogen temperature (77 K), which is a much cheaper and easily accessible quantity than liquid helium. Later in 1987, La was substituted with yttrium (Y) in the form of $YBa_{2-x}Cu_3O_{7-x}$ which showed a T_c of about 92 K. The materials show

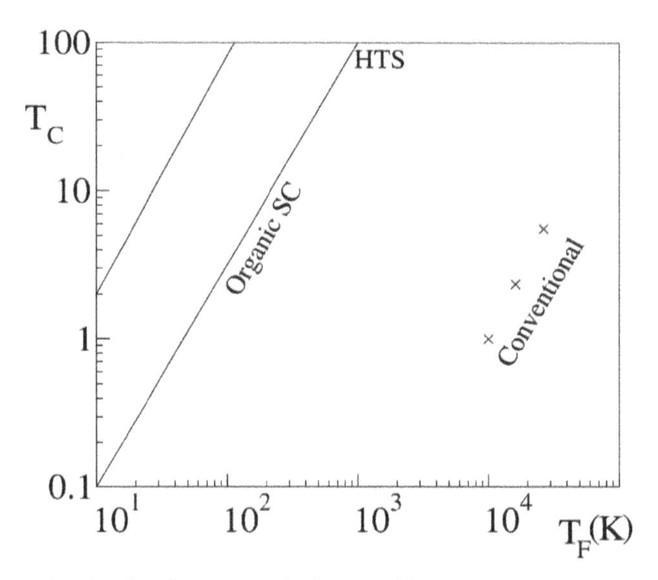

Figure 6.24. Uemura plot showing the superconducting transition temperature, T_c versus the Fermi temperature, T_F (both in log scale). The positions of different superconductors are shown. The conventional (BCS) superconductors lie far away from the unconventional superconductors.

highest T_c when they are slightly oxygen deficient, that is, when $x \simeq 0.15$. Superconductivity disappears at larger values of x, which is also accompanied by a structural phase transition when YBCO changes from an orthorhombic structure to a tetragonal structure. Subsequently thallium (Th) and mercury (Hg) based superconductors show even higher transition temperatures. These superconductors are generally type-II superconductors which show gradual change in temperature as a function of the external magnetic field.

A large volume of research has taken place since its discovery [17], however, owing to its ill understood normal state, a comprehensive understanding of the microscopic phenomena remained elusive. The electronic pairing mechanism is itself questionable and possibly other pairing symmetries, other than the conventional s-wave discussed in the context of BCS theory play crucial roles. Even after several decades of active research, several anomalies that questioned the established paradigm of superconductivity, were left without clarification. Similar unanswered questions exist for superconductivity found in heavy fermion compounds, iron based metals and more recently in twisted bilayer graphene.

The cuprate superconductors have a generic phase diagram (see figure 6.25) which itself poses a bunch of surprises. The proximity of the Mott insulating state and the superconducting phase, a pseudogap phase in the underdoped phase are among the few, while the overdoped regime is probably a familiar metal. We shall not prolong the discussion here, and rather motivate readers to look at the review mentioned in the footnote and the references therein. A few of the CuO_2 superconductors are listed here along with transition temperatures (table 6.2).

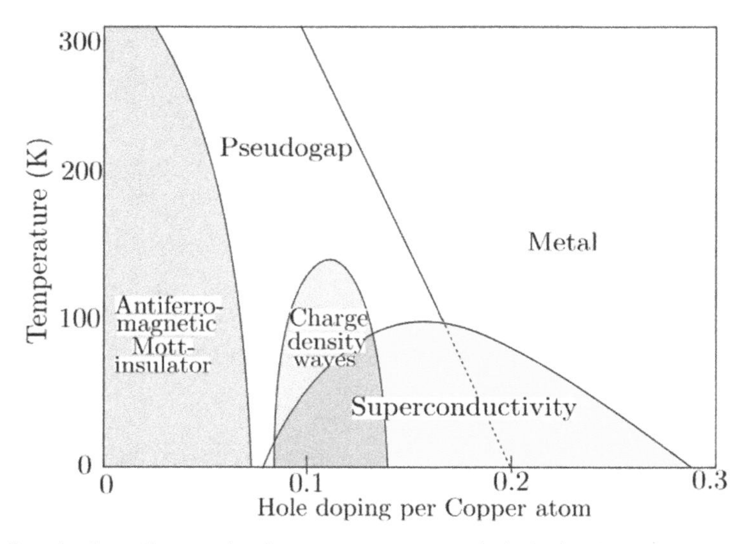

Figure 6.25. Generic phase diagram showing temperature versus hole doping (per copper atom) for cuprate superconductors. Different phases are shown. The superconducting dome extends up to a doping of 0.3. In the underdoped regime, at high temperatures, one gets a *strange*' metallic phase, which cannot be explained by the Fermi liquid theory, whereas the overdoped regime shows more familiar metallic features.

Table 6.2. T_cs of some typical cuprate superconductors are shown.

	T_c
$La_{2-x}Ba_xCuO_4$	30 K
$La_{2-x}Sr_xCuO_4$	39 K
$YBa_2Cu_3O_7$	92 K
$Bi_2Sr_2Ca_2Cu_3O_10$	110 K
$Te_2Ba_2Ca_2Cu_3O_10$	125 K
$HgBa_2Ca_2Cu_3O_8$	134 K

The compounds that are of special interest are $Bi_2CaSr_2Cu_2O_8$ (called Bi – 2122 compounds) and $Tl_2CaBa_2Cu_2O_8$ (called Tl – 2122 compounds). Because of their large T_c (>100 K), they have different numbers of cuprate layers and are denoted by a general formula $A_2Ca_nY_2Cu_nO_{2n+4}$ with A = Bi, Tl and Y = Sr, Ba, etc. The ones with Bi and Sr are more widely studied and are described by the formula $Bi_2Sr_2Ca_{n-1}Cu_nO_{2n+4+\delta}$ where $n = 1 \rightarrow 3$ and δ denotes the oxygen doping. These are called BSCCO (pronounced as BISKO). $n = 1$, $n = 2$ and $n = 3$ have a T_c of 33 K, 96 K, and 108 K, respectively. Similarly, for the Tl family (denoted by $TlBa_2Ca_{n-1}Cu_nO_{2n+3}$) $n=$ 1, 2 and 3 correspond to $T_c \sim$ 14 K, ~80 K and ~120 K, respectively.

There are numerous examples of anomalies noted in various experiments on the cuprate superconductors. For example, consider ultrasound attenuation measurements. In conventional (BCS) superconductors, the ultrasound attenuates exponentially due to loss of unpaired electrons (in favour of forming Cooper pairs). The coefficient of attenuation α is proportional to $e^{-\Delta/K_BT}$ where Δ denotes the superconducting energy gap. The scenario is not too different in the cuprate family. However, the ultrasonic attenuation below T_c is found to be lower than that of the BCS superconductors, possibly owing to the formation of pseudogap. However, the longitudinal sound velocity and the elastic moduli show discontinuities in most of the cuprate superconductors (except for the Tl-based family). The discontinuity points towards a structural change and '*lattice softening/hardening*' at the superconducting transition. The anomaly in the elastic moduli points toward a collective ordering of the oxygen atoms in the CuO_2 layers. There are many such examples which show significant digression compared to the behaviour observed for conventional superconductors. Apart from being beyond the scope of the present discussion, there are many good reviews on the subject, and interested readers are encouraged to look at some of these. In the following we review the most significant feature of the cuprates which eluded a microscopic theory so far, namely the formation of a pseudogap phase.

6.8 The pseudogap phase

The nature of the spectral gap in superconductors and how it vanishes (or changes) across a transition can be detected by angle resolved photoemission spectroscopy (ARPES) experiments. In the context of high T_c superconductors, high energy photons from a synchrotron source with an energy $\approx$20 eV made to incident on the surface of the single crystal. As a result, a photoelectron is ejected at an angle θ normal to the surface of the crystal. The kinetic energy of the ejected electron is measured via an electron spectrometer whose energy resolution is of the order of a few meV. The ARPES data for high T_c cuprates are thoroughly analysed and reviewed in [18, 19].

The parallel component of the momentum ($\mathbf{k}$) is balanced by the momentum of the excitations created inside the sample. By varying the detecting angle θ between the normal and the position of the detector it is possible to map out the momentum dependence of the filled states of the crystal. Thus the intensity of the photoelectron distribution will demonstrate a sharp peak corresponding to the parallel component of the momentum with an energy ω which is given by,

$$I(\mathbf{k}, \omega) = I_0(\mathbf{k})f(\omega)A(\mathbf{k}, \omega) \tag{6.177}$$

where the energy is measured with respect to the Fermi level [19] and $A(\mathbf{k}, \omega)$ denotes the spectral function of the excitations that corresponds to the hole that is left behind. $I_0(\mathbf{k})$ is the arbitrary intensity factor that preserves the dimensionality of the equation and $f(\omega)$ is the Fermi distribution function, $f(\omega) = \frac{1}{e^{\beta(\omega-\mu)}+1}$, $\hbar = 1$ and β is the inverse temperature.

The important quantity in the above equation (6.177) is the spectral function which contains the information about the electron–electron interaction present in the material via the self energy $\Sigma(\mathbf{k}, \omega)$, which is in general a complex quantity. The spectral function $A(\mathbf{k}, \omega)$ is expressed in terms $\Sigma(\mathbf{k}, \omega)$ as [20],

$$A(\mathbf{k}, \omega) = \frac{\Sigma''(\mathbf{k}, \omega)}{(\omega - \epsilon_k - \Sigma'(\mathbf{k}, \omega))^2 + \Sigma''(\mathbf{k}, \omega)} \tag{6.178}$$

Where Σ' and Σ'' represent the real and the imaginary parts of the self energy and ϵ_k denotes the single-particle energies.

The ARPES spectra measure $A(\mathbf{k}, \omega)$ and show a broad background as a function of ω that extends to lower energies which yields the energy loss of the photoelectrons within the crystal. For a recent review see reference [21]. However, as a result of strong interparticle interactions, which dominate in these cuprate superconductors, this broad background is attributed due to the electronic correlations. The energy of the emergent photoelectrons yields a sharp coherent peak at the single-particle energies given by ϵ_k and broadens rapidly for state below the Fermi level as the detection angle θ is varied. As said earlier, the loss of coherence is due to the presence of strong electronic correlations [21] and hints towards the breakdown of the Fermi liquid picture (see section 1.5). The optical conductivity data support the loss of coherence as well. The sharp peak at the Fermi level as a function of the detection angle θ maps out the Fermi surface as it yields the information on the momentum (and hence the energy) of the emitted photoelectron. The same experiment can be repeated at various doping levels. Mostly in cuprates, at optimal doping and at the overdoped regime, the presence of the sharp peak in the ARPES spectrum confirms a Fermi liquid picture to be valid. However, a noticeable broadening is noted in the underdoped regime at the Fermi level and that implies breakdown of Fermi liquid theory (FLT).

Both at $T < T_c$ (superconducting state) and $T > T_c$ (normal state), there are deviations from the FLT which are attributed to the formation of pseudogap in cuprates. The pseudogap is most likely related to and has the same symmetry as the superconducting gap (believed to be *d*-wave like). However, unlike a conventional superconducting gap, the magnitude of the pseudogap is temperature independent. ARPES and other experiments indicate that the magnitude of the gap does not change significantly as the temperature is lowered through T_c. Thus, in the case of the cuprates, unlike a conventional superconducting gap, the gap starts as a pseudogap in the normal state and evolves into the superconducting gap below T_c. In other words, $\Delta(T)$ does not go to zero at $T = T_c$.

The pseudogap appears in the underdoped regime of the phase diagram and weakens towards optimal doping. A weak pseudogap is still present at optimal doping[8], but rapidly disappears thereafter. The spectral gap is purely superconducting in the overdoped region. This is illustrated in the phase diagram shown in figure 6.25 where T^* approaches T_c just into the overdoped region. T^* is temperature dependent in the underdoped regime. Well into the overdoped region, the pseudogap merges with the superconducting gap, which means that T^* merges with T_c, and there remains no trace of the pseudogap in the normal state.

[8] Optimal doping corresponds to highest T_c.

To wind up the discussion, we note down a few salient features. They are:

(i) the pseudogap is different than the superconducting gap in the underdoped phase of cuprates, and gradually evolves into the superconducting gap at larger doping levels;

(ii) both the pseudogap and superconducting gaps are believed to have *d*-wave symmetry;

(iii) the crossover temperature T^* (shown in figure 6.25) merges with T_c in the overdoped region of the phase diagram.

6.8.1 Summary and outlook

Let us include a brief recap of the topics discussed. By and large we have confined ourselves to the study of conventional superconductors. We have highlighted several distinguishing properties of superconductors, such as Meissner effect, isotope effect, distinction between perfect conductors and superconductors, flux quantization, penetration depth, and the coherence length, type-I and type-II superconductors, etc. Hence we have discussed magnetic and thermodynamic properties of superconductors, where the former shows phase diagram as a function of the external magnetic field that encodes a phase transition, either directly from a superconductor to a metal, or the same intervened by a mixed phase that admits the magnetic flux lines (in unit of a constant flux quantum) to penetrate the sample. Most of the well-known conventional superconductors are of the latter variety. The thermodynamic properties, such as the specific heat, show a jump of fixed magnitude at the transition point and are a generic feature of a superconducting phase transition.

Hence we have embarked on the famous BCS theory, where we have discussed Cooper's instability that leads to the formation of bound pairs of electrons mediated via phonons and is known to provide the microscopic origin of the pairing phenomena. This was followed by a detailed description which yields an appropriate estimation of the temperature dependence of the spectral gap, and hence the transition temperature, where the latter provides enumeration of different superconductors. Among the crucial properties, we have outlined the calculation of the specific heat, computed the behaviour of the specific heat and the jump therein at the transition temperature, and discussed Meissner effect.

We have also included a brief recap of the Ginzburg–Landau (GL) theory which is a phenomenological theory of superconductivity proposed by Ginzburg and Landau in the early 1940s, and is broadly applicable to all second-order phase transitions. Apart from discussion of the two GL equations, we have estimation of two important length scales that characterize a superconductor, namely the coherence length and the penetration depth. By no means is the above discussion complete, and readers are encouraged to look up more detailed notes on the subject.

We have also described in brief a few experimental methods of estimating the energy gap in superconductors. These are electromagnetic, ultrasound absorption experiments, and measuring the tunnelling spectra in junction systems, involving a normal metal and a superconductor (N–S junction), or a metal between two superconductors (SNS junction), etc.

To wind up, we have mentioned very briefly about the unconventional superconductors that have been discovered over the last two and a half decades. There are a number of issues that necessitates going beyond the BCS paradigm. Once again readers are encouraged to look at literature, including some excellent review articles on the subject.

References

[1] Kamerlingh Onnes H *Research Notebooks* **56** 57 Kamerlingh Onnes H *Archive* (Leiden, the Netherlands: Boerhaave Museum); Kamerlingh Onnes H 1911 *Commun. Phys. Lab. Univ. Leiden* **122** 124**;** Kamerlingh Onnes H 1913 *Commun. Leiden* **I33atoI33d**
[2] Hulm J K and Matthias B T 1951 *Phys. Rev.* **82** 273
Hulm J K and Matthias B T 1953 *Phys. Rev.* **89** 439
Hulm J K and Matthias B T 1952 *Phys. Rev.* **87** 799
Hulm J K and Matthias B T 1953 *Phys. Rev.* **92** 874
[3] Becker R, Heller G and Sauter F 1933 Uber die Stromverteilung in einer supraleitenden Kugel *Z. Phys.* **85** 772–87
[4] London F and London H 1935 *Proc. R. Soc. A* **149** 71
[5] Deaver B S and Fairbank W M 1961 *Phys. Rev. Lett.* **7** 43
[6] Tinkham M 1973 *Introduction to Superconductivity* 2nd edn (Dover: New York)
[7] Ramakrishnan T V and Rao C N R 1999 *Superconductivity Today* (Hyderabad: Universities Press)
[8] Kittel C 2004 *Introduction to Solid State Physics* 8th edn (Hoboken, NJ: Wiley)
[9] Fröhlich H 1950 *Phys. Rev.* **79** 845
[10] Cooper L N 1956 *Phys. Rev.* **104** 1189
[11] Sakurai J J and Napolitano J 2011 *Modern Quantum Mechanics* 2nd edn (Reading, MA: Addison-Wesley)
[12] Reif F 1965 *Fundamentals of Statistical and Thermal Physics* (New York: McGraw-Hill)
[13] Ginzburg V L and Landau L D 1950 *Zh. Eksp. Teor. Fiz.* **20** 1064 English translation in: Landau L D 1965 *Collected Papers* (Oxford: Pergamon Press) p 546
[14] Fetter A L and Hohenberg P C 1969 *Superconductivity* 1st edn ed R D Parks (Boca Raton, FL: CRC Press)
[15] Tinkham M and Glover R E 1958 *Phys. Rev.* **110** 778
[16] See the Nobel lecture by I Giaever available at: http://www.nobelprize.org/prizes/physics/1973/giaever/lecture/
[17] Verma C M 2020 *Rev. Mod. Phys.* **92** 031001
[18] Shen Z-X *et al* 1993 *Phys. Rev. Lett.* **70** 3999
[19] Randeria M and Campuzano J C *Varenna Lectures* (arXiv:9709107)
[20] Fetter A L and Walecka J D 1971 *Quantum Theory of Many Particle Systems* (New York: McGraw-Hill)
[21] Timusk T and Statt B 1999 *Rep. Prog. Phys.* **62** 61

IOP Publishing

Condensed Matter Physics: A Modern Perspective

Saurabh Basu

Chapter 7

Superfluidity

7.1 Introduction

Superfluidity is a phenomenon that is identified with the ability of fluids to flow without friction. Superfluidity was first discovered in liquid helium in which the helium atoms were transported seamlessly through a capillary tube. However, in addition to the frictionless flow, there are more interesting properties observed in superfluid helium. For example, if a mass of liquid helium is rotated, '*vortices*' are formed. In fact the occurrence of vortices is an important property that encodes superfluid behaviour. There is a critical temperature for helium below which it becomes superfluid, and above which it remains as a normal fluid. Thus, there is a phase transition that occurs at the critical temperature. For helium, that is, ^{4}He, the superfluid phase is a Bose–Einstein condensate (BEC), where all the (Bosonic) atoms occupy the lowest lying ($k = 0$) quantum state, thereby forming a condensate.

Even though a superfluid transition for ^{4}He occurs at a temperature $T = 2.2$ K, for other atomic gases, such phase transitions occur at very different temperatures, such as, in the regime of a micro or nano Kelvin, that is, 10^{-6}–10^{-9} K. Also, such phase transitions occur at very large temperature scales in other systems. For example, in the astrophysical contexts, the critical temperature for the onset of the superfluid phase in neutron stars is about 10^8 K, while the corresponding temperature for quark matter is of the order of 10^{11} K. This shows that superfluidity occurs in a temperature range spanning over nearly 10^{20} K.

Let us come back to superfluid helium which would facilitate a discussion of the BEC. But before we do that, a historical overview is necessary. Helium comprises an even number of fermions, that is, two protons, two neutrons and two electrons, and hence is a boson with spin zero. Helium was first liquefied by H Kamerlingh Onnes in 1908 at a temperature 4.2 K. In fact, the discovery of superconductors, the other 'condensation' phenomenon occurring in fermionic systems, just followed this discovery, and set the stage for achieving temperatures that are low enough to study vanishing of the resistivity of mercury.

doi:10.1088/978-0-7503-3031-2ch7

About two decades later, in 1927 Wolfke and Keesom [1] found a second phase transition at even lower temperatures, namely around 2.17 K, which showed a discontinuity of the specific heat (along with a peak in the number density) at this temperature. In fact, because of the structure of the discontinuity at 2.17 K, which closely resembles the Greek letter λ, the transition is called the λ point transition. The two phases above and below λ point are termed as helium-I and helium-II, respectively. Helium-II has some remarkable superfluid properties, which were discovered by Kapitza [2], and independently by Allen and Misener [3] in 1938. Kapitza won a Nobel prize for the experimental demonstration of helium flowing smoothly, that is, without any viscous drag below the λ-point through a narrow orifice of about 0.5 μm.

It is important to note that the phenomenon of superfluidity occurs at very low temperatures and thus it is bound to be a quantum effect. It is also very important for sustenance of the superfluid behaviour that it remains a liquid at low temperature as most of the other elements tend to form solid at very low temperatures. It was not long after this discovery that London [4] proposed the connection between superfluidity and the BEC. A detailed account of this exciting discovery can be found in the articles below [5]. The theoretical proposal of a BEC in a gas of non-interacting bosons was put forward by S N Bose, who was apparently trying to reconcile Planck's quantum hypothesis with the experiments on the intensity of black body radiation. Einstein joined the collaboration when he read and translated Bose's paper on photon statistics which correctly explained the non-monotonic variation of the black body radiation intensity as a function of the wavelength of the emitted radiation. Both Bose and Einstein had realized that a certain fraction of the total number of the (massive) bosons would macroscopically occupy the lowest energy or the zero momentum state below a certain critical temperature. This is called the condensate fraction. For non-interacting particles, the condensate fraction comprises nearly 100% of the particles. However, for ^{4}He, the condensate fraction is barely 10% even at low temperatures. Even if there is an effort to explain superfluidity as a result of BEC, however, there is enough evidence for the following:

(i) ^{4}He is not a non-interacting system. It has very short-ranged repulsive interactions;
(ii) a BEC does not have the ability to superflow as seen for superfluids;
(iii) neither does it show a divergence in the specific heat at the transition point.

7.2 Bose–Einstein condensation

Let us briefly review the fundamentals of Bose–Einstein condensation (BEC) at the outset. Among the known states of matter, a BEC is probably the most mysterious one and could only be created more than seven decades in the lab after it was theoretically predicted by S N Bose (1894–1974). Bose communicated his research findings to A Einstein, and Einstein immediately saw the merit of Bose's research findings for them to get published. The formalism by Bose on the distribution function of particles with integer spin (including zero) was later identified as the Bose–Einstein (BE) statistics, and found to be equally applicable for light (photons) and atoms (matter waves).

The BE statistics allows for an affinity of the bosons to congregate in a particular quantum state. This along with a symmetric wavefunction upon exchange of a pair of particles, are the main differences between them and fermions, where the latter obeys Pauli exclusion principle, and it is hence forbidden for two of them with an identical set of quantum numbers to occupy the same state. A significant phenomenon occurs in the case of non-interacting bosons owing to the unrestricted occupancy in the lowest quantum state at very low temperatures (of the order of a micro Kelvin) which we denote as the BEC.

A detailed mathematical approach of condensation in ideal Bose systems is discussed in the book by Pathria [6]. In three dimensions with quadratic dispersion, the DOS, $N(\epsilon)$ as a function of the energy ϵ behaves as $\epsilon^{1/2}$, which demands a zero weight for the lowest energy states, namely, $\epsilon = 0$. However, for an ideal Bose gas at low temperature, this state ($\epsilon = 0$) is the most important one, and hence cannot be assigned a zero weight. In fact, it may have an infinite occupancy under certain conditions. If that happens, this state will house a macroscopically large number of particles, and the system will behave like a giant coherent matter wave.

There is a simple way to tackle the problem of zero weight (that is, zero DOS) for $\epsilon = 0$. It has to be made discernible from all other states ($\epsilon \neq 0$). One can now write the particle density comprising two terms, one which corresponds to $\epsilon = 0$, and the other for $\epsilon \neq 0$. Mathematically, this is written as,

$$\frac{N}{V} = \frac{1}{V}\frac{e^{\beta\mu}}{(1 - e^{\beta\mu})} + \int N(\epsilon)f_{\mathrm{B}}(\epsilon)d\epsilon = \frac{N_0}{V} + \frac{N_{\epsilon\neq0}}{V} \tag{7.1}$$

where $f_{\mathrm{B}}(\epsilon) = (e^{\beta(\epsilon-\mu)} - 1)^{-1}$ is the Bose distribution function, $\frac{N_0}{V}$ and $\frac{N_{\epsilon\neq0}}{V}$ are the particle densities corresponding to $\epsilon = 0$ and $\epsilon \neq 0$, respectively. The first term in equation (7.1), which is of interest to us, can be written as,

$$\frac{N_0}{V} = \frac{1}{V}\frac{z_f}{(1 - z_f)} \tag{7.2}$$

where $z_f = e^{\beta\mu}$ and is called the fugacity of the system of particles, with μ being the chemical potential. μ approaches zero from below at the phase transition, that is, at the onset of the Bose condensation, or equivalently, when $\frac{N_0}{V}$ diverges. The divergence is expressed in terms of the Riemann zeta function, $g_n(z_f)$. Stated more concretely, the excited state occupancy is expressed as,

$$\frac{N_{\epsilon\neq0}}{V} = \frac{N - N_0}{V} = \frac{1}{\lambda^3}g_{3/2}(z_f) \tag{7.3}$$

where,

$$g_{3/2}(z_f) = \frac{1}{\Gamma(3/2)}\int_0^\infty \frac{\sqrt{x}}{z_f^{-1}e^x - 1}dx.$$

Quite accidentally, and unlike in two dimensions, $g_{3/2}(z_f)$ has a finite value as $\mu \to 0$ from below, for which $g_{3/2}(1) \simeq \zeta(3/2) = 2.612$. Consequently, below a certain

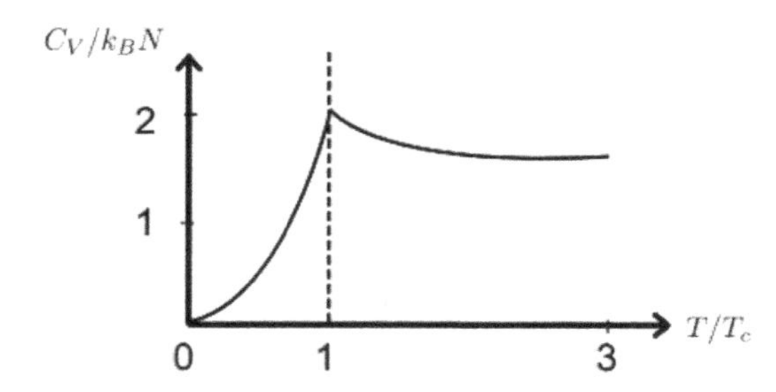

Figure 7.1. The variation of the specific heat of an ideal Bose gas as a function of temperature. At large temperatures, the plot approaches its classical value of 3/2.

temperature, T_c[1], the total number of particles in all the excited states put together is bounded. Thus, if the system possesses more particles, they will be accommodated in the ground state, which, as we saw has infinite occupancy. As the system can have infinitely large number of particles, below a temperature, T_c, all of them will accumulate in the ground state leading to a condensation.

Among many of the striking properties of this condensate, the pressure exerted by the particles at $T = T_c$ is remarkably low, and is half that of a classical ideal gas. The result is particularly interesting, since the pressure exerted by a Fermi gas at $T = 0$ is enormous owing to the exclusion principle. Further, the entropy, S (first derivative of the free energy) and the specific heat, C_v (second derivative of the free energy) are continuous (see figure 7.1), while $\frac{dC_v}{dT}$ shows a discontinuity at $T = T_c$. The specific heat of liquid ^{4}He as a function of temperature corroborates the above finding. In fact, the proof of the formation of a condensate is robustly demonstrated by computing the entropy of the system. The entropy is entirely contributed by the excited state occupancy, $N_{\epsilon \neq 0}$, while N_0 does not contribute anything.

7.3 Superfluidity

Bogoliubov developed microscopic theory which demonstrated that the excitation spectrum of a weakly interacting Bose gas is linear in momentum (such as that in the relativistic case). Moreover, the critical flow speed at which superfluidity gets destroyed is finite. Thus, weak inter-particle repulsive interactions do not destroy the Bose condensate. Also, the ideal Bose gas in the condensed phase has a vanishingly small critical velocity and hence cannot be termed as a superfluid. Thus formation of a condensate, that is a BEC, is distinct from a superfluid, where the latter is associated with a current that flows without dissipation and must involve inter-particle interaction.

Superfluid He transports mass (of the He atoms) without friction. It does so via forming a Bose–Einstein condensate which can carry mass, and flow without losing energy. However, the excitation spectrum of the condensate leads to dissipation and accounts for a poor condensate fraction of He. Landau proposed a general scheme to incorporate these excitations. Let us consider a superfluid is flowing through a

[1] Detailed calculations reveal $T_c = \frac{h^2}{2mk_B}\left[\frac{N}{V\zeta(3/2)}\right]^{2/3}$, m: mass, k_B: Boltzmann constant, V: volume.

capillary with velocity $\mathbf{v}_s$. In the frame of the fluid, the capillary moves with a velocity $-\mathbf{v}_s$. The excitation spectrum for such a situation is given by $\mathcal{E}_p > 0$ (say) corresponding to a momentum $\mathbf{p}$. In the rest frame of the capillary, the total energy of the fluid is the kinetic energy (ϵ_{kin}) plus the energy of the elementary excitations (ϵ_{exc}). Hence one can write combining the two conditions,

$$\epsilon_{\text{tot}} = \epsilon_{\text{kin}} + \epsilon_{\text{exc}}.$$

Here the excitation energy is,

$$\epsilon_{\text{exc}} = \epsilon_{\mathbf{p}} + \mathbf{p} \cdot \mathbf{v}_s.$$

The fluid is dissipative if $\epsilon_{\text{exc}} < 0$, that is,

$$\epsilon_{\mathbf{p}} + \mathbf{p} \cdot \mathbf{v}_s < 0.$$

The sum of the two quantities can only be negative if $\mathbf{p}$ is in opposite direction to $\mathbf{v}_s$ and $\epsilon_p - p\mathbf{v}_s = 0$. That defines a critical velocity below which He transports charge.

Landau's theory of superfluidity is based on the Galilean transformation equations for energy (E) and momentum ($\mathbf{p}$) of the fluid. Let us consider two frames O and O', where O is attached to the frame of fluid, and O' is attached with the capillary. Thus the momentum of the fluid in the O' frame is $\mathbf{p}' = \mathbf{p} - m\mathbf{V}$. Thus the energy is,

$$E' = \frac{|\mathbf{p}'|^2}{2m} = \frac{(\mathbf{p} - m\mathbf{V})^2}{2m}$$
$$= E - \mathbf{p}.\ \mathbf{V} + \frac{1}{2}m|V|^2$$

where $E = \frac{p^2}{2m}$ and V denote the relativistic velocity between the two frames. Suppose a viscous fluid is flowing through a capillary at a constant velocity $\mathbf{v}$. The system loses energy while flowing due to the friction between the fluid and capillary wall. Such dissipative process occurs through the elementary excitations. The excitations are known as Bogoliubov quasiparticles for the case of an interacting Bose gas.

Let us assume that the fluid is at rest in the frame O (that is, O moves with a velocity $\mathbf{v}$ with respect to the wall of the capillary[2].) Now if an elementary excitation with momentum $\mathbf{p}$ is created in the fluid, the total energy becomes the ground state energy (E_0) plus this energy of the excitations ($\mathcal{E}(\mathbf{p})$), that is $E_{\text{tot}} = E_0 + \mathcal{E}(\mathbf{p})$. Now shift to the O' frame (which moves with a velocity $-\mathbf{v}$ with respect to the frame of the fluid), where the energy and the momentum are given by,

$$\mathbf{p}' = \mathbf{p} + m\mathbf{v} \qquad (\text{putting} \quad \mathbf{V} = -\mathbf{v}) \tag{7.4}$$

and

$$E' = E_0 + \mathcal{E}(\mathbf{p}) + \mathbf{p}.\ \mathbf{v} + \frac{1}{2}m|v|^2. \tag{7.5}$$

[2] This may be confusing because there is another frame O' that moves with a velocity $\mathbf{v}$ with respect to O.

The above results confirm that there is the appearance of elementary excitation due to $\mathbf{p}$ (in equation (7.4)) and $\mathcal{E}(\mathbf{p}) + \mathbf{p}.\,\mathbf{v}$ (equation (7.5)). Dissipation of energy through the appearance of elementary excitations can occur if the phenomenon is energetically favourable, which happens when

$$\mathcal{E}(\mathbf{p}) + \mathbf{p} \cdot \mathbf{v} < 0.$$

The above condition gets satisfied when, $|\mathbf{v}| = \frac{\mathcal{E}(\mathbf{p})}{|\mathbf{p}|}$ and $\mathbf{p}.\,\mathbf{v} < 0$. The last condition signals that the elementary excitation corresponds to the velocity of the fluid being opposite to the direction of the momentum. This is certainly an unconventional scenario. It also says that the condition holds till a critical fluid velocity, namely, $v_c = \min_{\mathbf{p}} \frac{\mathcal{E}(\mathbf{p})}{|\mathbf{p}|}$. If the velocity of the fluid is smaller than this, then there will be no elementary excitations. This is known as Landau's criterion of superfluidity, which asserts that superfluidity occurs when the relative velocity between the fluid and the wall of the capillary is smaller than that of v_c. Weakly interacting Bose gas, or even strongly interacting Bosonic fluids, such as, ^{4}He fulfil Landau's criterion. The critical velocity is given by the sound velocity. In ^{4}He, the critical velocity is smaller than the sound velocity. It is easy to see that for a free gas of particles $v_c = \min_{\mathbf{p}} \frac{\mathcal{E}(\mathbf{p})}{|\mathbf{p}|} = 0$. Thus for a non-zero critical velocity, v_c, interaction effects are important for superfluidity to occur.

Let us recall that the energy and the momentum in the rest frame of the fluid (frame O), namely $\mathcal{E}(\mathbf{p})$ and $\mathbf{p}$ transform to $\mathcal{E}(\mathbf{p}) + \mathbf{p}.\,\mathbf{v}$ and $\mathbf{p} + m\mathbf{v}$, respectively, in the capillary frame (O' frame). Let us rename $\mathbf{v}$ as $\mathbf{v}_s$ where the subscript denotes superfluid. Thus at finite temperature, the population of quasiparticles follows Bose–Einstein distribution, namely,

$$f_{\mathrm{B}}(\mathcal{E}) = \frac{1}{e^{\beta(\mathcal{E}(\mathbf{p})+\mathbf{p}.\mathbf{v})} - 1}. \tag{7.6}$$

Thus, for $\mathcal{E}(\mathbf{p}) + \mathbf{p}.\,\mathbf{v} > 0$, or $|\mathbf{v}_s| < \min_{\mathbf{p}} \frac{\mathcal{E}(\mathbf{p})}{|\mathbf{p}|}$, $f_{\mathrm{B}}(\mathcal{E})$ is positive for all values of $\mathbf{p}$. Besides, the above condition is identical to Landau's condition.

It is quite neat to express the critical velocities for the weakly and strongly interacting Bose gases as shown in figure 7.2. For a weakly interacting gas, the critical velocity is the same as the sound velocity, while the same for strongly interacting gas is much lower than the sound velocity.

7.3.1 Gross–Pitaevskii equation

An alternate route to establish a connection between BEC and superfluid is via the relationship between the superfluid velocity and the order parameter of the condensate. The Gross–Pitaevskii equation (GPE) [7] is a useful tool in this regard. It describes the ground state of interacting bosons using the Hartree–Fock approximation and a pseudopotential that describes the inter-particle interaction. GPE is a non-linear Schrödinger equation for the single-particle wavefunction in a BEC. We present a derivation of the GPE in the appendix.

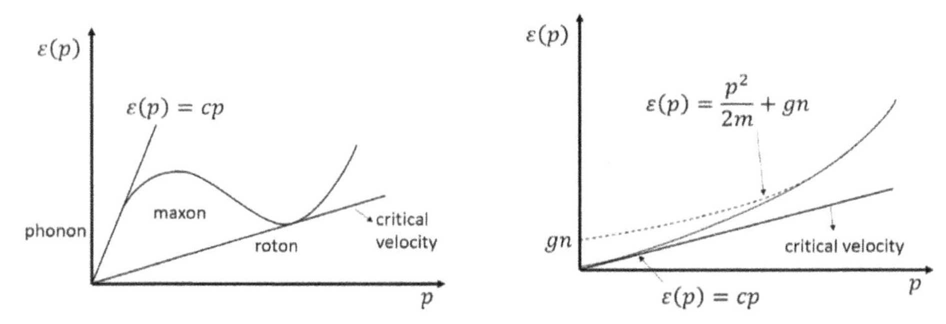

Figure 7.2. The excitation spectra of weakly (left) and strongly (right) interacting Bose gases. For the weakly interacting case, the critical velocity is identical to the sound velocity, while for the strongly interacting scenario, the critical velocity is smaller than the sound velocity.

In the following we review the formalism needed for our purpose in brief. Consider a quantum mechanical field operator,

$$\psi(\mathbf{r}) = \sum_i \phi_i(\mathbf{r}) a_i$$

where a_i ($a_i^\dagger$) denote annihilation(creation) operators corresponding to the single-particle states $\phi_i(\mathbf{r})$ which obey,

$$[a_i, a_i^\dagger] = \delta_{ij}; \qquad [a_i, a_j] = \left[a_i^\dagger, a_j^\dagger\right] = 0; \qquad \forall\, i, j.$$

The single-particle states obey the orthonormal relation,

$$\left\langle \phi_i(\mathbf{r}) | \phi_j(\mathbf{r}) \right\rangle = \delta_{ij}$$
$$\left\langle \phi_i(\mathbf{r}) | \phi_i(\mathbf{r}') \right\rangle = \delta(\mathbf{r} - \mathbf{r}')$$

where $\delta(\mathbf{r} - \mathbf{r}')$ denotes the Kronecker delta function. Further, it is easy to see that the field operators $\psi(\mathbf{r})$ obey,

$$[\psi(\mathbf{r}), \psi(\mathbf{r}')] = \delta(\mathbf{r} - \mathbf{r}').$$

Suppose the ground state of a system of particles is macroscopically populated, we can separate the corresponding field operator into a condensate part ($i = 0$) and a non-condensate part ($i \neq 0$), such that,

$$\psi(\mathbf{r}) = \phi_0(\mathbf{r}) a_0 + \sum_{i \neq 0} \phi_i(\mathbf{r}) a_i. \tag{7.7}$$

In the Bogoliubov approximation [8], the single-particle operators are separated by c-numbers using

$$a_0 = a_0^\dagger = \sqrt{N_0}$$

where $N_0 = \langle a_0^\dagger a_0 \rangle$ is the condensate electron density. If we define,

$$\psi_0 = \sqrt{N_0}\,\phi_0, \quad \text{and} \quad \delta\psi = \sum_{i\neq 0}\phi_i a_i,$$

the Bogoliubov ansatz can be written as,

$$\psi(\mathbf{r}) = \psi_0(\mathbf{r}) + \delta\psi(\mathbf{r}).$$

Since $\psi_0(\mathbf{r})$ denotes the condensate wavefunction, it has a magnitude and a phase. For example,

$$\psi_0(\mathbf{r}) = |\psi_0(\mathbf{r})|e^{i\phi(\mathbf{r})}.$$

$\psi_0(\mathbf{r})$ is considered as the order parameter for the condensed state, that is, a condensate phase has a finite $\psi_0(\mathbf{r})$, while vanishing $\psi_0(\mathbf{r})$ denotes a non-condensate phase. $\psi_0(\mathbf{r})$ yields the particle density of the condensate via, $n(\mathbf{r}) = |\psi_0(\mathbf{r})|^2$, while the phase $\phi(\mathbf{r})$ describes the coherent nature of the wavefunction and hence characterizes the superfluid phenomena.

The field operator $\psi(\mathbf{r}, t)$ obeys the Heisenberg equation of motion (EOM), namely,

$$i\hbar\frac{d}{dt}\psi(\mathbf{r}, t) = [\psi(\mathbf{r}, t), \mathcal{H}]. \tag{7.8}$$

It may be noted that the field operator is assumed to be a function of both $\mathbf{r}$ and t. The Hamiltonian for the system can be denoted by the field operators $\psi(\mathbf{r})$. Using,

$$\begin{aligned}\mathcal{H} = &\int \frac{\hbar^2}{2m}\nabla\psi^\dagger(\mathbf{r})\nabla\psi(\mathbf{r})\, d\mathbf{r} \\ &+ \frac{1}{2}\int \psi^\dagger(\mathbf{r})\psi^\dagger(\mathbf{r}')V(\mathbf{r}' - \mathbf{r})\psi(\mathbf{r})\psi(\mathbf{r}')\, d\mathbf{r}'\, d\mathbf{r}.\end{aligned} \tag{7.9}$$

The first term is the kinetic energy and $V(\mathbf{r})$ is Coulomb term. In real systems, $V(\mathbf{r})$ always contains a short-range term which, however, makes the solution of the Schrödinger's equation different. Thankfully, the actual form of $V(\mathbf{r})$ is unimportant for describing the macroscopic properties of the system. It is in fact convenient to replace the actual potential by an effective potential $V_{\text{eff}}(\mathbf{r})$ which is smooth and yields a correct description for the low momentum component of the Fourier transformed potential, $V_{\mathbf{q}\simeq 0}$ given by,

$$V_{\mathbf{q}\simeq 0} = V_0 = \int V_{\text{eff}}\, d\mathbf{r}$$

which yields a reasonable answer for the complicated many-body problem. Further, these may also be an external potential (such as a trapping potential), namely $V_{\text{ext}}(\mathbf{r}, t)$, which is a one-body potential. The EOM now yields,

$$\begin{aligned}i\hbar\frac{d}{dt}\psi(\mathbf{r}, t) = &\left[-\frac{\hbar^2}{2m}\nabla^2 + V_{\text{ext}}(\mathbf{r}, t)\right]\psi(\mathbf{r}, t) \\ &+ \psi(\mathbf{r}, t)\int \psi^\dagger(\mathbf{r}', t)V_{\text{eff}}(\mathbf{r}')\psi(\mathbf{r}', t)d\mathbf{r}'.\end{aligned} \tag{7.10}$$

The idea is to replace $\psi(\mathbf{r}, t)$ by $\psi_0(\mathbf{r}, t)$ (a c-number) which has negligible variation over the range of the interaction $V_{\text{ext}}(\mathbf{r})$ and thus can be taken as a constant. This permits replacing $\mathbf{r}'$ by $\mathbf{r}$ in equation (7.10) which finally can be written as,

$$i\hbar\frac{d}{dt}\psi_0(\mathbf{r}, t) = \left[-\frac{\hbar^2}{2m}\nabla^2 + V_{\text{ext}}(\mathbf{r}, t) + \int V_{\text{eff}}(\mathbf{r})|\psi_0(\mathbf{r}, t)|^2\, d\mathbf{r}\right]\psi_0(\mathbf{r}, t). \quad (7.11)$$

Let us substitute $\int V_{\text{eff}}(\mathbf{r})\, d\mathbf{r}$ by g (the interaction term), and write the well-known GPE as,

$$i\hbar\frac{d}{dt}\psi_0(\mathbf{r}, t) = \left[-\frac{\hbar^2}{2m}\nabla^2 + V_{\text{ext}}(\mathbf{r}, t) + g|\psi_0(\mathbf{r}, t)|^2\right]\psi_0(\mathbf{r}, t). \quad (7.12)$$

This has the form of a non-linear Schrödinger's equation owing to the presence of the last term in the RHS of equation (7.12). Thus the dynamics of the condensate is described by the GPE. GPE plays an important role in the study of BEC and non-linear optics.

Here we shall concentrate on the physics of BEC, where the density of particles is related to the square of the order parameter, namely,

$$n(\mathbf{r}, t) = |\psi_0(\mathbf{r}, t)|^2.$$

The total number particles, N may be obtained using,

$$N = \int n(\mathbf{r}, t)\, d\mathbf{r} = \int |\psi_0(\mathbf{r}, t)|^2\, d\mathbf{r}. \quad (7.13)$$

The corresponding continuity equation can be written as,

$$\frac{d}{dt}n(\mathbf{r}, t) + \nabla \cdot [j(\mathbf{r}, t)] = 0 \quad (7.14)$$

where the probability current density can be expressed in terms of the order parameter as,

$$j(\mathbf{r}, t) = \frac{i\hbar}{2m}(\psi_0^*\nabla\psi_0 - \psi_0\nabla\psi_0^*). \quad (7.15)$$

Replacing $\psi_0(\mathbf{r}, t) = \sqrt{n(\mathbf{r}, t)}\, e^{i\phi(\mathbf{r}, t)}$ in equation (7.14), the current density assumes a form,

$$j(\mathbf{r}, t) = n(\mathbf{r}, t)\frac{\hbar}{m}\nabla\phi(\mathbf{r}, t). \quad (7.16)$$

The superfluid velocity, v_s is related to this gradient of the phase via, $v_s = \frac{\hbar}{m}\nabla\phi(\mathbf{r}, t)$. This is how the superfluid velocity, v_s encodes the superfluid behaviour of the Bose gas as mentioned by us before. Since $\mathbf{v}_s$ is related to the gradient of a phase, the flow must be strictly irrotational, that is, $\nabla \times \mathbf{v}_s = 0$. It is for the same reason that the circulation of $\mathbf{v}_s$ obeys the Bohr–Sommerfeld quantization condition, namely,

$$k = \oint \mathbf{v}_s \cdot d\mathbf{l} = \frac{nh}{m}$$

where n is an integer and m denotes the mass of the atoms, for example, He atom in ^{4}He

Inserting the order parameter in the GPE (equation (7.12)), one may get an EOM for ϕ,

$$\hbar\frac{d}{dt}\phi(\mathbf{r}, t) + \left(\frac{1}{2}mv_s^2 + V_{\text{ext}} + gn - \frac{\hbar^2}{2m\sqrt{n}}\nabla^2\sqrt{n}\right) = 0. \tag{7.17}$$

This equation along with the continuity equation is equivalent to the GPE. In fact, the phase and the number of the particles are conjugate variables. They obey Heisenberg's uncertainty relation, namely, $\Delta n \Delta\phi \simeq \frac{\hbar}{2}$.

The stationary state solution of the GPE can be written as,

$$\psi_0(\mathbf{r}, t) = \psi_0(\mathbf{r})\, e^{-i\mu t},$$

where $\mu = \frac{1}{\hbar}\frac{\partial E}{\partial n}$ is the chemical potential. The total energy of the system is,

$$E = \int \left(\frac{\hbar^2}{2m}|\nabla\psi_0|^2 + V_{\text{ext}}(r)|\psi_0(r)|^2 + \frac{g}{2}|\psi_0(r)|^4\right) dr. \tag{7.18}$$

Plugging in the stationary state solution into GPE gets rid of the time dependence. We arrive at the time independent GPE,

$$\left[-\frac{\hbar^2}{2m}\nabla^2 + V_{\text{ext}}(r) - \hbar\mu + g|\psi_0(r)|^2\right]\psi_0(r) = 0. \tag{7.19}$$

The above equation can be solved numerically, where the lowest energy solution is real, while the excited states are complex. Quantized vortices realized in superfluid denote such excited states. We include a brief description of the quantized vortices below.

7.3.2 Quantized vortices

The dynamical properties of an ideal fluid become interesting in the presence of vortices. We are familiar with vortices in fluids, which arise when the water in the river becomes turbulent as it passes through an obstacle. In a similar fashion, a superfluid circulates around the vortices. The velocity of the fluid near the vortices is larger than that far away from them. In fact, the velocity around each vortex is determined by h/m and the presence of h (Planck's constant) in the expression implies that the vortices are quantized. Hence superfluidity (similar to superconductivity) is a quantum phenomenon occurring at the microscopic level.

It was first proposed by L Onsager and later confirmed by R P Feynman that such vortices should be present in the superfluid ^{4}He. In accordance with the

Bohr–Sommerfeld quantization condition, the line integral of the superfluid velocity should be quantized,

$$\oint \mathbf{v}_s \cdot d\mathbf{l} = \frac{nh}{m} \tag{7.20}$$

where $n = 0, 1, 2, \ldots$ and m denotes the mass of He atom. In reality, only $n = 1$ is observed. $\mathbf{v}_s$ can be determined in the same way as computing the magnetic field for a straight current carrying wire at a certain distance using the Biot–Savart law. This explains the quantization condition where the magnetic field (here it is superfluid velocity) falls of as the inverse of distance from the wire (here the vortex filament). As the distance from a given reference grows linearly, the value of the integral yields a constant. Incorporating the exact structure of the vortex (unlike that for a simple geometry of the straight wire) is a complicated problem, and has received wide attention. However, in most of the cases, the radius of the core is too small (of the order of a few nm) to bother about the detailed structure of the vortex.

There exist a few experimental techniques by which these quantized vortices can be detected. In one of them, the charged ions are injected into the superfluid which are hence trapped by the vortices. Tracking the motion of the ions, brings out the existence of the vortices. Besides, an indirect method constitutes investigating the dissipation that they cause, which eventually modifies the dynamical properties of the vortices.

It is also important to explore the creation of the vortices in the superfluid. Consider extremely low temperatures such that thermal effects can be ignored. Now let us consider a cylinder, containing a superfluid, that is set into rotation. Unlike a classical fluid where rigid body rotation occurs, in a superfluid, such rotation is forbidden owing to $\nabla \times \mathbf{v}_s = 0$. Instead the vortex lines appear parallel to the axis of rotation which, at large rotational speeds, are arranged in the form of a triangular lattice. This yields an interplay between a triangular order (for the vortices) and a circular symmetry imposed by the rotation, thereby resulting in a long-range, logarithmic interaction between the vortices.

Let us consider a superfluid contained in a cylinder of radius R and length L (see figure 7.3). The solution of the GP equation corresponding to a rotation about the z-axis of the cylinder can be written as,

$$\psi(r_0) = e^{is\phi}|\psi(r_0)| \tag{7.21}$$

where the amplitude is given by, $\psi(r) = \sqrt{n(r)}$ and s is an integer to satisfy. Substituting the ansatz equation (7.21) in the GPE (equation (7.19)), we arrive at an equation for the order parameter $|\psi(r_0)|$,

$$-\frac{\hbar^2}{2m}\frac{1}{r}\frac{d}{dr}\left(r\frac{d}{dr}|\psi(r_0)|\right) + \frac{\hbar^2 s^2}{2mr^2}|\psi(r_0)| + g|\psi(r_0)|^3 - \hbar\mu|\psi(r_0)| = 0 \tag{7.22}$$

Let us introduce new variables, namely, $\eta = r/\xi$ with $\xi = \hbar/\sqrt{2mgn}$. With these, the solution can be written as,

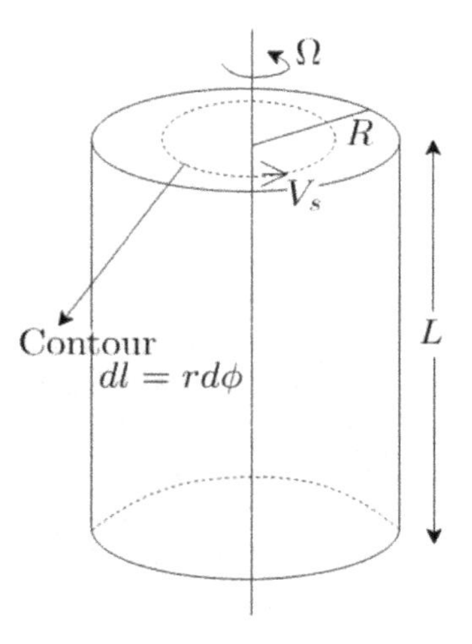

Figure 7.3. Schematic diagram of a vortex is shown.

$$|\psi(r_0)| = \sqrt{n}\, g(\eta),$$

which yields an equation for $g(\eta)$, namely,

$$\frac{1}{\eta}\frac{d}{d\eta}\left(\eta\frac{dg}{d\eta}\right) + \left(1 - \frac{s^2}{\eta^2}\right)g - g^3 = 0. \tag{7.23}$$

For η (or r) to assume large values, ($\eta \to \infty$), $v_s \to 0$, as a result of which the density approaches its unperturbed value ($|\psi_0| = \sqrt{n}$) demanding $g(\infty) = 1$. On the other limit, that is $\eta \to 0$, $g(\eta) \to 0$, which also implies that the local density, $n(r)$ $(=|\psi_0(r)|^2)$ tends to vanish. The energy of the vortices can be obtained by comparing with the ground state solution ($s = 0$) using,

$$\begin{aligned} E_v &= E(s \neq 0) - E(s = 0) \\ &= \frac{Ln\pi\hbar^2}{m}\int_0^{\frac{R}{\xi}}\left[\left(\frac{dg}{d\eta}\right)^2 + \frac{s^2}{\eta^2}g^2 + \frac{1}{2}(g^2 - 1)^2\right]\eta\, d\eta. \end{aligned} \tag{7.24}$$

The energy corresponding to $s \neq 0$ is,

$$E_v = L\pi n\frac{\hbar^2 s^2}{m}\ln\left(\frac{R}{r_c}\right)$$

where r_c denotes certain core radius. For $s = 1$, it is comparable to η. Two things may be noted about E_v here, namely, the s^2 dependence and the logarithmic dependence on r_c. The former implies that a single vortex is not stable for $s \geqslant 2$. This is so since the z component of the angular momentum, L_z behaves as s (since superfluid velocity, $v_s \sim s$), while the energy goes as s^2. Thus the stability of the vortices is limited to $s = 1$. Further the logarithmic dependence on the core radius implies the weak dependence of energy on the core size.

7.4 Many-body physics with cold atomic systems

BEC was realized in the lab about seven decades later after its proposition by Bose and Einstein, on June 5, 1995, by E Cornell and C Wieman of University of Colorado, at Boulder NIST-JILA lab in a gas of Rb atoms cooled down to a temperature of about 170 nK. Shortly afterwards, W Ketterle at MIT produced a BEC in a gas of Na atoms. This feat of achieving a condensate in dilute atomic gases (possibly along with evolving technologies to cool down a system to few hundred nK) earned a Nobel prize for Cornell, Wieman and Ketterle in 2001. There are several research groups around the world who have created BEC in dilute vapours and even in systems comprising atoms, molecules, photons in the labs [9].

For physicists ranging from disciplines such as, condensed matter physics and atomic physics, this discovery marked the commencement of a new field of research, where 'synthetic' systems were created with inter-atomic distances much larger than those in crystal lattices (~300 nm and this value can be tuned), typical energy scales determined by hundreds of nano Kelvin (or micro Kelvin), and most importantly, in an environment of tunable inter-particle interactions (which can never be controlled in crystals) and being able to perform experiments without disorders, defects, impurities, etc (again not under control in solid state physics).

Let us stress the inter-particle interaction aspect stated above. A typical atomic scattering length is of the order of tens of nm, which is certainly an order of magnitude larger than inter-atomic distances. This facilitates reliable calculations of the physical properties under controlled approximations. The scenario is in stark contrast to superfluid ^{4}He, where owing to strong inter-particle interactions, detailed and systematic studies are unavailable. In the following subsection we shall focus on BEC of weakly interacting systems which store a large number of interesting and unanswered questions.

7.4.1 BEC in weakly interacting systems

In dilute atomic systems, the range of the inter-atomic forces, r_0 is usually much smaller than the average distance between the constituent particles, l_0, that is,

$$r_0 \ll l_0 = \frac{1}{n^{1/3}} = \left(\frac{N}{\mathcal{V}}\right)^{-1/3}. \tag{7.25}$$

The interaction potential is non-zero only when the two particles come within a distance r_0. In the following we assume an infinitesimal value for r_0, so that the interaction becomes a contact interaction. The microscopic Hamiltonian for a system of atoms (Bosons) with contact interaction is written as,

$$\mathcal{H} = \sum_p \epsilon_p b_p^\dagger b_p + \frac{V_0}{2\mathcal{V}} \sum_{p,\, p',\, q} b_{p+q}^\dagger b_{p'-q}^\dagger b_{p'} b_p - \mu \sum_p b_p^\dagger b_p \tag{7.26}$$

where $b_p(b_p^\dagger)$ is the annihilation(creation) operator, ϵ_p denotes single-particle states, μ is chemical potential which fixes the number of particles and $\mathcal{V}$ is the volume. Here $V_0 = \int V(\mathbf{r})\, d\mathbf{r}$, which, owing to the diluteness condition stated above (equation

(7.25)) can be expressed in terms of the s-wave scattering length, a_s. Thus, $V_0 = \frac{4\pi\hbar^2}{m}a_s$. In general, the scattering length can either be positive or negative depending on the short-range part of the interaction potential. The scattering length tending to infinity implies the formation of bound states. The scattering length can be tuned using a Feshbach resonance via an external magnetic field. In fact for some particular values of the magnetic field, the scattering length can be made comparable or even larger than the inter-particle distances [10].

We shall incorporate the inter-particle interaction perturbatively into the problem. Thus the zeroth approximation consists of all non-interacting atoms condensed in the $k = 0$ state, which yields the following N_0 particle state,

$$| \psi_N\rangle = \frac{1}{\sqrt{N!}}(b^\dagger_{p=0})^N \mid 0\rangle,$$

where $| 0\rangle$ denotes a vacuum. The unperturbed energy is obtained as,

$$\langle\psi_N|\mathcal{H}|\psi_N\rangle = E(N_0) = -\mu N_0 + \frac{V_0}{2\mathcal{V}}N_0^2. \tag{7.27}$$

Minimizing the energy with respect to the ground state occupancy, N_0,

$$\frac{\partial E(N_0)}{\partial N_0} = 0 \qquad \text{yields,}$$

$$\frac{\mu}{V_0} = \frac{N_0}{\mathcal{V}} = n \quad \text{(density).} \tag{7.28}$$

Now the effect of interaction has to be included. The standard approach is to use a mean field decoupling procedure. One can replace $b^\dagger_{p=0}$ by their expectation values in the ground state, that is, $<b^\dagger_{p=0}> = \sqrt{N_0}$. This reduces the $p = 0$ contribution to a number, leaving the $p \neq 0$ contribution to be taken incorporated in a mean field manner (where the quartic operators are reduced to quadratic ones, the rest of which carries a factor of $\sqrt{N_0}$).

$$\begin{aligned}\mathcal{H}_{\text{MF}} = &-\frac{N_0^2 V_0}{2\mathcal{V}} + \sum_{p\neq 0}(\epsilon_p + 2n_0V_0 - \mu)\left(b^\dagger_p b_p + b^\dagger_{-p}b_{-p}\right)\\ &+ n_0V_0\sum_{p\neq 0}\left(b^\dagger_p b^\dagger_{-p} + b_p b_{-p}\right)\end{aligned} \tag{7.29}$$

where $n_0 = \frac{N_0}{\mathcal{V}}$. The mean field Hamiltonian can be diagonalized using a Bogoliubov transformation,

$$b_p = u_p\alpha_p + v_p\alpha^\dagger_{-p}$$
$$b_{-p} = u_p\alpha_{-p} + v_p\alpha^\dagger_p$$

subject to the condition[3],

$$u_p^2 - v_p^2 = 1$$

where u_p and v_p are the amplitudes for an empty state and a filled state, respectively, with momentum p. It may be noted that the above formalism has a perfect analogy with mean field studies of superconductors, except that we now deal with bosons instead of fermions. If we now replace $\mu = n_0 V_0$ from equation (7.29), the $\mathcal{H}_{MF}$ becomes,

$$\begin{aligned} \mathcal{H}_{\mathrm{MF}} = &-\frac{N_0^2 V_0}{2\ V} + \sum_{p\neq 0}\left(\alpha_p^{\dagger}\alpha_p + \alpha_{-p}^{\dagger}\alpha_{-p}\right)\left[(\epsilon_p + n_0 V_0)(u_p^2 + v_p^2)\right. \\ &\left.+ n_0 V_0(2u_p v_p)\right] \\ &+ \sum_{p\neq 0}\left(\alpha_p^{\dagger}\alpha_{-p}^{\dagger} + \alpha_p\alpha_{-p}\right)\left[(\epsilon_p + n_0 V_0)(2u_p v_p) + n_0 V_0(u_p^2 + v_p^2)\right]. \end{aligned} \quad (7.30)$$

Demanding the cancellation of the off-diagonal terms, that is, the last term in equation (7.30) (since the transformations in equation (7.30) are supposed to diagonalize $\mathcal{H}_{\mathrm{MF}}$),

$$(\epsilon_p + n_0 V_0)(2u_p v_p) + n_0 V_0(u_p^2 + v_p^2) = 0. \quad (7.31)$$

Further, the normalization condition in equation (3) can be satisfied if we choose,

$$\begin{aligned} u_p &= \cosh\theta_p \\ v_p &= \sinh\theta_p. \end{aligned} \quad (7.32)$$

Solution of equation (7.31) yields,

$$u_p = \left(\frac{\epsilon_p + n_0 V_0 + E_p}{2E_p}\right)^{1/2}$$

$$u_p = -\left(\frac{\epsilon_p + n_0 V_0 - E_p}{2E_p}\right)^{1/2}$$

for the amplitudes. Alongside we have,

$$\cosh\theta_p = -\frac{(\epsilon_p + n_0 V_0)}{E_p}$$

$$\sinh\theta_p = -\frac{n_0 V_0}{E_p}$$

with the quasi-particle energies E_p given by,

[3] This condition is in contrast with the corresponding scenario for superconductors where the squares of the amplitudes add up to unity.

$$E_p = \sqrt{\epsilon_p^2 + 2n_0 V_0 \epsilon_p}.$$

Thus the quasi-particle energies differ from the bare energies ϵ_p due to the presence of the second term inside the square root which incorporates the inter-particle interactions.

Finally a characteristic length, called the healing length, ξ, can be defined via,

$$\xi = \frac{1}{\sqrt{mn_0 V_0}}, \quad \text{where} \quad n_0 = \frac{N_0}{\mathcal{V}}$$

which yields a long wavelength behaviour ($p\,\xi \ll 1$) for the quasi-particle energies, $E_p \sim v_s|q|$. This is the so-called sound modes with $v_s = \left(\frac{n_0 V_0}{m}\right)^{1/2}$. The sound modes arise with spontaneously broken $U(1)$ symmetry arising out of the conservation of the total number of particles. For the other limit, namely $p\,\xi \gg 1$, one recovers the free-particle dispersion, $E_p = \frac{p^2}{2m}$.

7.5 Strongly correlated systems

Having studied the weakly interacting systems, we may explore the regime of strong interactions in many-body systems in atomic systems. The popular methods are based on Feshbach resonances, optical lattice, etc. As said earlier, there are a number of excellent reviews on Feshbach resonances; while we include a brief introduction of it in the subsequent discussion, we mainly focus on the optical lattices.

7.5.1 Optical lattice

The understanding of correlated many-body phenomena in usual solid state physics are extremely difficult due to the complicated band structure, complexity in interactions between the constituent particles, the presence of disorder and impurities, etc. Sometimes it is quite impossible to take care of all the underlying consequences in a single theoretical model, and often hard to gauge out the relevance of a particular effect while synthesizing the outcome of an experiment in an interacting system. An alternative way to bypass such restrictions is the usage of an artificial, known as optical lattice potential [11] and thus provide a gateway to explore the phenomena at the interface of solid state physics and (ultracold) atomic physics.

An optical lattice is formed due to the interference of counter propagating laser beams from all directions that renders a periodic potential which replicates lattice structures as perceived by an electron in a real crystal. To create a one- dimensional optical lattice potential, consider two counter propagating laser beams each of amplitude E_0, and linearly polarized in the x-direction with a wave vector $\mathbf{k} = (\pm k, 0, 0)$ which yields a resultant electric field that can be written as,

$$\mathbf{E}(x) = E_0 \hat{e}_z [e^{ikx} + e^{-ikx}]. \tag{7.33}$$

The field intensity is given by,

$$I(x) = |\mathbf{E}(x)|^2 = 4|E_0|\cos^2(kx). \tag{7.34}$$

Thus the optical lattice potential becomes,

$$V_{\text{op}}(x) = V_0\cos^2(kx) \tag{7.35}$$

where V_0 is the depth of the optical lattice[4], and the lattice periodicity is $a = \lambda/2$. Using further two pairs of laser beams in y and z directions, a three-dimensional optical lattice potential is formed as shown in figure 7.4 [11],

$$V_{\text{op}}(\mathbf{r}) = V_{0x}\cos^2(kx) + V_{0y}\cos^2(ky) + V_{0z}\cos^2(kz). \tag{7.36}$$

The height of the optical lattice potential, V_{0i} is often expressed in terms of recoil energy, via, $E_R = \hbar^2k^2/2m$. The superiority of the optical lattice potential originates from the vast control of the lattice parameters as well as the potential height by smoothly calibrating the laser intensity only. More so, it is free from defects, impurities and the lattice vibrations, etc. Thus the tunability of different parameters that are agents of the phase transition yields an edge over the experiments performed in the context of usual condensed matter physics. Furthermore, a host of phenomena, such as, metal-insulator, topological features demonstrated through the edge modes, etc can be designed in optical lattices.

An optical lattice allows the neutral atoms to cool down to extremely low temperatures using various cooling mechanisms, such as laser cooling [13–16], sympathetic and evaporative cooling [17] techniques, etc. Thermal velocity of an atom is directly related to the temperature and hence slowing down their motion will help in lowering the temperature. The central idea of laser cooling is to decelerate an

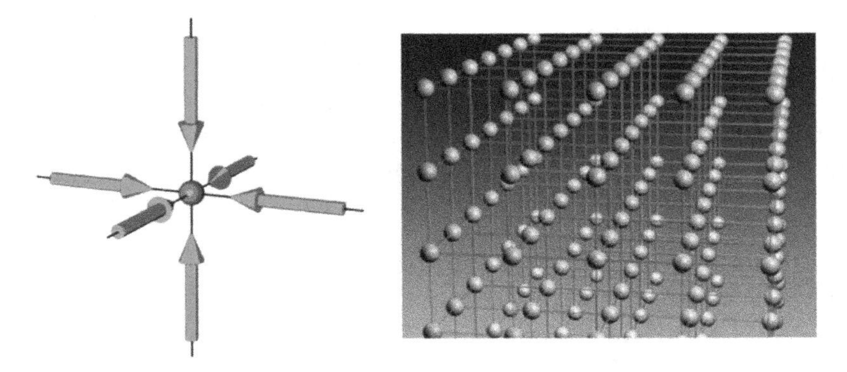

Figure 7.4. A three-dimensional optical lattice potential is formed by superimposing three orthogonal standing waves. In the 3D case, the optical lattice can be approximated by a 3D simple cubic array of tightly confining harmonic oscillator potentials at each lattice site. Figure courtesy of I Bloch [12], reprinted/adapted by permission from Nature Physics: Springer Nature, copyright 1969.

[4] Here V_0 has distinct meaning with respect to the discussion in the preceding section.

atom by using a radiation pressure from a laser beam near the atomic resonance. In this process, an atom in the ground state absorbs a photon from an incident laser beam. After absorbing a photon of energy $h\nu$, the atom acquires a momentum impulse of $h\nu/c$ along the incoming direction. In order to absorb a photon again, the atom has to return to the ground state by emitting a photon. After emitting a photon, the atom recoils in the opposite direction which results in its slowing down for a span of a few microseconds, a phenomenon responsible for the origin of 'optical molasses' [13] and subsequently producing low temperature. Apart from the laser cooling, a few other techniques such as polarization gradient cooling [18], Raman cooling [18], Sisyphus cooling [14, 16, 19], etc have been developed to reach ultra low temperatures.

7.5.2 Atom–atom interaction: Feshbach resonance

In optical lattices, experimental navigation from a weak to a strong interaction region is possible by simply changing the optical lattice potential. However, without disturbing the lattice geometry, a strong interaction region can be achieved by controlling the atom–atom interaction using a Feshbach resonance [20, 21], which in turn determines the on-site interaction potential. For ultracold gases at extremely low temperatures, the two-body interaction potential is essentially short-ranged in nature and hence entirely depends upon the s-wave scattering length, a_s. The Feshbach resonance was first introduced in the context of nuclear physics, which was later extended for the ultracold atomic gases to tune the scattering lengths via using a magnetic field.

Apart from the Feshbach resonance, a few other resonances such as, a 'shape resonance', etc which occurs in a potential barrier, where the scattering cross-section is a function of the angular momentum and a 'potential resonance' that happens in a single channel where the s-wave phenomenon is dominant [22]. Unlike the shape and potential resonances, a Feshbach resonance involves a two-particle collision process between different channels and hence occurs when a bound state in a closed channel is resonantly coupled to the scattering state of an entrance or an open channel [20, 21]. Here the open and the closed channels may correspond to the two different spin configurations of the atom. Further, by simply changing the magnetic field, the energy difference and hence the interaction potential can be controlled over a wide range of values.

Instead of a magnetically induced resonance, a Feshbach resonance can be obtained by inducing an optical transition by detuning the light from the atomic resonance [21]. In a magnetically tuned Feshbach resonance, the s-wave atomic scattering length, a_s, is now a function of the magnetic field, B, as [23],

$$a_s(B) = a_{\text{bg}}\left(1 - \frac{\Delta}{B - B_0}\right). \tag{7.37}$$

Here a_{bg} is the background scattering length which arises as a result of the background collision in the open channel and represents an off-resonant value. B_0 is the resonance field at which scattering length diverges ($a \to \pm\infty$) and Δ is called

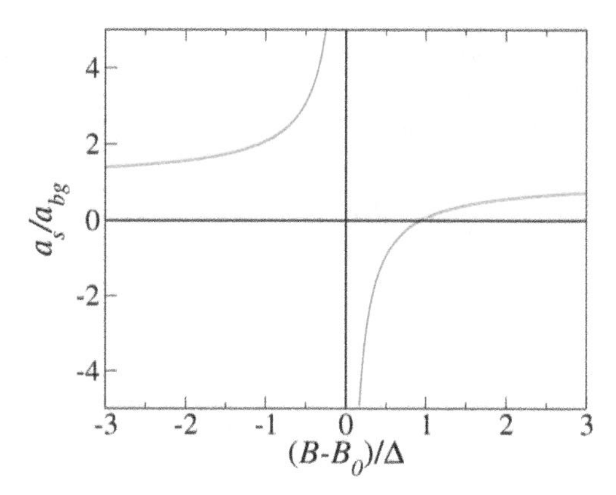

Figure 7.5. The variation of the scattering length, $a_s/a_{\rm bg}$ in a magnetically tuned Feshbach resonance. Figure courtesy of Chin *et al* [21], copyright (2010) by the American Physical Society..

resonance width over which the scattering length changes sign. Figure 7.5 shows a schematic plot of the atomic scattering length as a function of magnetic field, B. Both a and $a_{\rm bg}$ can have positive as well as negative values which implies repulsive and attractive interactions, respectively. Such tunability of the interaction by using Feshbach resonances has resulted in successful observation of BEC [24], physics of the BEC–BCS crossover [25], and the existence of Effimov states [26], etc.

7.5.3 Ultracold atoms on optical lattice and Bose–Hubbard model

Let us review the physics of the Bose–Hubbard model (BHM) which has enriched our understanding of bosons in an interacting environment. In 1963, J Hubbard introduced a simple Hamiltonian, commonly known as Fermi–Hubbard model to describe the electronic structure of strongly correlated systems which also aided in understanding of high-temperature superconductivity to a certain extent. In 1989, Fisher *et al* [27] first extended the idea of a Fermi–Hubbard model for a correlated bosonic system in a lattice, leading to the birth of the Bose–Hubbard model (BHM) to study the Mott insulator (MI) and the superfluid (SF) phases. Later, in 1998, Jacksh *et al* showed that, ultracold atoms cooled and trapped in optical lattices can also be mapped into similar BHM consisting of a hopping term as well as an on-site interaction potential [28].

In 1989 a simple model was proposed by Fisher *et al* [27]. for interacting bosons in a lattice potential. The model is known as BHM and is capable of predicting a quantum phase transition from SF to a Mott insulating phase (MI) at zero temperature [27]. The SF (MI) phase is characterized by a non-zero (zero) SF order parameter and finite (vanishing) compressibility.

The Hamiltonian in its second quantization formulation reads,

$$\mathcal{H} = -t\sum_{\langle i,j\rangle}(b_i^\dagger b_j + \text{h. c. }) + \frac{U}{2}\sum_i \hat{n}_i(\hat{n}_i - 1) + \sum_i(\epsilon_i - \mu)\hat{n}_i \tag{7.38}$$

where the angular bracket indicates i and j as nearest neighbouring sites, $b_i^\dagger$ (b_i) creates (annihilates) a boson on site i, and $\hat{n}_i = b_i^\dagger b_i$, is the local density operator for bosons. The first term is the kinetic energy, describing the hopping with strength t between the neighbouring sites. The second term determines the on-site interaction energy between the bosons. The parameter U measures the strengths of interaction at a lattice site i. In the last term, the chemical potential energy, μ controls the number of particles in the system and ϵ_i is the random on-site energy that may be used to model disorder. At this stage there is no way to ascertain the explicit presence of an optical lattice in the system under consideration, however, the connection can be understood in the following sense (that also paves the way for our dealing with the interacting Bose gas on a lattice).

In 1998, Jaksch *et al* [28] experimentally realized that the dynamics of cold atoms in optical lattices can be described by BHM with an excellent control on the Hamiltonian parameters by tuning the properties of the laser field. The Hamiltonian can be simulated with a great accuracy by loading bosonic atoms in an optical lattice to study SF–MI transition at $T = 0$ by varying the depth of the optical potential [28]. The atoms are thus held together by a trapping potential, V_T along with an optical lattice potential, V_0 thereby confining the atoms to individual lattice sites. As mentioned earlier, the interaction between the ultracold atoms is mostly determined by the s-wave scattering length a_s. The resulting many-body Hamiltonian is given by,

$$\begin{aligned}\mathcal{H} = &\int d^3x \psi^\dagger(x)\left(-\frac{\hbar^2}{2m}\nabla^2 + V_0(x) + V_T(x)\right)\psi(x)\\ &+\frac{1}{2}\frac{4\pi a_s \hbar^2}{m}\int d^3x \psi^\dagger(x)\psi^\dagger(x)\psi(x)\psi(x).\end{aligned} \tag{7.39}$$

Expanding the bosonic field operators ψ and $\psi^\dagger$ in site-localized Wannier basis as,

$$\psi(x) = \sum_i w(x - x_i) b_i \tag{7.40}$$

where $w(x - x_i)$ represents the Wannier orbitals localized at site x_i. Considering the lowest vibrational states (assuming the energy scale involved is sufficient up to the excitation energy of the first band), equation (7.39) reduces to the BMH in equation (7.38), where the Hamiltonian parameters are defined as,

$$t = -\int d^3x \; w^*(x - x_i)\left[-\frac{\hbar^2}{2m}\nabla^2 + V_0(x)\right]w(x - x_j) \tag{7.41}$$

$$U = \frac{4\pi a_s \hbar^2}{m}\int d^3x \; |w(x)|^4. \tag{7.42}$$

In the limit $V_0 \gg E_r$ (E_r is the recoil energy of the atom), where a single band approximation is valid, the hopping parameter, t can be obtained as,

$$t = \frac{4}{\sqrt{\pi}} E_r \left(\frac{V_0}{E_r} \right)^{\frac{3}{4}} e^{-2\sqrt{V_0/E_r}}. \tag{7.43}$$

Also, the form of the on-site interaction parameter, U in terms of V_0 and E_r is given by the equation,

$$U = \frac{8}{\sqrt{\pi}} E_r k a_s \left(\frac{V_0}{E_r} \right)^{\frac{3}{4}} \tag{7.44}$$

a_s being the s-wave scattering length and k is the wave vector of the laser field.

After such a connection being established between the BHM and atoms in an optical lattice, the signature of SF–MI quantum phase transition was observed experimentally by Greiner *et al* [47] in 2002 in a gas of ultracold 87^{Rb} atoms in a three-dimensional optical lattice. This observation of the quantum phase transition from an SF to a MI phase has offered a first glimpse of the plethora of physical phenomena that were awaiting. Later on, the SF–MI phase transition was found also in 1D [48] and 2D [49]. It must be realized at this stage that engineering such a phenomenon in crystal lattices (where electrons are the key players) is extremely difficult owing to the lack of control over the parameters. Thereafter, the theoretical studies of cold atomic gases have dramatically increased. These include the theoretical studies of disordered bosonic system [29] multi-component Bose mixture [30] Bose-Fermi mixture [31]. Besides, a large number of analytical and computational methods are proposed to study the BHM such as strong coupling expansion, projection method, mean field theories (MFTs), Gutzwiller approach, quantum Monte Carlo (QMC) method, etc.

Recent progress in the experiments with cold chromium atoms [32] suggested that an extension to the BHM is experimentally relevant. Unlike the alkali atoms, which have been used in most of the optical lattice experiments so far, ^{52}Cr atoms have a large dipole moment which leads to long-range dipolar interactions. This makes ^{52}Cr atoms in an optical lattice a potentially ideal system for experimentally exploring the phase diagram of the BHM with long-range interactions, known as the extended Bose–Hubbard model (EBHM). To understand the effect of longer range repulsive interactions, $\mathcal{H}_{\mathrm{LR}}$ among the atoms, an inter-site potential term of the form,

$$\mathcal{H}_{\mathrm{LR}} = V \sum_{\langle i,j \rangle} n_i n_j$$

is added to the Hamiltonian in equation (7.38) that takes into account the interaction between nearest neighbours, a scenario that is capable of giving rise to a rich quantum phase diagram, which should be experimentally accessible. A large number of theoretical studies have been carried out to investigate the possibility of different exotic phases in the EBHM.

7.6 Various aspects of ultracold atoms in optical lattices

7.6.1 Disorder optical potential

Although the correlated bosons described by the BHM successfully demonstrate the presence of an incompressible MI and a compressible SF phase in a disorder-free scenario, some kinds of impurities, such as disorder, defects, etc are always present in an interacting system. Thus, understanding the interplay between disorder and interaction still remains a focused topic from theoretical as well as experimental point of view and is expected to show a plethora of physics at the boundary between a 'clean' and 'dirty' bosons problem.

Motivated by the experimental observation of the superfludity in a weakly interacting ^{4}He absorbed in a porous Vycor glass [33], the effects of a random potential, 'disorder' were first studied in a correlated Bose system by Giamarchi *et al* [34] and Fisher *et al* [27]. They found that apart from the usual SF and MI phases in a clean state, an additional phase, known as a Bose glass (BG) phase, which is compressible, but gapless yet insulating in nature appears due to the boson localization in presence of disorder. They showed that a BG phase is sandwiched between SF and MI phases, that is, a direct transition from the MI to the SF phase is interrupted by the presence of such a glassy phase, however, a direct transition is not fundamentally impossible.

While the theoretical understanding of the location and appearance of the BG phase still remains a contradictory subject, the possible scenarios at which the BG phase intervenes between the SF and the MI phases are the following: (i) a BG phase appears in between the MI and the SF phases (figure 7.6 (top panel)); (ii) a BG phase appears between the MI and the SF phases except at the tip of the MI lobes where a direct transition from an MI to an SF phase is possible; (iii) a BG phase appears only in between the MI lobes and a direct transition from MI to SF phases is possible. Several numerical studies, such as, QMC and renormalization group (RG) [35, 36] show that for a weakly disordered system, the disorder is insufficient to introduce a BG phase and thus paves the way for a direct transition from the MI to the SF phase.

Recently, Pollet *et al* gave rigorous proof which forbids possible direct transition from the MI to the SF phase in scenarios (ii) and (iii) and pointed out that for a generic bounded disorder, a BG phase always intervenes in between the MI and the SF phases (scenario (i)) [37]. They formalized a theorem (Theorem 1) which was introduced earlier in [27] and further postulated another theorem, known as, *theorem of inclusions* in support of scenario (i). Theorem 1 states that when the critical strength of a bounded disorder (Δ_c) is larger than half of the width of the energy gap ($E_g/2$) in the MI phase then the transition inevitably occurs in a compressible and gapless phase, that is, the BG phase. This theorem is based on the fact that for an infinite system, the chemical potential is homogeneously shifted upwards or downwards by an amount Δ since there exists an arbitrarily large 'Lifshitz' region where no energy gap exists for particle transfer in that region [37]. The '*theorem of inclusions*' says that for a generic bounded disorder, there exists a rare but arbitrary large region on either side of a transition line. This implies that if

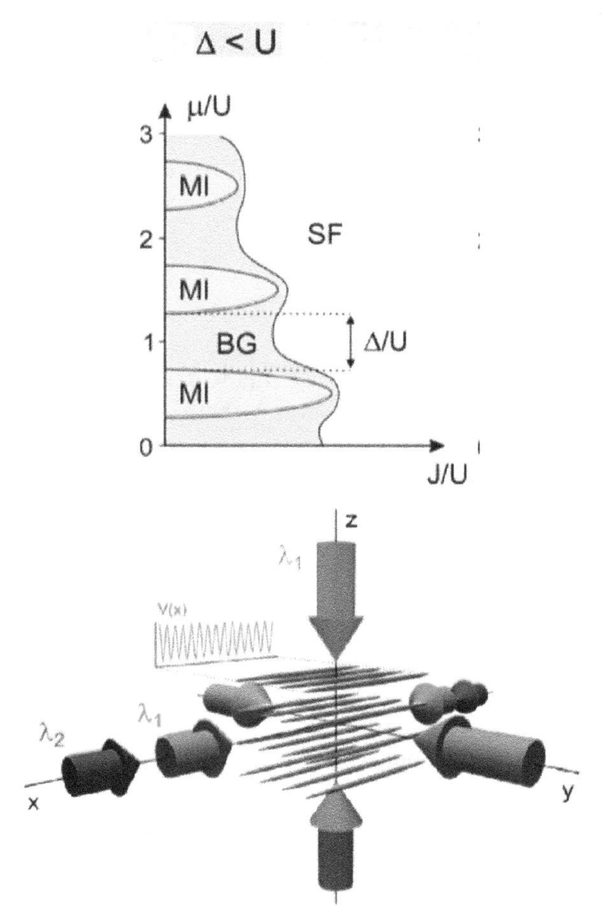

Figure 7.6. The appearances of a BG phase in between MI and SF phases in scenario (i) in the presence of disorder, Δ (top). A disordered optical lattice potential is created using an auxiliary laser beam with slightly different wavelength along with a primary laser beam to create an optical lattice (main) potential (bottom). Figure courtesy of Fallani *et al* [38], copyright (2007) by the American Physical Society.

one phase across the transition line is gapless, then the other phase across the transition line is automatically gapless because of arbitrarily large rare regions which locally look exactly the same as the previous phase [37].

Study of the BG phase in the context of ultracold Bose gas in optical lattices has attracted much attention due to the ability to introduce disorder in a controlled manner. Although an optical lattice is free from disorder or defects, one can easily create a disordered optical lattice potential by several methods, such as using speckle laser beams, two-colour superlattices, etc.

Recently Fallani *et al* designed an experiment to create a disordered optical lattice potential using different laser beams and thus observed a transition from MI to BG phase by loading the ^{87}Rb atoms in optical lattices [38]. A primary laser beam of wavelength, λ_1, was used to create an optical lattice (main) along with an auxiliary laser beam with slightly different wavelength, λ_2 ($<\lambda_1$), to introduce the disorder and thus controlling the auxiliary laser beam, a transition from weak to strong disorder strength is possible (figure 7.6 (bottom)). They initially prepared the system in the MI phase by turning off the disorder potential ($\lambda_2 = 0$) and measured the excitation spectra in different phases via time of flight (TOF) at different disorder strengths. The peaks in the excitation spectra show that the system is in MI phase in the clean state but with increasing disorder potential (λ_2), the disappearances of such peaks in the excitation spectra indicate a transition to the Bose glass phase.

7.6.2 Synthetic magnetic field

Ultracold atoms suffer a serious drawback due to the charge neutrality despite the fact of their unchallenging ability in tuning the Hamiltonian parameters which simulates the quantum behaviour. Unlike a charge particle which rotates around a field axis, it is quite difficult to know the true response of an ultracold Bose gas in the presence of a magnetic field because of their charge neutrality.

One possible way to overcome such a limitation is to put these neutral atoms in rotational motion so that they can mimic the effects of a charge particle. They use the equivalence between the Lorentz force experienced by a charge particle and the Coriolis force on the neutral atoms due to rotation. Thus the neutral atoms in rotation generate an artificial gauge field, known as synthetic magnetic vector potential ($\mathbf{A}^*$) which was observed by the appearances of quantized vortices in BEC [39, 40] (figure 7.7 (top)). However large synthetic vector potential requires large rotation of the quantum gases which may cause the atoms to fly apart.

Later, Lin *et al* engineered a large synthetic magnetic field $\mathbf{B}^*=\nabla \times \mathbf{A}^*$ by introducing a Berry phase which arises due to the coupling between the different hyperfine levels of an atom and does not depend explicitly on the rotational motion [41]. For that purpose, they created a spatially varying synthetic vector potential by illuminating ^{87}Rb BEC ($F = 1$) by a pair of Raman laser beams with momentum difference and a real magnetic field which varies in a single direction. At first the BEC was formed in a uniform motion at the Raman resonance and then by tuning the laser beams and applying a magnetic field gradient (b'), a non-uniform motion is created. Such non-uniform motion where atoms swirl in a whirlpool-like motion generates quantized vortices which are the signature of the synthetic vector potential. The number of vortices which determine the strength of the synthetic vector potential depends upon the detuning gradient ($\delta \propto b'$) and it is thus possible to create large vector potential without the restriction of large rotation [41] (figure 7.7 (bottom)). After the observation of vortices in the BEC, study of Abrikosov vortex lattice which shows a structural transition due to the pinning effects and observation of Dirac mono-pole, etc are now of great importance.

7.6.3 Dipole–dipole interaction

For correlated Bose systems, the interaction between the constituent particles, which can be either short or long range, plays an important role in determining the ground state properties of the system. Although most of the BHM have been analysed by

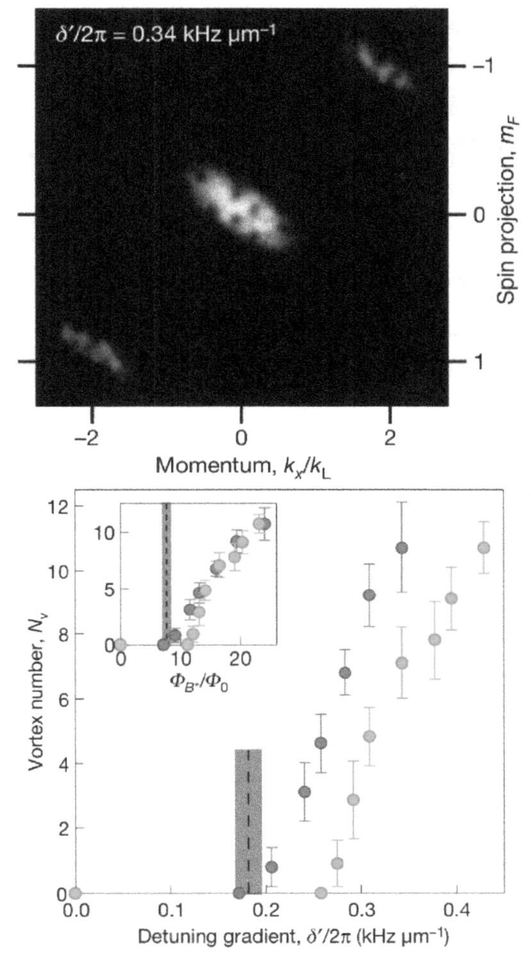

Figure 7.7. Dressed BEC image of three hyperfine states of ^{87}Rb atoms after a TOF of 25.1 ms in a crossed optical dipole trap with a magnetic field, $B = (B_0 - b'y)\hat{y}$, that is, with a detuning gradient, $\delta' \propto b'$ (top). Vortex number, N_v versus δ' at Raman coupling $\hbar\Omega_R = 5.85E_L$ (blue circles) and $\hbar\Omega_R = 8.20E_L$ (red circles) (bottom). The inset displays N_v versus the synthetic magnetic flux, ϕ_{B^*}/ϕ_0. Figure courtesy of Lin *et al* [41] adapted by permission from Nature: Springer (Springer Nature), copyright (2009).

taking care the effects of two-body short-range interaction which certainly raises the concern to revisit such quantum phase properties by considering the long-range interaction such as a dipole–dipole interaction (DDI). In contrast to the *s*-wave-dominated two-body pseudopotential which is isotropic in nature, the DDI is highly anisotropic and has direct consequences on the stability of the BEC, that is, it leads to the *d*-wave collapse in the dipolar condensates.

Consider two dipoles that are separated by a relative distance r and their dipole moments are pointed along two different directions with the unit vectors $\mathbf{e}_i$ then the interaction potential of the two dipoles is given by [42]

$$U_{\mathrm{DDI}}(\mathbf{r}) = \frac{C_{\mathrm{DDI}}}{4\pi}\frac{(\mathbf{e}_1.\ \mathbf{e}_2)r^2 - 3(\mathbf{e}_1.\ \mathbf{r})(\mathbf{e}_2.\ \mathbf{r})}{r^5} \tag{7.45}$$

where C_{DDI} is the strength of the DDI which is $C_{\mathrm{DDI}} = \mu_0\mu_m^2$ for particles with magnetic dipole moments, μ_m and $C_{\mathrm{DDI}} = d^2\epsilon_0$ for electric dipole moments, d. Such dependency of $1/r^3$ shows that the DDI is a long-range interaction and the anisotropy which depends upon the relative angles between two dipoles is now responsible for the phase shift where all of the partial wave contributes to the scattering phenomena [42] (figure 7.8).

It was found that for polarized molecules, the electric nature of the DDI is dominant, while for spinor particles, the DDI is essentially magnetic origin. The strengths of magnetic DDI of few spinor particles are listed in table 7.1 and till now the maximum value of the experimentally observed DDI corresponds to the ^{52}Cr atoms which has a hyperfine spin $F = 3$ [32]. The effects of such long-range interaction in the context of the BHM can be studied by considering the off-site interaction term to be the interaction between the density imbalanced nearest neighbour sites which is often referred to as an extended interaction potential [43]. Thus the inclusion of such extended interaction in the ultracold Bose gas shows a plethora of interesting phases such as solid as well as crystalline along with the usual quantum phases of the system.

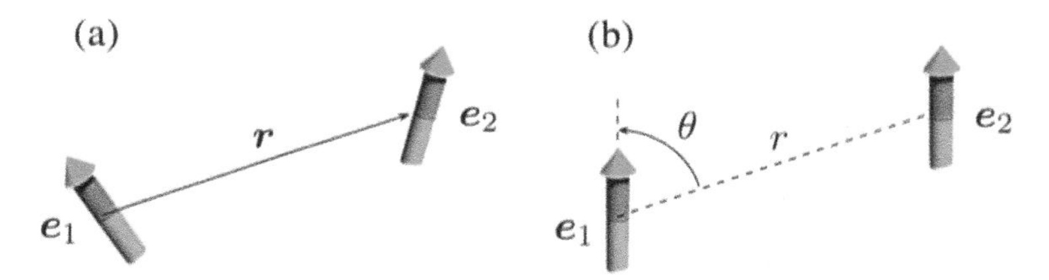

Figure 7.8. Two particles interacting via dipole–dipole interaction for non-polarized atoms in (a) and polarized atoms in (b). r and θ are the relative distance and angle between the two dipoles. Figure courtesy of Lahaye *et al* [42], copyright IOP Publishing. Reproduced with permission. All rights reserved..

Table 7.1. Magnetic dipole moment of various species where μ_{B}: Bohr magneton. Source: Lahaye *et al* [42].

Species	I	F	Dipole moment
^{7}Li	3/2	1	$0.94\mu_{\mathrm{B}}$
^{23}Na	3/2	1	$0.91\mu_{\mathrm{B}}$
^{39}K	3/2	1	$0.95\mu_{\mathrm{B}}$
^{87}Rb	3/2	1	$0.73\mu_{\mathrm{B}}$
^{52}Cr	0	3	$6\mu_{\mathrm{B}}$

7.6.4 Bose glass phase

Motivated by the experimental results on superfluidity in ^{4}He on Vicor glass [33], Fisher *et al* [27] discussed the behaviour of bosons with short-range interactions moving in random external potentials. Unlike the SF and MI phase for pure system, a BG phase intervening to SF and MI phases, was predicted in presence of disorder [37]. The BG phase is characterized by a gap in the particle–hole excitations, however, finite compressibility, but the system is still an insulator because of the localization effect of the random potential [27]. Since then, a large number of analytical, computational and experimental studies have been carried out for finding the correct answer about the location and nature of the BG phase and in many cases reaching contradictory conclusions.

The emergence of the BG phase along with the SF and the MI phases described by the disordered BHM in its parameter space can be understood as follows. When the interaction is sufficiently strong and the total number of bosons is an integral multiple of the number of lattice sites, the model should describe a Mott insulator. In this phase, the bosons are strongly localized in the potential wells of the optical lattice. When both the interaction and the disorder strengths are weak, the model should describe an SF phase. In the presence of disorder, depending on the system parameters, the BG phase appears and can be described either by localized single-particle states or as an isolated SF region. The BG does not allow phase coherence to extend over the entire lattice, and as mentioned before, is characterized by a finite compressibility and gapless particle–hole excitations.

7.6.5 Methods of solution of the BHM

This section is mainly to familiarize readers about different techniques that are employed to solve the BHM under controlled approximations. The techniques involve a moderate to advanced level of numerical expertise and are extensively used in the research on ultracold atoms. We briefly review a few of the techniques, however, it is by no means complete, and would require a significant amount of further work to be able to apply them to study interacting bosons in an optical lattice.

The techniques that are applied to study the BHM (or equivalent model, such as model in the presence of disorder, long-range interaction, trapping potential, etc) have included a large number of approaches. For example exact diagonalization (ED) method [28], MFT [44], QMC schemes [45], etc.

In this section, we only review the most common mean field technique, namely a single-site MFT that is used to solve and find the ground state of the BHM for a minimal characterization of the system by evaluating certain correlation functions. A site decoupled mean field scheme based on the decoupling of the hopping term succeeds in providing an SF–MI phase transition. The method was first proposed by Sheshadri *et al* [44] to study a homogeneous system. Later it was extended to the inhomogeneous cases [46].

7.6.6 Single-site MFT

The method consists of decoupling the kinetic energy operator which can be shown as,

$$b_i^\dagger b_j \rightarrow \left\langle b_i^\dagger \right\rangle b_j + b_i^\dagger \left\langle b_j \right\rangle - \left\langle b_i^\dagger \right\rangle \left\langle b_j \right\rangle = \psi_i b_j + \psi_j b_i^\dagger - \psi_i \psi_j \tag{7.46}$$

where $\psi_i = \langle b_i^\dagger \rangle = \langle b_i \rangle$ is the order parameter which for the number conserving case, is real. In the single-site mean field techniques, the couplings between the sites are completely replaced by average fields as shown in figure 7.9(a). Now the Hamiltonian in equation (7.38) can be written as a sum of single-site Hamiltonian as,

$$H = \sum_i \mathcal{H}_i^{\mathrm{MF}}(\psi_i, \phi_i) \tag{7.47}$$

where $\phi_i = \sum_{j=1}^{z} \psi_j$, z being the coordination number of the lattice. $\mathcal{H}_i^{\mathrm{MF}}$ is hence given by,

$$\mathcal{H}_i^{\mathrm{MF}}(\psi_i, \phi_i) = \left[\frac{U}{2}\hat{n}_i(\hat{n}_i - 1) - \mu\hat{n}_i\right] - zt\left[\phi_i(b_i + b_i^\dagger) + \psi_i\phi_i\right]. \tag{7.48}$$

For a homogeneous system, the observables are independent of the lattice sites and we can set $\phi_i = \psi_i = \psi$. Hence the Hamiltonian becomes site independent and can be written as,

$$\mathcal{H}^{\mathrm{MF}} = \left[\frac{U}{2}\hat{n}(\hat{n} - 1) - \mu\hat{n}\right] - \psi(b + b^\dagger) + \psi^2. \tag{7.49}$$

In the above equation the interaction parameter, U and the chemical potential, μ are scaled by the energy zt. The matrix representation of the above Hamiltonian can be obtained in the occupation number basis, $|n\rangle$ (truncated at some maximum value $n = n_{\max}$) by computing the matrix elements as

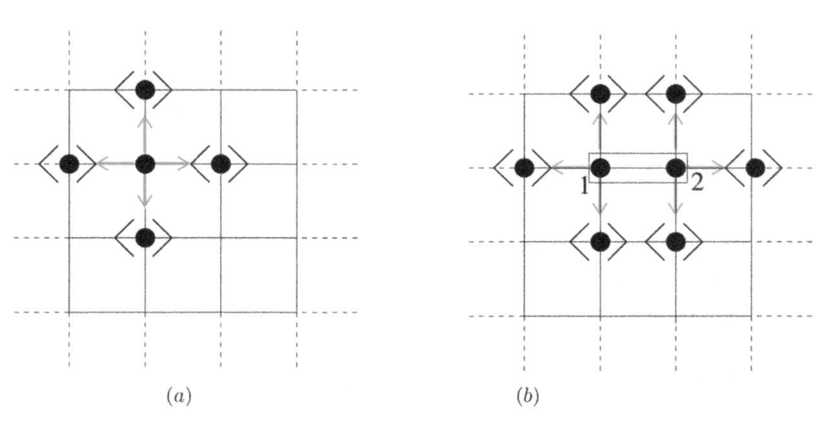

Figure 7.9. The decoupling scheme of a single site in (*a*) and for a cluster of two site in (*b*) is presented. The angular brackets are used to indicate average field of the bosonic operators.

$$\langle n'|\mathcal{H}^{\mathrm{MF}}|n\rangle = \left[\psi^2 + \frac{U}{2}n(n-1) - \mu n\right] \delta_{n,n'} - \left[\psi(\sqrt{n+1}\,\delta_{n+1,n'} + \sqrt{n}\,\delta_{n-1,n'})\right] \tag{7.50}$$

where the operators $b_i^\dagger$ and b_i are the raising and lowering operators of the quantum harmonic oscillator. Thus,

$$\begin{aligned} b_i^\dagger|n\rangle &= \sqrt{n+1}\,|n+1\rangle \\ b_i|n\rangle &= \sqrt{n}\,|n-1\rangle. \end{aligned} \tag{7.51}$$

Suppose $n_{\max} = 3$, then the order of the matrix becomes 4×4 and the matrix representation of the Hamiltonian is,

$$\begin{bmatrix} \psi^2 & -\psi & 0 & 0 \\ -\psi & \psi^2 - \mu & -\sqrt{2}\psi & 0 \\ 0 & -\sqrt{2}\psi & \psi^2 - 2\mu + U & -\sqrt{3}\psi \\ 0 & 0 & -\sqrt{3}\psi & \psi^2 - 3\mu + 3U \end{bmatrix}.$$

The eigenvalues and eigenstates for a homogeneous system can be found by diagonalizing the above matrix. The order parameter now needs to be obtained self-consistently as, $\psi = \langle b\rangle$ ($= \langle b^\dagger\rangle$ for a real ψ). In figure 7.10, the ground state energy (E_g) variation as function of $n_{\max}$ is shown which yields the following. When $n_{\max} < 6$, E_g depends on the choice of $n_{\max}$. So to attain a stable ground state, the value of $n_{\max}$ should be at least 6. The results on the phase diagram are shown for $n_{\max} = 8$.

For an inhomogeneous system (which can result from disorder, harmonic potential), the order parameter no longer remains site independent. Hence the Hamiltonian at each site will have to be treated individually and since each site is connected to z neighbours, a simultaneous self-consistent solution of the problem at all sites is necessary.

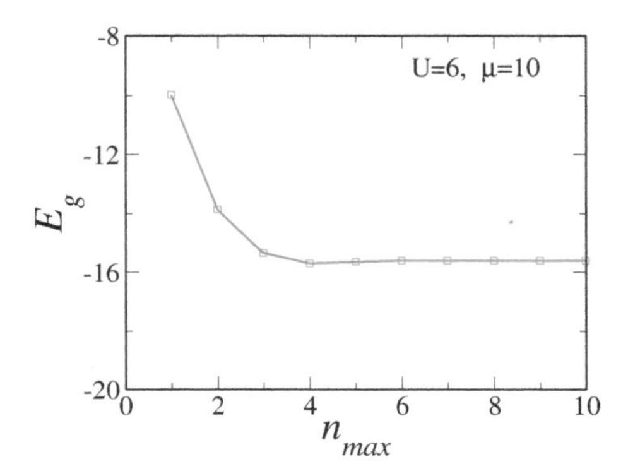

Figure 7.10. The plot shows the variation of ground state energy, E_g as a function of truncated occupation number, $n_{\max}$. For $n_{\max} < 6$, E_g depends on $n_{\max}$. But after those values, E_g is independent of the choice of $n_{\max}$.

7.6.7 Superfluid–Mott insulator (SF–MI) transition

As emphasized in the preceding discussion, at zero temperature, bosons in a periodic potential exhibit two types of phases: an SF phase and an MI phase [27]. In the context of the BHM, let us see what it means. The model is characterized mainly by two energies. One is the tunnelling energy, t which determines the probability amplitude of an atom to hop from a lattice site to another neighbouring site and the on-site interaction energy, U, which is the interaction energy between the atoms located in the same site. Depending upon the strengths of these parameters, the physics described by the BHM can be divided into two separate regimes. One is the kinetic energy dominated regime, when the on-site interaction is small compared to the hopping amplitude, where the ground state is an SF as it tries to delocalize the bosons over the entire lattice. In the opposite limit, where the on-site interaction supersedes the tunnelling energy, the ground state is in the MI phase in which the bosons are localized and each lattice site has an integer occupation density.

When the kinetic energy term dominates the Hamiltonian, that is, when the tunnelling amplitude is sufficient to delocalized the bosons, quantum correlations can be ignored and the system can be described by a macroscopic wavefunction (since the many-body state is a product over single identical particle wavefunctions). There is a macroscopic well-defined phase and the system is an SF. This is because when the atoms are delocalized over the lattice with equal relative phases between the adjacent sites, they exhibit an interference pattern from an array of phase coherent matter wave sources. As interaction increases, the average kinetic energy required for an atom to hop from one site to the next becomes insufficient to overcome the potential energy cost. Thus atoms tend to get localized at the individual lattice sites and subsequently the fluctuations in the number density are suppressed. Hence in the MI phase, the ground state of the system, instead consists of localized atomic wavefunctions with a fixed number of atoms per site. The lowest lying excitations that conserve the particle number (namely the $U(1)$ symmetry) are the particle–hole excitations. This phase is characterized by the existence of an energy gap. The gap is determined by the energy necessary to create one such particle–hole pair.

In experiments, the SF–MI transition was observed by Greiner *et al* [47] by loading ^{87}Rb atoms from a Bose–Einstein condensate into a three-dimensional optical lattice potential. The condensate is an SF, where the bosons can tunnel from one lattice site to the next. When the lattice potential is turned on smoothly, the system remains in the SF phase, until the interaction between the atoms are small compared to tunnelling strength. If the lattice potential depth is increased, the repulsive interaction between the atoms becomes large compared to the tunnelling term, which signals the onset of the SF–MI transition. In the MI phase, the total energy is minimized when each lattice site is occupied by a fixed number of bosons. Hence the fluctuation in the number of atoms reduces on each site, leading to enhanced fluctuation in the phase of the wavefunction[5]. Thus the phase coherence is

[5] This is due to Heisenberg's uncertainty relation $\Delta n \Delta \phi \simeq \frac{\hbar}{2}$.

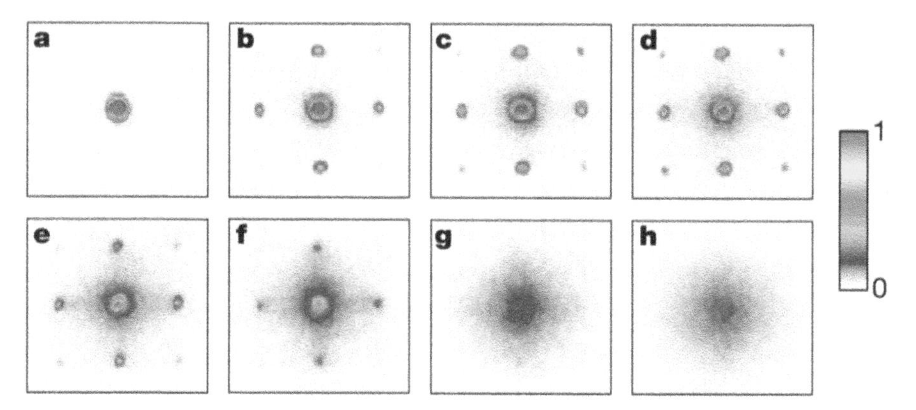

Figure 7.11. Figure from reference [11] of absorption images of multiple matter wave interference patterns of ultracold atoms released from an optical lattice across the superfluid to Mott insulator transition. As one goes from a to h, the depth of the lattice potential is increased. The time interval of imaging is fixed at $\tau = 15ms$. The SF phase yields to sharp interference peaks, the vanishing of pattern indicating transition to MI phase (g and h, etc). Reprinted/adapted by permission from Nature: Springer Nature [11], copyright (2002).

lost. These features were observed in the experiment by recording absorption images after suddenly releasing the atoms from the lattice potential at an interval of time $\tau = 15$ms. The respective images corresponding to the different values of the lattice potential depth are presented in figure 7.11. The absorption images confirm the signature of the SF–MI phase transition. When the depth of the potential is relatively small, the atoms have a reasonable kinetic energy. Hence the system is in SF state which signals the emergence of coherent matter waves during the expansion, and appears as Bragg-like interference pattern provided by the absorption images (figure 7.11). With increasing the potential depth, the sharp peaks in the absorption pattern vanish, indicating the transition to an MI phase. Further, the critical ratio $(U/t)_c$ (see figure 7.12) at which SF–MI transition occurs obtained from the experiment is in good agreement with the theoretical prediction.

7.6.8 Limitations of MFT

It may be noted that the agreement between the MFT and the other more refined techniques is only qualitative in one dimension, but it becomes progressively better in two and three dimensions. It should be stressed that the MFT results agree with those obtained from other techniques in most parts of the boundary that separates the compressible (SF) and incompressible (MI) phases, however, the agreement becomes poor at the tip of the Mott lobes which is the most sensitive region. This can be understood as follows. The fluctuation in the order parameter is ignored by MFT and hence it yields a continuous phase boundary with smooth boundaries separating the SF–MI phases, while more refined methods, such as a field theoretic method and a QMC may miss a real solution at certain discrete values of the chemical potential near the tip of the lobe.

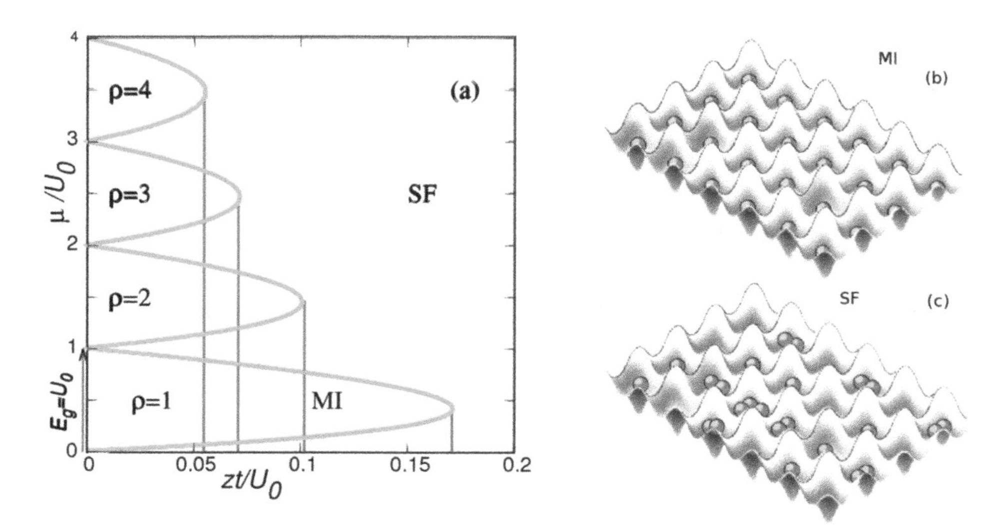

Figure 7.12. The zero temperature phase diagram of a homogeneous BHM in (a). The lobes indicate MI phase with occupation densities, ρ. The vertical line corresponds to the critical tunnelling strength, zT_c/U_0 from the MI to SF phase. Pictorial representations of the occupation densities in the MI lobe ($\rho = 1$) in (b) and in the SF phase in (c) of bosons on an optical lattice. Figure courtesy of M Greiner [50].

7.6.9 Optical dipole trap (ODT)

A pre-requisite of the discussion of spinor gases is a discussion of optical dipole trap. A neutral atom can be cooled and trapped using various mechanisms such as a magneto-optical trap (MOT) [18, 50] and purely optical dipole trap (ODT) [50] techniques. In MOT, ultracold atoms are exposed in a simple Helmholtz coil arrangement where the applied magnetic field in opposite direction orients the ultracold atoms in a single magnetic sublevel despite the presence of different hyperfine states, m_F. To unlock all the hyperfine states, an optical dipole trap can be employed which works on the principle of energy shifts induced by the radiation on the atomic internal energy levels. In ODT, the interaction between the electric field of a laser beam with the induced dipole moment on a neutral atom paves the way in retaining the hyperfine degrees of freedom and makes them a spinor Bose gas. In order to see how a dipole force arises, consider a spatially varying electric field of a laser beam which oscillates with frequency ω_l of the form,

$$\mathbf{E}(\mathbf{r}, t) = 2E_0 \cos(\mathbf{k} \cdot \mathbf{r} - \omega_l t) \tag{7.52}$$

and the induced dipole moment on the atom at a position $\mathbf{r}$ is $\mathbf{d} = -e\sum \mathbf{r}$ which oscillates with the same frequency of the laser field. Thus the Hamiltonian for the atom–light interaction is given by,

$$H_{AL}(t) = -\mathbf{d} \cdot \mathbf{E}(\mathbf{r}, t). \tag{7.53}$$

Using atom–light Hamiltonian, the energy shift after including the second-order correction is given by,

$$\Delta E(\mathbf{r}) = -\frac{1}{2}\alpha_{m_{\mathrm{F}}}(\omega_l)\overline{\mathbf{E}(\mathbf{r},\, t)^2} \tag{7.54}$$

where $\overline{\mathbf{E}(\mathbf{r},\, t)}$ denotes the time average of the electric field and $I(\mathbf{r}) = \overline{\mathbf{E}(\mathbf{r},\, t)^2}$ is the intensity of the laser field. $\alpha_{m_{\mathrm{F}}}(\omega_l)$ is the atomic polarizability which is given by,

$$\alpha_{m_{\mathrm{F}}}(\omega_l) = \sum_{\gamma} |\langle \gamma|\mathbf{d}\cdot\mathbf{e}|g\rangle|^2 \left(\frac{1}{E_\gamma - E_g + \hbar\omega_l} + \frac{1}{E_\gamma - E_g - \hbar\omega_l}\right). \tag{7.55}$$

Here $\mathbf{e}$ is the unit vector in the direction of the electric field and $|g\rangle = |F,\, m_{\mathrm{F}}\rangle$ is the ground state of the atom and the summation is taken over all the excited states, $|\gamma\rangle$. For a far off resonance laser light, the polarizability, $\alpha_{m_{\mathrm{F}}}(\omega_l)$ takes the form,

$$\alpha_{m_{\mathrm{F}}}(\omega_l) \simeq -\frac{|\langle \gamma|\mathbf{d}\cdot\mathbf{e}|g\rangle|^2}{\hbar\Delta} \tag{7.56}$$

where Δ is the detuning parameter from the resonance, $\hbar\Delta = \hbar\omega_l - (E_\gamma - E_g)$. Thus the effective dipole potential and hence the force experienced by an atom are given by,

$$\begin{aligned} U_{\text{dipole}}(\mathbf{r}) &= -\frac{1}{2}\alpha_{m_{\mathrm{F}}}(\omega_l)\overline{\mathbf{E}(\mathbf{r},\, t)^2} \propto \frac{I(\mathbf{r})}{\Delta} \\ \mathbf{F}_{\text{dipole}}(\mathbf{r}) &= -\nabla U_{\text{dipole}}(\mathbf{r}). \end{aligned} \tag{7.57}$$

For a red detuned light, $\Delta < 0$ ($\omega_l < \omega_0$), the atoms are attracted towards potential minima, that is, the regions of maximum field intensity, $I(\mathbf{r})$, while for a blue detuned light, $\Delta > 0$ ($\omega_l > \omega_0$), the atoms are attracted towards potential maxima, that is, the regions of minimum field intensity, $I(\mathbf{r})$ [50].

For a system with hyperfine spin, $F = 1$, the detuning parameter, Δ involves different nS to nP transition where $\Delta = \omega_l - \omega_{PS}$ and the atomic polarizability, $\alpha_{m_{\mathrm{F}}}(\omega_l)$ can be approximated as [52],

$$\alpha_{m_{\mathrm{F}}}(\omega_l) = -\frac{1}{\omega_{PS}^2 - \omega_l^2}\left(\sum_{\gamma\in P} |ed_{\gamma,g}|^2\omega_{\gamma,g}\right) + O\left(\frac{1}{\Delta}\right) \tag{7.58}$$

where $ed_{\gamma,g}$ are the dipole matrix elements between the ground state $|g\rangle$ and the excited state $|\gamma\rangle$.

7.6.10 Spin-1 Bose gas: an era of quantum magnetism

Although the commonly used MOT during the early days of experiment successfully show the BEC as well as the quantum phase transition (QPT) of ultracold atoms in optical lattices. But such a trapping technique comes at the cost of dropping most of the rich phase properties due to the suppression of the internal degrees of freedom. In contrast, experiments involving the optical dipole trap (ODT), which purely uses an electric field instead of a magnetic field, have opened up a new window of opportunity in studying the quantum degenerate Bose gases with atomic hyperfine (that is, nuclear) spin states and hence treat them as spinor Bose gases.

Remarkably, a spin-1 ($F = 1$) condensation [51] was experimentally observed using ^{23}Na atoms where all three hyperfine states are populated by transferring the spin-0 BEC into a far off-resonant optical trap (figure 7.13), initiated following theoretical studies by Ho [52] and Machida [53] in the low energy limit. Later, spinor condensates of ^{87}Rb atoms corresponding to hyperfine spin state $F = 1$ as well as $F = 2$ and ^{23}Na atoms with $F = 2$ were observed and it was found that the higher spin manifolds ($F = 2$) of ^{23}Na atoms are very much unstable compared to those for ^{87}Rb atoms.

The vector nature of the order parameter of a spin-1 condensate which is invariant under the spin-gauge rotation provides new insights in exploring various spin textures, and quantum magnetization properties present therein [52, 53]. The application of an external magnetic field which breaks the rotational symmetry of the system shows a transition between easy-axis and easy-plane ferromagnetism. The transition was observed by applying an *in situ* magnetization of a ^{87}Rb ($F = 1$) spinor condensates (figure 7.14).

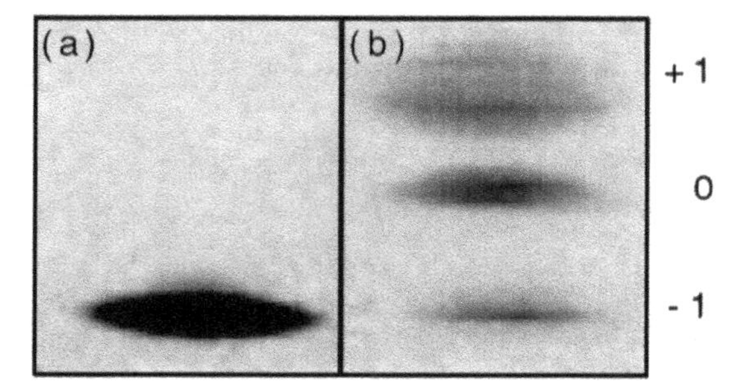

Figure 7.13. Optical trapping of a spin-1 ($F = 1$) BEC using ^{23}Na atom where all three hyperfine states are populated. Absorption images are captured after 250 ms in (a) where atoms remains polarized. After 340ms of optical confinement in (b), hyperfine states are separated by a magnetic field gradient of pulse duration 40 ms. Reprinted with permission from [51], copyright (1998) American Physical Society.

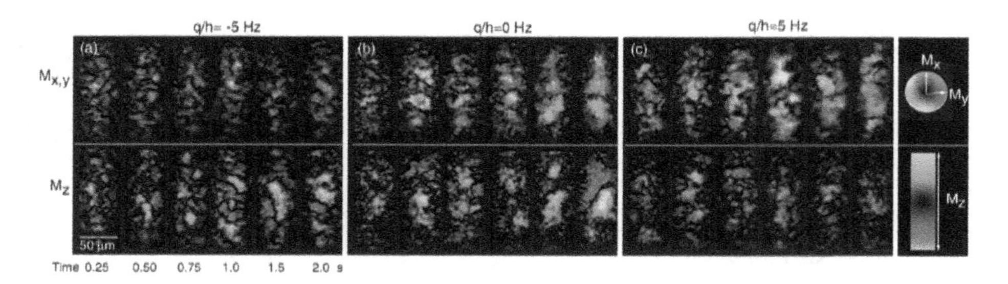

Figure 7.14. Transverse (top) and longitudinal (bottom) magnetization of a $F = 1$ ^{87}Rb spinor Bose gas that was prepared initially in a non-degenerate incoherent (1/3, 1/3, 1/3) population mixture of the Zeeman states and then cooled to quantum degeneracy. The degenerate spin texture evolves for a variable time at a quadratic shift of (a) $q/h = -5$, (b) $q/h = 0$ and (c) $q/h = 5$ before column density of the vector magnetization is measured. Reprinted with permission from [54], copyright (2013) American Physical Society.

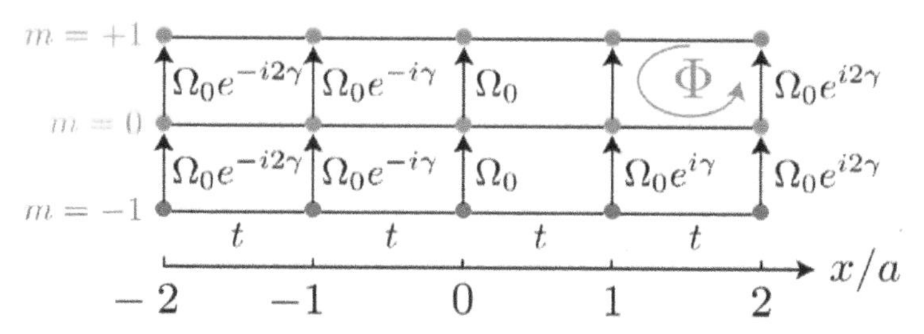

Figure 7.15. A synthetic dimension is created by illuminating a $F = 1$ ^{87}Rb spinor Bose gas in a pair of Raman laser beam of wavelength λ_R. A synthetic 2D lattice with magnetic flux $\Phi = \gamma/2\pi$ per plaquette ($\gamma = 2k_R a$, a: lattice spacing). Reprinted with permission from [55], copyright (2014) American Physical Society.

Apart from such intriguing aspects of spinor condensates, the ground state of a spin-1 Bose gas in optical lattices is determined completely by the magnetic sublevels. Recently, it was shown that the different hyperfine spin states can be coupled sequentially to form an extra dimension, known as a synthetic dimension, similar to the spatial dimension of an optical lattice (figure 7.15). For ^{87}Rb ($F = 1$) atoms, a Raman transition which splits the three spin states $m_F = \pm 1, 0$, thus providing the synthetic dimension, can be used along with an optical lattice dimension to create a synthetic magnetic flux due to Peierl's coupling. Moreover, a synthetic dimension can be used to generate high synthetic magnetic field of the order of unit flux per plaquette to overcome the rotational limitation of the neutral atoms in optical lattices.

7.6.11 A comparison between spin-0 (spinless) and spin-1 Bose gases

The above section shows that the spin-1 Bose gas exhibits an edge in the direction of quantum magnetism compared to spin-0 Bose gas due to its spin degrees of freedom. Particularly, the spin dependent interaction plays an important role in determining different kinds of spontaneous symmetry breaking of the ground state which in turn directly affects the properties of the MI and SF phases. Thus a comparison between the spin-0 and spin-1 Bose gases with regard to different quantities will help us to get a comprehensive overview on these two systems which are listed in the following table 7.2. We have included a few properties of a spin-1 Bose gas here but this table does not include the complete list of phase properties that a spinor Bose gas exhibits.

7.6.12 Phase diagrams

The zero temperature phase diagrams computed based on the self-consistent values of ψ and ρ in the homogeneous case for both values of the spin dependent interactions are shown in figure 7.16. Although the two phase diagrams look similar but the major intriguing aspects appear in the AF case.

We discuss various phases for some representative values of the interaction parameters. For example, corresponding to $U_2/U_0 = 0.05$ (figure 7.16(a)), the even MI lobes ($\rho = 2, 4, \ldots$) expand by encroaching into the SF phase compared to the

Table 7.2. A comparison between the spin-0 and spin-1 Bose gases.

Parameters	Spin-0	Spin-1
Interaction potential	U_0	$U_0 \propto a_0 + 2a_2$ $U_2 \propto a_2 - a_0$
Nature of interaction		Antiferromagnetic (AF): $U_2/U_0 > 0$
Ferromagnetic: $U_2/U_0 \leqslant 0$		
SF order parameter	scalar quantity: ψ	vector quantity: $\boldsymbol{\psi} = (\psi_{+}, \psi_0, \psi_{-})$
Symmetry of BHM	$U(1)$	$U(1) \times S^2$: AF $SO(3)$: ferromagnetic
MI phase	Insulating phase $\rho = 1, 2, \ldots$	Spin singlet: (ρ= even) spin nematic: (ρ= odd)
SF phase	Atomic SF	Transverse or longitudinal polar, ferromagnetic, broken axis symmetry
Order of MI–SF phase transition	Second order	First order: spin singlet MI–SF phase second order: spin nematic MI–SF phase

odd MI lobes ($\rho = 1, 3, \ldots$). Such asymmetry is due to the formation of the spin singlet (nematic) phase corresponding to the even (odd) MI lobes and hence leads to the stability of the even MI phase. While the outer boundary of each odd MI lobe is suppressed but for the even MI lobes, it remains constant at ρ = even which is also expected from the atomic limit. This shows that the chemical potential width and hence the energy gap, E_g for the odd MI lobes corresponds to $U_0 - 2U_2$, while for the even MI lobes, it is $U_0 + 2U_2$. Further, the dashed line represents a first-order phase transition from spin singlet MI to the SF phase and solid line indicates a second-order transition from spin nematic MI to SF phase.

In the ferromagnetic case (figure 7.16(b)), the phase diagram is similar to the spin-0 (scalar) case and the MI lobes gradually decrease in size with increasing of the chemical potential. The phase transition from the MI to the SF phase is a second-order transition, and the width of the chemical potential for all MI lobes is $U_0 + U_2$. Since any further discussion may get too technical for general readers, we would like to stop here and present a summary of the discussion held so far.

7.7 Summary and outlook

We have described Bose–Einstein condensation and superfluidity, and have emphasized the similarities and the differences between them. We have further introduced the Gross–Pitaevskii equation, which yields the SF velocity that encodes the superfluid behaviour of a Bose gas. Subsequently, the utility of the cold atomic gases in the domain of quantum many-body systems is underscored, especially to realize phase transitions, say from an insulator to an SF in a controlled manner. In this connection, we have described the optical lattices formed by counter propagating laser beams, where the ultracold atomic particles are loaded, and in such an environment one can access different quantum phases by carefully varying the laser parameters. A microscopic Hamiltonian, known as the Bose–Hubbard model is

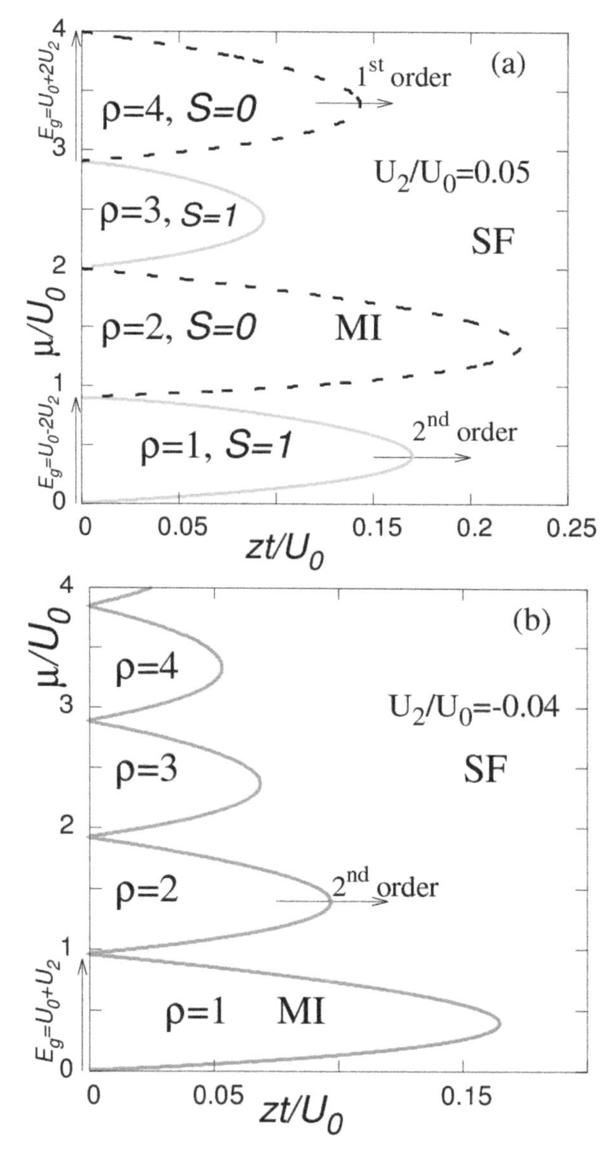

Figure 7.16. The mean field phase diagram corresponding to AF case ($U_2/U_0 = 0.05$) in (a) and ferromagnetic case ($U_2/U_0 = -0.04$) in (b). The solid line represents a second-order and the dashed line represents a first-order transition.

hence studied within a single-site MFT. The approach is shown to reproduce an SF to an insulator transition, which are, respectively, characterized by finite and vanishing SF order parameter, besides having fractional and integer particle densities, respectively, at a particular site in the optical lattice. Such a phase transition has been experimentally observed in a controlled environment. Further, disorder is seen to have important consequences on the phase diagram of these cold atomic systems loaded on the optical lattice, where a new phase, known as the Bose glass phase is seen to intervene in a direct insulator to an SF transition. Finally, generalization of the scenario to spinor particles has been discussed, which hosts a much richer phase diagram.

7.8 Appendix

7.8.1 Derivation of the Gross–Pitaevskii equation

A generic Hamiltonian for a many-particle system with a two-body interaction is written as,

$$\mathcal{H} = \sum_{i=1}^{N}\left(\frac{\mathbf{p}_i^2}{2m} + V_{\text{ext}}(\mathbf{r}_i) - \mu\right) + \frac{1}{2}\sum_{i=1}^{N}\sum_{j\neq i} V(|\mathbf{r}_i - \mathbf{r}_j|) \tag{7.59}$$

where $V_{\text{ext}}(\mathbf{r}_i)$ is a confinement potential and μ is the chemical potential. It is further assumed that the potential V in the second term depends on $|\mathbf{r}_i - \mathbf{r}_j|$, and not separately on $\mathbf{r}_i$ and $\mathbf{r}_j$. An usual approach to find the ground state energy is to minimize $\langle\mathcal{H}\rangle$ with respect to the space of the single-particle states, $|\phi\rangle$, which define the many-particle wavefunction, Ψ via,

$$|\Psi\rangle = |\phi(\mathbf{r}_1)\rangle \otimes |\phi(\mathbf{r}_2)\rangle \otimes |\phi(\mathbf{r}_3)\rangle \cdots \otimes |\phi(\mathbf{r}_N)\rangle. \tag{7.60}$$

Computing $\langle\Psi|\mathcal{H}|\Psi\rangle$, one can write various terms (after some algebra) sequentially as,

(i) $\langle\Psi|\frac{\mathbf{p}_i^2}{2m}|\Psi\rangle = -\frac{N\hbar^2}{2m}\int \phi^*(\mathbf{r})\nabla^2\phi(\mathbf{r})d^3r.$

(ii) $\langle\Psi|V_{\text{ext}}(\mathbf{r}_i)|\Psi\rangle = N\int \phi^*(\mathbf{r})V_{\text{ext}}\phi(\mathbf{r})d^3r$

(iii) $\mu\langle\Psi|\Psi\rangle = \mu(\int \phi^*(\mathbf{r})\phi(\mathbf{r})d^3r)^N$

(iv) $\frac{1}{2}\langle\Psi|V(|\mathbf{r}_i - \mathbf{r}_j|)|\Psi\rangle = \frac{N(N-1)}{2}\int d^3r \int d^3r' \phi^*(\mathbf{r})\phi^*(\mathbf{r}')V(|\mathbf{r}_i - \mathbf{r}_j|)\phi(\mathbf{r})\phi(\mathbf{r}').$

We shall now consider the energy, E (a sum of terms in (i) to (iv)) to be functions of ϕ and ϕ^* which are to be considered as independent variables, and hence minimize E with respect to ϕ or ϕ^*. A little algebra yields,

$$\begin{aligned}\frac{\delta E}{\delta\phi^*} = N\int \Bigg[&\frac{\hbar^2}{2m}\nabla^2\phi(\mathbf{r}) + V_{\text{ext}}(\mathbf{r})\phi(\mathbf{r}) - \mu\phi(\mathbf{r}) \\ &+ (N-1)\int (|\phi(\mathbf{r})|^2 V(|\mathbf{r} - \mathbf{r}'|)d^3r')\phi(\mathbf{r})\Bigg]\delta\phi^* d^3\mathbf{r}.\end{aligned} \tag{7.61}$$

Putting the above to zero yields,

$$-\frac{\hbar^2}{2m}\nabla^2\phi(\mathbf{r}) + V_{\text{ext}}\phi(\mathbf{r}) - \mu\phi(\mathbf{r}) + NV(|\mathbf{r} - \mathbf{r}'|)|\phi(\mathbf{r})|^2\phi(\mathbf{r}) = 0 \tag{7.62}$$

where we have approximated $N - 1 \simeq N$ which is reasonable for large N. This is the time independent Gross–Pitaevskii equation. Including a time dependence in ϕ yields a dynamics of the BEC.

A few comments are in order. ϕ is a complex quantity which can be expressed as,

$$\phi = \sqrt{\rho}\, e^{i\theta}$$

where ρ is the density of the condensate and the phase θ is related to the SF velocity,

$$\mathbf{v} = \frac{\hbar}{m}\nabla\theta.$$

The trap potential is usually an externally applied one which can be approximated to a harmonic form of the type[6],

$$V_{\text{ext}}(\mathbf{r}) = \frac{1}{2}m\omega^2 r_0^2$$

where m denotes the mass of the particles and ω is the angular frequency that characterizes the trapping potential. Finally, the two-body potential can be taken to be a contact type, namely,

$$V(|\mathbf{r} - \mathbf{r}'|) = V_0\delta(\mathbf{r} - \mathbf{r}') = \frac{4\pi\hbar^2}{m}a_s\delta(\mathbf{r} - \mathbf{r}')$$

where a_s denotes the s-wave scattering length.

It may be useful to quote a few values that are relevant to experiments. For example, in ^{87}Rb, $m = 1.45 \times 10^{-25}$ kg, $\omega = 674.2$ rads^{-1}, $N \simeq 10^5$, $a_s = 109a_0$ (a_0: Bohr radius) and $r_0 \simeq 1\mu$m.

References

[1] Wolfke M and Keesom W H 1936 *Physica* **3** 823
[2] Kapitza P 1938 *Nature* **141** 74
[3] Allen J F and Misener A D 1938 *Nature* **141** 75
[4] London F 1938 *Nature* **141** 643; *Phys. Rev. B* **54** 947
[5] Balibar S 2007 *J. Low Temp. Phys.* **146** 441; Griffin A 2009 *J. Phys.: Condens. Matter.* **21** 164220
[6] Pathria R K and Beale P D 2011 *Statistical Mechanics* 3rd edn (Amsterdam: Elsevier)
[7] Gross E P 1961 *Il Nuovo Cimento* **20** 454; Pitaevskii L P 1961 *Sov. Phys. - JETP* **13** 451
[8] Bogoliubov N N 1947 *J. Phys. (USSR)* **11** 23
[9] Cornell E 1996 *J. Res. Natl. Inst. Stand. Technol.* **101** 419 ; Bloch I, Dalibard J and Zwerge W 2008 *Rev. Mod. Phys.* **80** 885
[10] Chin C, Grimm R, Julienne P and Tiesinga E 2010 *Rev. Mod. Phys.* **82** 1215; Giogini S, Pitaevskii L and Stringari S 2008 *Rev. Mod. Phys.* **80** 1215
[11] Greiner M, Mandel O, Esslinger T, Hansch T W and Bloch I 2002 *Nature* **415** 39
[12] Bloch I 2005 *Nat. Phys.* **1** 23
[13] Chu S 1998 *Rev. Mod. Phys.* **70** 685
[14] Phillips W D 1998 *Rev. Mod. Phys.* **70** 721
[15] Cohen-Tannoudji C N 1998 *Rev. Mod. Phys.* **70** 707
[16] Hänsch T and Schawlow A 1975 *Opt. Commun.* **13** 68

[6] Anharmonic trapping potentials are considered as well in literature.

[17] Ketterle W and Van Druten N 1996 *Adv. At. Mol. Opt. Phys.* **37** 181
[18] Grimm R, Weidemüller M and Ovchinnikov Y B 2000 *Adv. At. Mol. Opt. Phys.* **42** 95
[19] Wineland D J, Dalibard J and Cohen-Tannoudji C 1992 *J. Opt. Soc. Am.* B **9** 32
[20] Bloch I, Dalibard J and Zwerger W 2008 *Rev. Mod. Phys.* **80** 885
[21] Chin C, Grimm R, Julienne P and Tiesinga E 2010 *Rev. Mod. Phys.* **82** 1225
[22] Kukulin V I, Krasnopolsky V and Horácek J 2013 *Theory of Resonances: Principles and Applications* vol 3 (Dordrecht: Springer)
[23] Moerdijk A J, Verhaar B J and Axelsson A 1995 *Phys. Rev.* A **51** 4852
[24] Abraham E R I, McAlexander W I, Sackett C A and Hulet R G 1995 *Phys. Rev. Lett.* **74** 1315
[25] Zwierlein M W, Abo-Shaeer J R, Schirotzek A, Schunck C and Ketterle W 2005 *Nature* **435** 1047
[26] Viverit L 2002 *Phys. Rev.* A **66** 023605
[27] Fisher M P A, Weichman P B, Grinstein G and Fisher D S 1989 *Phys. Rev.* B **40** 546
[28] Jaksch D, Bruder C, Cirac J I, Gardiner C W and Zoller P 1998 *Phys. Rev. Lett.* **81** 3108
[29] Sanchez-Palencia L, Clément D, Lugan P, Bouyer P, Shlyapnikov G V and Aspect A 2007 *Phys. Rev. Lett.* **98** 210401
[30] Roati G, Zaccanti M, D'Errico C, Catani J, Modugno M, Simoni A, Inguscio M and Modugno G 2007 *Phys. Rev. Lett.* **99** 010403
[31] Albus A, Illuminati F and Eisert J 2003 *Phys. Rev.* A **68** 023606
[32] Griesmaier A, Werner J, Hensler S, Stuhler J and Pfau T 2005 *Phys. Rev. Lett.* **94** 160401
[33] Crooker B C, Hebral B, Smith E N, Takano Y and Reppy J D 1983 *Phys. Rev. Lett.* **51** 666
[34] Giamarchi T and Schulz H J 1987 *Europhys. Lett.* **3** 1287
[35] Svistunov B V 1996 *Phys. Rev.* B **54** 16131
[36] Lee J-W, Cha M-C and Kim D 2001 *Phys. Rev. Lett.* **87** 247006
[37] Pollet L, Prokof'ev N V, Svistunov B V and Troyer M 2009 *Phys. Rev. Lett.* **103** 140402
[38] Fallani L, Lye J E, Guarrera V, Fort C and Inguscio M 2007 *Phys. Rev. Lett.* **98** 130404
[39] Schweikhard V, Coddington I, Engels P, Mogendorff V P and Cornell E A 2004 *Phys. Rev. Lett.* **92** 040404
[40] Abo-Shaeer J, Raman C, Vogels J and Ketterle W 2001 *Science* **292** 476
[41] Lin Y-J, Compton R L, Jimenez-Garcia K, Porto J V and Spielman I B 2009 *Nature* **462** 628
[42] Lahaye T, Menotti C, Santos L, Lewenstein M and Pfau T 2009 *Rep. Prog. Phys.* **72** 126401
[43] Dutta O, Gajda M, Hauke P, Lewenstein M, Lühmann D-S, Malomed B A, Sowiński T and Zakrzewski J 2015 *Rep. Prog. Phys.* **78** 066001
[44] Sheshadri K, Krishnamurthy H R, Pandit R and Ramakrishnan T V 1993 *Europhys. Lett.* **22** 257
[45] Sengupta P, Rigol M, Batrouni G G, Denteneer P J H and Scalettar R T 2005 *Phys. Rev. Lett.* **95** 220402
[46] Pai R V, Kurdestany J M, Sheshadri K and Pandit R 2012 *Phys. Rev.* B **85** 214524
[47] Greiner M, Mandel O, Esslinger T, Hänsch T W and Bloch I 2002 *Nature* **415** 39
[48] Stöferle T, Moritz H, Schori C, Köhl M and Esslinger T 2004 *Phys. Rev. Lett.* **92** 130403
[49] Spielman I B, Phillips W D and Porto J V 2007 *Phys. Rev. Lett.* **98** 080404
[50] Greiner M 2003 Ultracold quantum gases in three-dimensional optical lattice potentials *PhD Dissertation* LMU München
[51] Stamper-Kurn D M, Andrews M R, Chikkatur A P, Inouye S, Miesner H-J, Stenger J and Ketterle W 1998 *Phys. Rev. Lett.* **80** 2027

[52] Ho T-L 1998 *Phys. Rev. Lett.* **81** 742
[53] Ohmi T and Machida K 1998 *J. Phys. Soc. Jpn.* **67** 1822
[54] Stamper-Kurn D M and Ueda M 2013 *Rev. Mod. Phys.* **85** 1191
[55] Celi A *et al* 2014 *Phys. Rev. Lett.* **112** 043001

Printed in the USA
CPSIA information can be obtained
at www.ICGtesting.com
JSHW060737120224
56719JS00007B/31

9 780750 330299